西北农林科技大学
数学建模
优秀论文集

（第一辑）

郑立飞　解小莉　胡小宁　主编

西北农林科技大学出版社

图书在版编目（CIP）数据

西北农林科技大学数学建模优秀论文集. 第一辑 /郑立飞，解小莉，胡小宁主编. —杨凌：西北农林科技大学出版社，2020.11

ISBN 978-7-5683-0860-1

Ⅰ. ①西… Ⅱ. ①郑… ②解… ③胡… Ⅲ. ①数学模型—文集 Ⅳ. ①O141.4-53

中国版本图书馆CIP数据核字（2020）第171179号

西北农林科技大学数学建模优秀论文集. 第一辑

郑立飞　解小莉　胡小宁　主编

出版发行　西北农林科技大学出版社
地　　址　陕西杨凌杨武路3号　　邮　编：712100
电　　话　总编室：029-87093195　发行部：029-87093302
电子邮箱　press0809@163.com
印　　刷　西安浩轩印务有限公司
版　　次　2020年11月第1版
印　　次　2020年11月第1次印刷
开　　本　889mm × 1194mm　1/16
印　　张　19.25
字　　数　377千字

ISBN 978-7-5683-0860-1

定价：75.00 元

PREFACE
前　言

西北农林科技大学自1999年开展数学建模活动以来，无论是数学建模教学，还是数学建模竞赛，都取得了较好的成绩。1999年，学校首次举行大学生数学建模竞赛活动，并同时参加国际大学生数学建模竞赛和全国大学生数学建模竞赛。2016年底新理学院成立以来，数学建模教学和数学建模竞赛发展快速，每年参加数学建模竞赛的人数累计1500余人。截止2020年，西北农林科技大学在国际数学建模竞赛中共获得国际提名奖（F奖）4项，国际一等奖（M奖）22项，国际二等奖（H奖）62项；在全国大学生数学建模竞赛中共获得全国一等奖4项，二等奖6项。另外，学生们自发组织参加了“电工杯”数学建模竞赛、Mathorcup高校数学建模挑战赛和APMCM亚太地区大学生数学建模竞赛，都取得了很好的成绩。

西北农林科技大学自2017年起，每年“五一”期间举办校级大学生数学建模竞赛，目前已经连续举办三届，参赛人数2000余人，参赛队员来自我校所有招生学院，真正做到了全覆盖——覆盖各个专业、覆盖各级学生。

西北农林科技大学的数学建模课程分为多个层次：理学院信息与计算科学系专业选修课、面向全校学生开设的选修课、赛前培训课、专题讲座、专家讲座、优秀学生报告等内容。这些选修课有效普及了数学建模基本内容，激发了学生参加数学建模竞赛的积极性，成为培养学生的数学建模创新教育平台。

为了对学校数学建模的成果进行总结，进一步提高数学建模水平以及提高所有参加数学建模活动学生的参赛水平，在有关领导的关心支持下，我们搜集整理了近三年优秀的获奖论文，出版了此书。

全书收录了2018—2020年西北农林科技大学参加国际大学生数学建模竞赛全国大学生数学建模竞赛的部分获得国际提名奖论文4篇，国家一等奖论文1篇，国家二等奖论文4篇。本书对收录的论文进行了统一的编排整理，对每篇论文的主体内容、建模方法、文章结构、计算结果等都保持了原来的面貌，这样可使读者真实地看到获奖者在三天（国际大学生数学建模竞赛为四天）比赛期间的论文成果，借鉴参赛论文的写作风格和方式，提高自己的写作水平。本书适合于那些初次参加数学建模竞赛（包括全国大学生数学建模竞赛和国际大学生数学建模竞赛）的队员作为问题求解和论文写作的参考借鉴，对老队员在写作中进一步提高水平也十分有益，另外也可以供从事相关教学和应用研究的工作人员参考。

本书的出版得到西北农林科技大学教务处、西北农林科技大学理学院的关心和全力支持，同时得到吴养会、任争争、罗虎明、宋怀波、魏宁等各位指导老师的支持。在此，还要感谢多年以来同我们一起奋斗过的同事，以及所有参加过全国和国际数学建模竞赛的同学们，感谢他们的辛勤努力！愿本书的出版能够给大学生数学建模活动带来积极的推动作用。

由于水平有限，书中难免有不妥之处，诚望批评指正。

编　者

2020年7月

CONTENTS
目　录

1.Out of Gas and Driving on E (for electric, not empty)

2018

MCM/ICM

Summary Sheet

(Your team's summary should be included as the first page of your electronic submission.)

Type a summary of your results on this page. Do not include the name of your school, advisor, or team members on this page.

Abstract

Electric vehicles (EVs) have a rapid growth for both environmental and economic reasons. Many governments (especially EVI members) adopt policy support to facilitate EVs uptake. However, a swift from petrol and diesel car to electric car needs transition time, during which the location and convenience of charging stations is critical.

We were tasked to design a plan of optimizing the distribution of charging stations that promotes all-electric cars. And we considered the final network of charging stations as well as the evolution of the network over time. We developed a **supply and demand network model** to measure the demand intensity of charging stations in a certain region. In order to determine the potential demand of charging stations, we forecast the EVs based on the improved **Bass diffusion model**. We evaluated the supply capacity of charging stations based on extended **penalty function**. Based on the **fish-swam algorithm**, we built a **network optimization model** to meet the annual growth of demand for charging.

In task1, cities were used as nodes to establish the US supply and demand network, and the optimal distribution of newly-invested charging stations were simulated each year. We found that the number of charging stations that the United States eventually need is 9.58×10^6. By comparing Tesla's current allocation of charging stations, we thought Tesla is on the track to allow a complete switch to all-electric in the US.

In task2, we discussed the best charging station distribution plan in Korea under transitional and non-transitional conditions. Results showed that it takes about 20 years for Korea to realize an all-electric vehicle. In the early stage, the stations should be built mainly in urban area. By changing the different factors and counting the time required for a full EVs, we find that the two main key factors are population density and total personal consumption expenditures.

In task3, we created a classification system based on the countries with different economic development. Taking China as an example, we analyzed the development of all- electric vehicles. Comparing China and the United States, we analyzed the reasons for the differences.

In task4, we analyzed the promotion of the development of EVs such as car-share and self-driving cars.

Finally, we conducted a sensitivity analysis and discussed the strengths and weaknesses of the model.

作　者：屈敏直，邢寇祯，荆朝

指导教师：宋怀波

奖　项：2018年美国大学生数学建模竞赛F奖

I Introduction

1.1 Background

The use of EVs has gained considerable attention over the last decade as an environmentally friendly alternative to conventional cars that consume fossil fuels and emit greenhouse gases. However, the development of EVs is hindered by many factors. One of the most critical barriers to widespread the adoption of EVs is the lack of charging station infrastructure. It is important to propose a reasonable distribution method of charging stations to meet the needs of people in different periods.

1.2 An Overview of the Problem

With the development of EV technology, more and more countries are paying attention to the potential of rapid growth in the adoption of EVs. The migration from gasoline and diesel cars to EVs, however, is not simple and can't happen overnight. There are limited resources, and it will take time for consumers to make the switch. Location and convenience, among many factors, were considered as the two critical factors for mainstream consumers to switch voluntarily.

There are several major challenges in this issue:

- There is a lack of an effective method to measure the **demand of charging station** to meet the future needs of potential customers.
- During transition period, the interaction between the **distribution of charging station** and demand for EVs is obscure.
- **Multi-phase optimization** of charging station distribution is lack, and it considers more realistic factors.

Tasks at hand

The following is our main tasks to be solved,

- Find the key factor that shaped the development of the plan.
- Explore the impact on the optimal distribution plan with or without concerning about transition.
- Analysis of the optimal development process and propose an invest plan.
- Develop a classification system and develop a reasonable distribution plan for different countries.

1.3 Analysis and Method

Related work

The optimization charging stations location model is widely studied by many research. The macro-aspect study is mainly based on the cost minimization (Li, Min, et al. 2009), while in micro-level researchers concentrate on the specific location of charging station individual (Ceylan, H. 163). However, there is less literature on the issues to be studied - the charging station network optimization process and plan aiming to eventually reach full coverage of EVs, and the different distribution schemes between urban and rural areas.

Method

There are two core challenges in the design of EV' s charging stations network: (1) How do the distribution of charging station demand changes? (2) Given the demand distribution, how to optimize the charging stations network to meet current demand for charging stations in different regions?

In order to figure out the change of charging station demand, we developed the **demand forecasting model** based on Bass model. The improved Bass model was put forward to predict the EV ownership, then the EV charging station demand was calculated in a region based on the EV ownership.

The distribution of additional charging stations depends on the changes of the supply and the demand network. Based on the fish-swarm algorithm, we proposed a **network optimization model** of charging station, which optimally allocates the newly added charging stations to the existing network of charging stations to meet EV' s needs in different regions.

The sketch map of this paper was shown in **Fig. 1**.

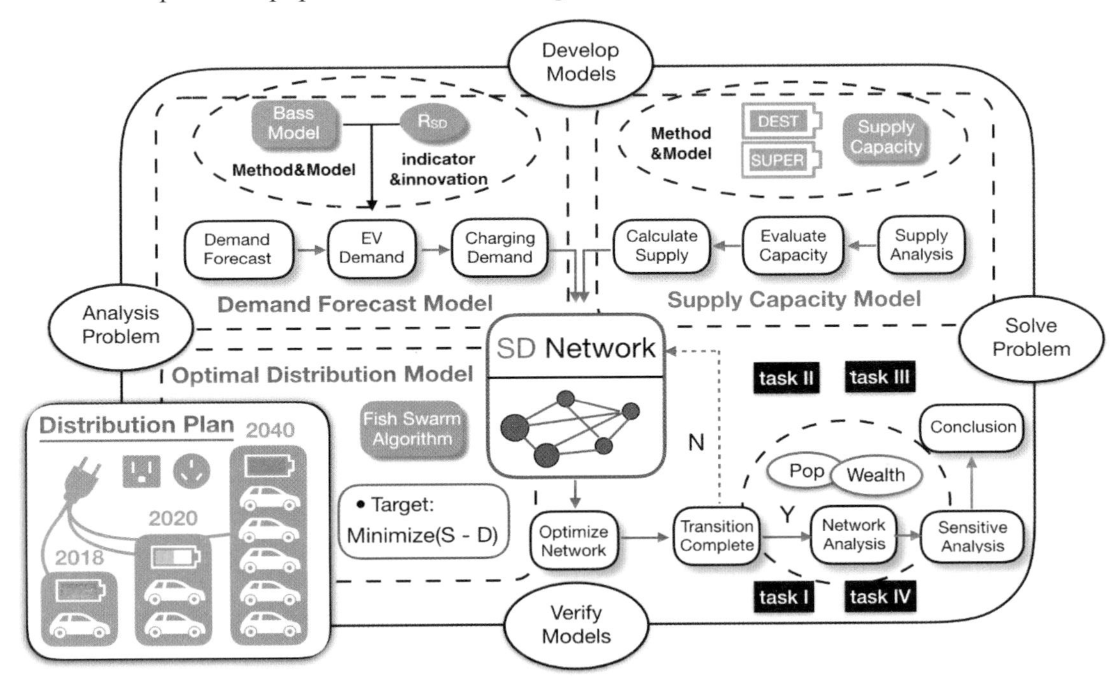

Fig. 1 Sketch map presented in this research

We developed a supply and demand network model to measure the demand intensity of charging stations in a certain region. In order to determine the potential demand of charging stations, we forecast the EVs based on the improved Bass diffusion model. We evaluated the supply capacity of charging stations based on extended penalty function. Based on the fish-swam algorithm, we build a network optimization model to meet the annual growth of demand for charging.

If the charging station coverage did not meet the standards of completing transition (for example 85%), we will forecast the amount of demand increase and estimate new-added supply. The latter will be allocated to the network based on Optimal Distribution Model in order to meet the needs of the distribution of the following year. If we meet the standard, we will analyze the status of the network. Finally, we conducted a sensitivity analysis of the network and made a conclusion.

II Model Assumptions and Notations

2.1 Assumptions and Justifications

The model we built was based on the following assumptions:

• **The number of buyers determines the unit sales of the product.** The basic Bass model assuming that there are no repeat buyers and only one unit per buyer in the planned product planning cycle.

• **Policies do not change.** We assume that current policies remain in place until they are set to expire, but we do not assume any fresh policies are introduced.

• **No substitution.** Assuming that under the government's policy support, EVs are marketed as a new product, regardless of demand analysis caused by the substitution with gasoline and diesel vehicles.

• **There is a transitional period in every country**: the number of charging stations increases each year is limited and less than the total number of charging stations at the completion of the transition. That is, the growth rate μ will not be too large.

• The maximum charging station power is not affected by the actual power of the surrounding substation.

• All charging stations in the world reference Tesla's charging station parameters. In our model, we don't consider about the differences of the type of charging ways from Tesla.

• The charge ability of the charging station will not increase over the next few years.

2.2 Notation

We list in **Tab. 1** the symbols and notations used in this paper. Some will be defined later.

Tab. 1 Indicators description

Variables	Definitions
p	innovation coefficient in Bass model
q	imitation coefficient
m	maximum potential market
R_{SD}	the ratio of supply and demand for charging stations
R_{SC}	the ratio of service coverage
μ	the growth rate
k	node (a city or a county) number.
l	edge (highway between one city to another) number
j	the number of the node that has the charging station
Q_k^D	the charging demand of node k.
Q_{jk}^S	the charging supply of node j to node k.

III Basic Model of Distribution

Firstly, the **Bass model** was improved to predict the future EV ownership, and then the demand for charging stations was got. After that, a **supply capacity model** was developed to quantify the current charging stations supply. Based on the quantified analysis of supply and demand, a **supply and demand network model** of charging station was set up, which describes its supply and demand distribution in a certain area. With the increase of EVs, more charging stations will be added into the network. We distributed these newly-invested stations through a **network optimization model** based on fish-swarm algorithm to meet the growth of demand in different regions.

3.1 Design of the Model

The model consists of the following four parts:

- **Demand forecast model:** forecast the potential demand for charging stations through forecasting EV ownership in the next phase. Get specific demand for charging stations at certain region.
- **Supply Capacity Model:** evaluate and quantify power supply capacity of charging stations and coverage area.
- **Supply-Demand Network Model:** describe the supply and demand distribution in different periods
- **Optimal Distribution Model:** optimize the distribution of new charging stations.

3.2 Submodels

3.2.1 Demand Forecast Model

Forecasting of EV ownership

In order to determine the optimal number and distribution of charging stations, we need to analyze the demand for charging stations through predicting the future electric vehicle(EV) ownership. In this paper, we improved the Bass model to calculate the ownership of the EVs at a certain moment.

The Bass model uses conditional likelihood functions to predict the share of a potential adopters of new industrial, durable products between the market introduction and market saturation. Market adoption mode follows the S-shaped pattern. (Thomas A. Becker, 30)

The discrete form of the Bass model is expressed as:

$$\frac{f(t+1)}{1-F(t)} = p \cdot R_{SD} + q \cdot \frac{N(t)}{m} \tag{1}$$

where,

$f(t+1)$ is the ratio of new customers and all potential customers at $t+1$moment; $F(t)$ is the ratio of cumulative customers and all potential customers at t;

R_{SD} is the ratio of supply and demand for charging stations;

$N(t)$ is the cumulative number of EVs purchased at t; $n(t)$ is the number of purchases at t.

There are two critical parameters:

p-innovation coefficient, which reflects the influence of external factors, indicates that the probability of purchase is affected by the media and other external environment (such as advertisements), and the value of p is usually within the range of 0.01-0.03.

q-imitation coefficient, which reflects the impact of internal factors, captures the behavior of those consumers for whom the purchasing decisions of others are important. The range of q values is usually within 0.3-0.7.

The spread of this new technology for EVs is restricted to environment. Namely, sufficient charging stations are critical to promote EVs smoothly. The inconvenience due to the lack of charging stations will reduce people's demand for charging cars, while adequate and convenient charging stations will promote the spread of charging cars. Therefore, we improved the Bass model for the diffusion characteristics of the charging car: we considered the impact of the ratio of charging station supply and demand(R_{SD})on the demand of EV. When the supply of the charging station is greater than demand, the diffusion rate of the innovator is greater, vice versa.

The relationship between $n(t)$ and $f(t)$ could be described as Equation (2):

$$n(t) = m \cdot f(t) \tag{2}$$

Thereby, we could get Equation (3) and Equation (4):

$$n(t) = \frac{dN(t)}{dt} = p \cdot R_{SD}[m - N(t)] + q \cdot N(t)[m - N(t)] \tag{3}$$

$$f(t+1) = \frac{dF(t)}{dt} = p \cdot R_{SD}[1 - F(t)] + q \cdot F(t)[1 - F(t)] \tag{4}$$

In Equation (4), we only need to know the cumulative amount of EVs purchased at a given point to get the probability of the next purchase.

In the prediction of their EV holdings, researchers in the United States considered that the p value is 0.015 and the q value is 0.35. (Li, Min, et al., 1010-1012)

Our improved Bass model also increased the flexibility of EV ownership forecasting in different regions. The maximum potential market (m) varies in different countries with specific very different geographies, population density distributions, and economic. For some specific regions, we increased the applicability of the model by changing the p and q parameters.

Forecasting of charging station demand

From Formula (2) and (4), we can get the total number of EVs in a region for the next period:

$$N(t+1) = m \cdot f(t+1) + m \cdot F(t) \tag{5}$$

Here, we only focus on personal passenger vehicles, not include commercial vehicles. According to the literature we have searched, average annual mileage is 15,000 kilometers. Therefore the average annual charging demand in a region as follows:

$$Q^D = N(t+1) \cdot 15000 \tag{6}$$

3.2.2 Supply Capacity Model

In this Section, we mainly established a model of the power supply capacity of charging stations. We hope that the actual supply capacity of the area around the charging station will be obtained, through the type and number of charging devices.

We defined Q_{jk}^S as the supply capacity of the charging station at location j and location k. The measure of Q_{jk}^S was the distance that EV can run when receiving energy from the charging station in a year, its unit is km/year, the same as the demand for EVs.

Penalty Function for Destination Charging

Joana Cavadas et al developed a kind of distance penalty function for Destination Charging . The penalty function was adopted in our model and it fits well with our model. The destination penalty function for Destination Charging was shown in Equation (7):

$$\Gamma_{jk}^d = \frac{(-(d_{jk})^4 + 30^4)}{30^4 \cdot \exp(\frac{d_{jk}}{30})} \qquad 0 \le d_{jk} \le 30 \tag{7}$$

where,

Γ_{jk}^{d} is the penalty of Destination Charging;

d_{jk} is the distance between charging station j and location k in km.

Penalty Function for Superchargers

Based on the data from on the TESLA website(https://www.tesla.com/supercharger). We expanded the supercharger penalty function Γ_s^{ij} as Equation (8):

$$\Gamma_{jk}^{s} = \frac{(-(d_{jk})^4 + 100^4)}{100^4 \cdot \exp(\frac{d_{jk}}{100})} \qquad 0 \le d_{jk} \le 30 \tag{8}$$

where,

Γ_{jk}^{s} is the penalty of Superchargers;

d_{jk} is the distance between charging station j and location k in km.

Supply Capacity Model

Based on the penalty function and configuration of two kind of charging station given by TESLA, the supply capacity of Q_{jk}^{S} was modeled as Equation (9):

$$Q_{jk}^{S} = 438000 \cdot num_d^j \cdot \Gamma_{jk}^{d} + 2978400 \cdot num_s^j \cdot \Gamma_{jk}^{s} \tag{9}$$

where,

num_d^j is the number of destination chargers;

num_s^j is the number of superchargers.

3.2.3 Supply-Demand Network Model

There are two basic network models in the model we built, that is the demand network and charging stations network (supply network).

Demand Network

We define a group within a certain distance as a node (e.g. a city or a county), the weight of the node represents the total demand of this group. There are edges between different nodes (e.g. highway between one city and another), The weight of the edge is defined as the average demand on the edge (e.g. the average demand for charging stations on a road). The demand graph consists of points and edge sets. It is an undirected graph since the highway is universally bi-directional.

The demand network is defined as $G = (V, E)$ with edge weights $w_d: E \to R^+ \ by \ (G, w_d)$, and node weights $q_d: E \to R^+ \ by \ (G, q_d)$.

Charging Station Network (or Supply Network)

The definition of a charging station network is similar to that of a demand network. The node is defined as a group of charging stations within a certain distance, the weight of which represents the total supply capacity of this group. Edges with different weights connect to different points, and the weight of the edge represents the average supply capacity on the edge.

We define the Charging Station Network $G = (V, E)$, with edge weights $w_s: E \to R^+ \ by \ (G, w_s)$, and node weights $q_s: E \to R^+ \ by \ (G, q_s)$.

Since the definitions of the two network models are similar, we replaced them with a single graph, named **Supply-Demand Network**.

Supply-Demand Network is defined as $G=(V,E)$, with edge weights $w_s: E \rightarrow R^+ \; by \; (G, w_d, w_s)$, and node weights $q_s: E \rightarrow R^+ \; by \; (G, q_d, q_s)$.

Important Parameters of the Network

a. **supply-requirement ratio R_{SD}:** is used to assess the supply capability of a node. The expression is defined as Equation (10),

$$R_{SD} = \frac{q_s}{q_d} \tag{10}$$

When $R_{SD} > 1$, the node is oversupplied; When $R_{SD} < 1$, the node lacks the supply capacity to meet the demand.

b. **Service radius of the node r_S:** represents the scope of the service impact. The linear fitting method was used to quantify the relationship between service radius and node supply-demand ratio based on the data from on the TESLA website: (https://www.tesla.com/supercharger). r_S could be described as Equation (11),

$$r_S = \alpha R_{SD} \tag{11}$$

where, α is the coefficient to represent the service influence. It depends on the property and population density of the node.

c. **Network service coverage R_{SC}:** is used to evaluate the extent of service coverage of the network. It could be calculated by Equation (12).

$$R_S = \sum_{i=1}^{n} \frac{\pi {r_{si}}^2}{A_{net}} \tag{12}$$

where, n is the number of node; A_{net} is the area of network.

We **divide the phase** of the development of EVs based on network service coverage.

3.2.4 Optimal Distribution Model

Artificial fish swarm algorithm is an animal-based intelligent optimization algorithm. This method has the characteristics of fast convergence in finding the global optimal value. Our goal was to build the charging stations for different nodes via artificial fish swarm algorithm to meet the electricity needs of EVs on different nodes as much as possible. The model is shown in Equation (13) and (14):

$$\min \sum_{k=1}^{m} \left(\frac{Q_k^S}{Q_k^D} - \mu\right)^2 + \sum_{l=1}^{n} \left(\frac{Q_l^S}{Q_l^D} - \mu\right)^2 \tag{13}$$

$$s.t. \quad \frac{\sum_{k=1}^{m} Q_k^S}{\sum_{k=1}^{m} Q_k^D} = \mu.$$

$$\sum_{j=1}^{m} Q_{jk}^S = Q_k^S \tag{14}$$

$$\sum_{j=1}^{n} Q_{jl}^S = Q_l^S$$

where,

$Q^S{}_k$ is the sum of energy received by node k around the power plant;

$Q^D{}_k$ is the demand of node k for electric energy;

$Q^S{}_l$ is the sum of the energy received by edge k around the power plant;

$Q^D{}_l$ is the demand of node edge for electric energy;

μ is the growth rate.

Equation (14) shows that the total energy supplied by the charging station is μ times of the total demand. Let μ=120%, we hope that the total supply is more than the total demand to promote the number of EVs. Our objective function is that the distribution of the charging stations can make certain power supply at any node and the edge of the network to make up for the power demand. Network cannot be oversupplied at a node, nor too low.

3.3 Example for Los Angeles

Take Los Angeles as an example, **Fig. 2** describes the overall circulation of our entire model, as the legend shows, the blue circles indicates that the supply of charging stations in the area is greater than the demand, namely, the ratio of the supply and demand (R_{SD}) is greater than 1. Oppositely, the red circles represent the supply of charging stations in the area are less than demand. And the orange circles, an intermediate state of influence, represent the original supply and demand unbalanced areas are affected by the new charging stations in the region. The size of the circle represents the population density in the area. Lastly, the light red area represents the scope affected by the charging station.

Fig2. a represents the initial stage of the model. Charging station supply and demand network shows the relationship between current charging station supply situation and demand (including the forecasting of EVs in a certain future period). According to the gap of supply and demand, a certain amount of charging station (lower right corner of **Fig2. a**) will be added into the network.

As is shown in **Fig2. b**, the charging stations are randomly placed in Los Angeles, which is an inefficient state.

Fig2. c shows an optimized distribution of the charging stations through the network optimization model. More charging stations should be allocated to areas with larger demands. The orange area represents the area affected.

Fig2. d shows the optimized supply and demand network. Most of the orange regions changed to blue from red after new stations are installed, indicating that the supply and demand in the region have been greatly improved. (Red did not change to blue represents the area has very inadequate supply originally.)

The potential demand of the charging station in the next phase was then forecasted on the basis of network shown in **Fig2. d**, so that we can obtain the new supply and demand network, as is shown in **Fig2. e**. After that, we can allocate newly additional charging stations again to optimize the allocations, and we can enter the next phase of the cycle.

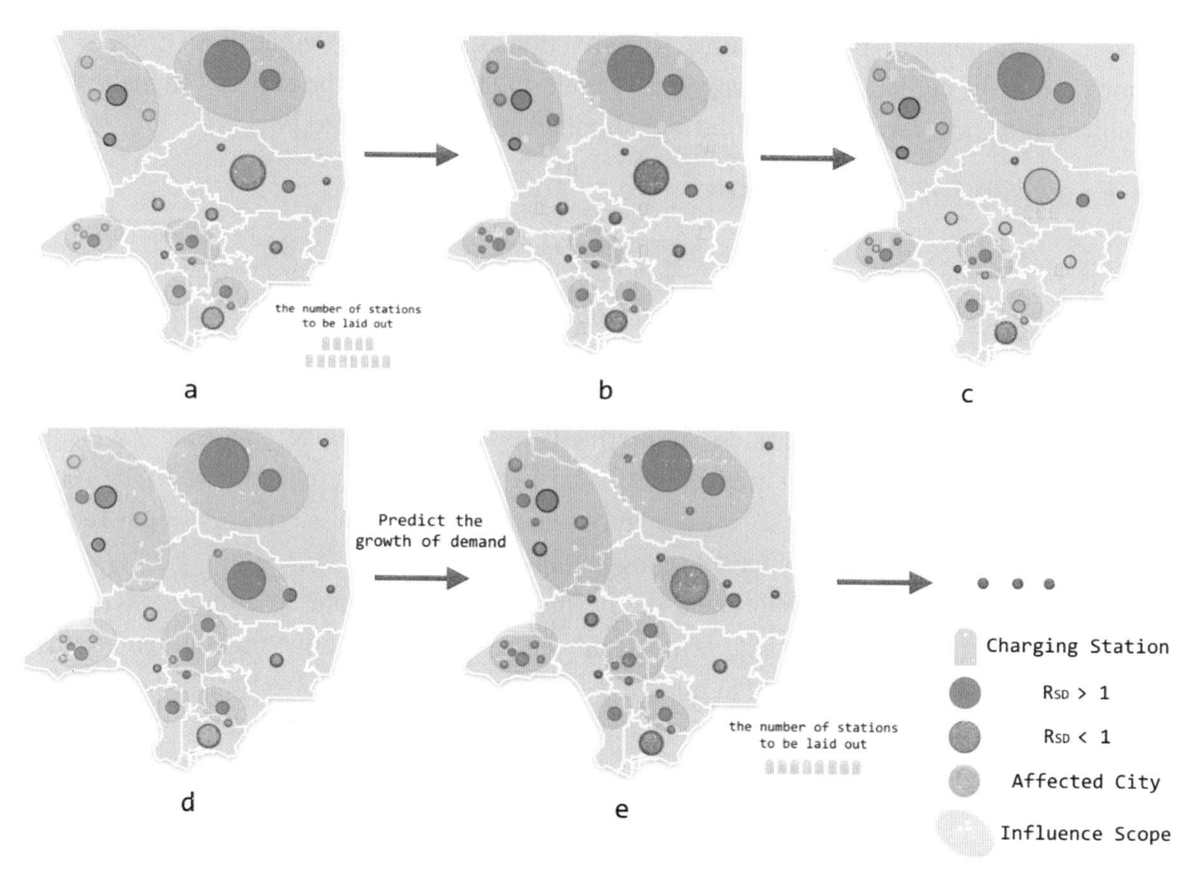

Fig. 2 Optimization process for Los Angeles

Through the optimization and recycling in multi-stages, the impact scope of the charging stations becomes more influential. When the coverage of the charging station reaches a certain level, we can further realize the full coverage of EVs through the ban.

IV Optimizing the Distribution Network

We simulated the distribution process in different environment given by the tasks. Our research object is the country network which is consisted of different regions. We specially studied three countries, the United States, South Korea and China. Through the analysis of national conditions, we developed the optimal plan of distribution of charging stations for them. The program is executed on MATLAB.

Task 1: Changes of EV network for US

Parameter Analysis:

In the demand forecasting model for EVs, the maximum market potential is related to the population in a certain area as well as economic conditions such as spending power. For cities, suburbs, and rural areas, where the three types of areas differ greatly, we calculated the parameters for each of them separately (the data is attached at the Appendix), we can predict the demand of the charging stations. $(p = 0.015,\ q = 0.35,\ \mu = 1.2)$

Discussion of result:

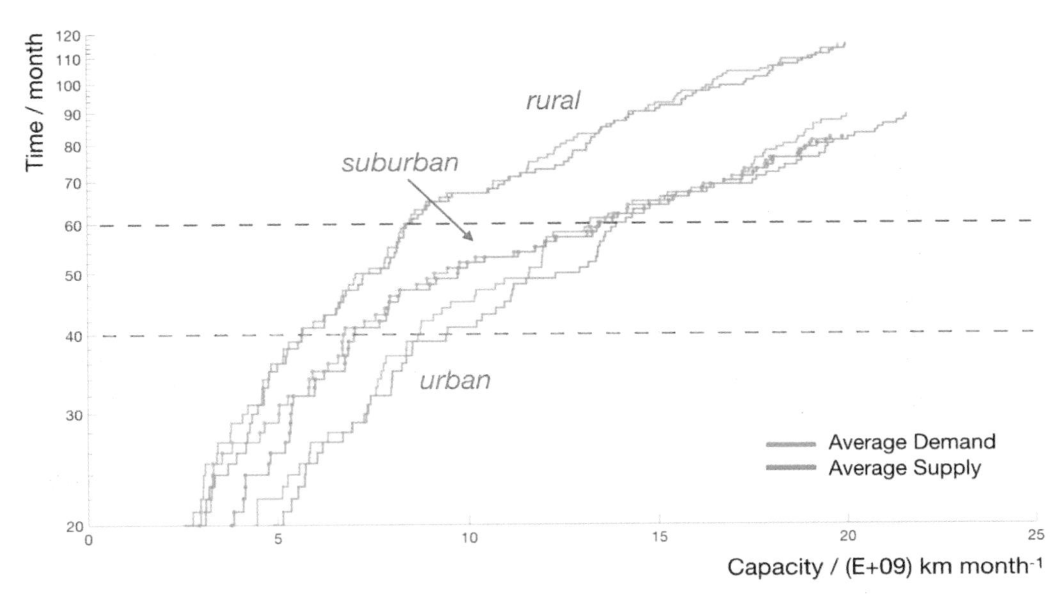

Fig. 3 Comparison of different areas

Comparison of different areas is shown in **Fig. 3**. From top to bottom, the lines show the monthly supply and demand map of suburban and urban regions. In order to avoid confusion, we thickened the line of supply and demand in the suburbs area. The distance between the blue and green lines at the same area in the same time shows supply and demand difference.

Fig. 3 also shows that, in the next 20 to 40 months, the demand for EVs is highest in the suburbs areas, followed by cities, despite the demand for EVs in all three regions. According to the idea of optimal distribution network, suburbs regions got more of the distribution of new charging stations, so the supply capacity of power plants has been greatly improved.

In the next 40-60 months, the supply capacity and demand in all three regions are all on rise tendency. In urban areas, the demand for EVs has increased more than the other two areas. However, from the experimental results, we find that the growth of suburban supply capacity is comparable to that of cities, that is, we still need to allocate comparable charging stations in the suburbs and cities according to the optimal distribution of networks. At this stage, the development of suburban EVs has been comparable to that of cities. The supply and demand in rural areas still showed a slow growth trend.

In the next 60-120 months, the development in urban and suburban areas is comparable, and rural areas have been given more power station allocation than ever before, at rates comparable to those of the other two regions.

In general, based on the optimal distribution plan provided by the supply and demand network, in the early stages of development (about 60 months), we prefer to allocate charging stations to cities and suburbs. In the medium period, the demand in rural areas will increase dramatically, and we should pay more attention to the distribution of rural supply capacity. And the Tesla station allocation plan (website) is consistent with the results of the optimal distribution network.

We summarized the total annual S&D in the United States, the result was shown in **Fig. 4**. Under the optimal distribution conditions, the growth of S&D across the United States can be divided into three stages. The first stage (the next 1-5 years), the demand growth rate will be higher than the supply capacity. The second phase (the next 5-15 years), with the rapid growth of S&D, EVs at this stage are in rapid popularity. The third stage (15 years and more), the demand is becoming saturated.

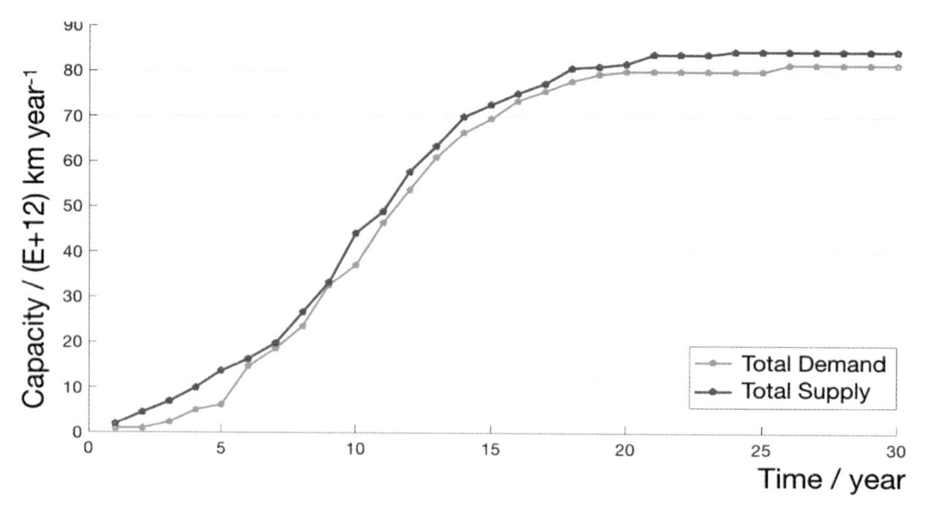

Fig. 4 Change of total S&D each year

Solution: When the demand tends to be saturated, the total demand is about $8.2 \times 10^{12} km/year$. If all destination stations are used, the total number is approximately 9.58×10^{6}. Through the experimental results, we calculated the distribution of charging pile in the next 20 years, for cities, the ratio is 0.4, 0.3, and 0.3 for urban, suburban, and rural areas, respectively.

Task 2: Implement on South Korea

In this section, South Korea, which is one of the few developed countries in Asia, was selected to conduct this research. According to data provided by the Korea Administration and Autonomy Department in August 2017, the current population of South Korea is about 51.75 million, with the GDP of 1.410 trillion US dollars. South Korea is one of the most densely populated countries in the world. The population distribution is very uneven in South Korea, most of the population live in cities, the urbanization rate was 82.5%. In 2015, the total population in the metropolitan area, including Seoul, Incheon and Gyeonggi Province, totaled 25.14 million, accounting for 49.7% of the total population. We gathered some data about the main cities of South Korea (see attached appendix), Limited to the time of the contest, some data of small cities and regions is not mentioned. Based on the available data, we estimate the cities without data before applying the model. So, in our model, the parameters are set as $p = 0.015$, $q = 0.35$, $\mu = 1.2$.

2a. Distribution of charging stations with no transition required

Because we do not need to consider the time transition, we do not consider the iterations over time variables in our model. So, this problem can be explained in the model as: In one calculation, the demand in each region increases directly to the highest value in a very short period of time. So, we can get the demand changes, as well as the number of charging pile numbers at different locations. After model analysis, the results obtained are shown in **Tab. 2.**

Tab. 2 Distribution of chargers in major cities of South Korea (100% EVs)

Area	Destination Charging	Superchargers	Supply (km/year)
Seoul	268349	74469	1. 40E+11
Busan	26200	11010	14754180501
Incheon	14806	6599	8450456639

Daegu	12393	5880	7179433371
Daejeon	4289	2523	2629737540
Gwangju	3773	2312	2341261089
Suwon	2983	1907	1874545461
Goyang-si	2190	1401	1376488634
Seongnam-si	1929	1418	1267264265
Ulsan	1397	1618	1093984817
Bucheon-si	1071	1244	839694924
Jeonju	1013	426	570373816
Ansan-si	822	366	469177411
Cheongju-si	760	373	443929509
Anyang-si	675	497	443576007
Changwon	407	486	322864114
Pohang	313	414	260260312
Uijeongbu-si	270	396	236109262
Hwaseong-si	250	414	232906158

As **Tab. 2** shows, we found that Seoul City has a very large number of charging piles. Some other cities have built fewer charging piles as well. According to our model, the distribution of charging stations is directly affected by the demand distribution.

The **key factor** influencing the demand is the population and per capita income in different cities. Thus, the distribution of population and income is critical in our plan to develop the charging station plan.

2b. Distribution concerning transition

In this section, we consider the impact of the transitional period on the distribution plan. We mainly consider two network parameters to help us to develop a reasonable distribution plan: (1) Service coverage rate, namely RSC, was used to assess the current network service capabilities, and to divide the important basis for different stages. (2) The transition of the total supply capacity of the network changes, was used to describe the network changes in the total amount of EVs.

The parameter μ, that is, the ratio of the amount of distribution at the next moment to the amount of demand growth at the current moment, was used to simulate the charger first or response to car purchase. As is shown in **Fig. 5**, $\mu>1$ means over-demand allocation, and $\mu<1$ means lower than the demand distribution.

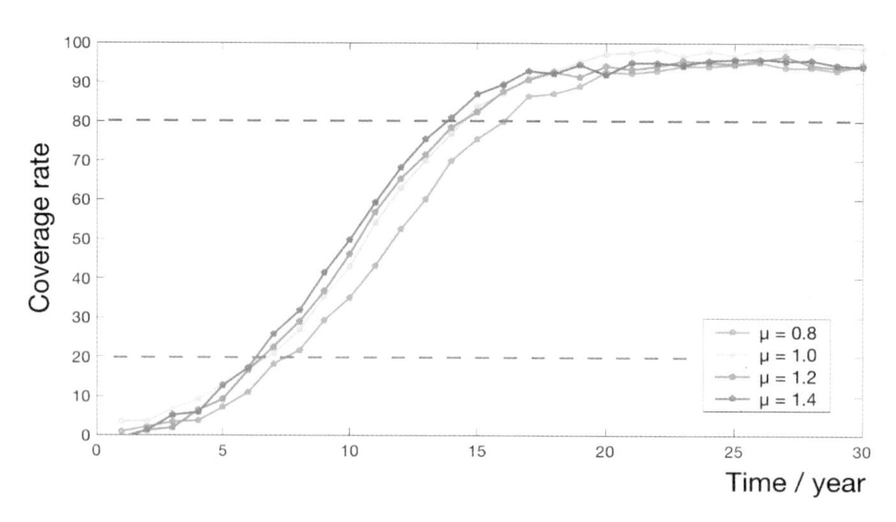

Fig. 5 Change of Coverage rate

If we take 80% coverage as the time to ban petrol car, we find that the bigger the μ is, the shorter it takes to issue an injunction. In other words, from the entire transfer process point of view, when the larger the μ is, the popularity of EVs will be faster. Therefore, we suggest build **the charger first**.

We got the similar results compared with three different regions of the United States (Task 1 **Fig. 3**). We do not repeat analysis details because the similar process has been shown in Task 1. Therefore, we suggest that the charging stations should be allocated in **city-based** regions in the early stages of the development of EVs. In the mid-term, there tends to be **a mixture of suburbs and cities**.

The **main factor** we considered in a step-by-step distribution is the **PD** (Population density) and **PCE** (Total personal consumption expenditures). We use the coverage rate of 85% as the judgment criteria of implementing the ban. As the experimental results shown in **Fig. 6**, with the increase of PD and PCE, the time required to popularize EVs will be shortened.

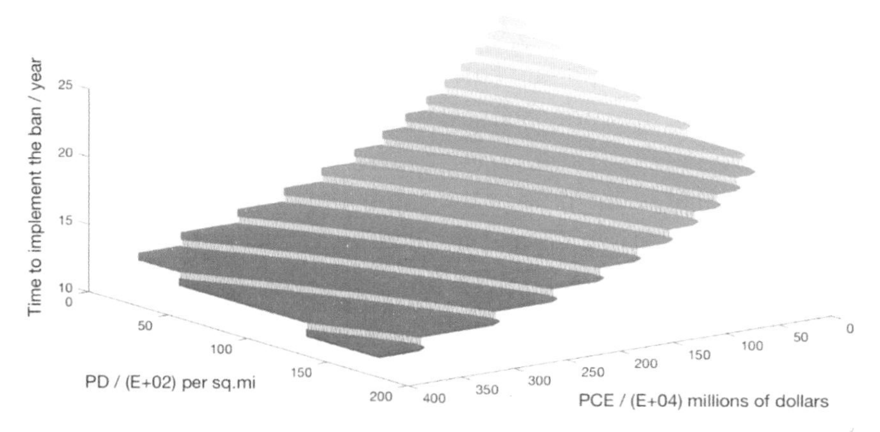

Fig. 6 the change of time consumption

2c. Distribution concerning transition

In order to better describe the changes in the network, we normalize the total network demand and total supply capacity at different times. And the normalized results and network coverage was used to assess the development of supply and demand networks, the results were shown in **Fig. 7**.

Solution: The supply capacity at different times was used to indicate the number of EVs. The number of EVs in Korea reaches 10%, 30%, 50% and 100%, respectively, which take about 6 years, 13 years, 18 years, and 50 years respectively.

We provide that when the coverage rate is higher than 85%, the ban can be implemented. So, if the optimal allocation is made annually to 120% of the increment in demand ($\mu = 1.2$), the ban can be enacted about 30 years later. And, the main factors that affect the distribution by stages are the urban population density and the wealth value.

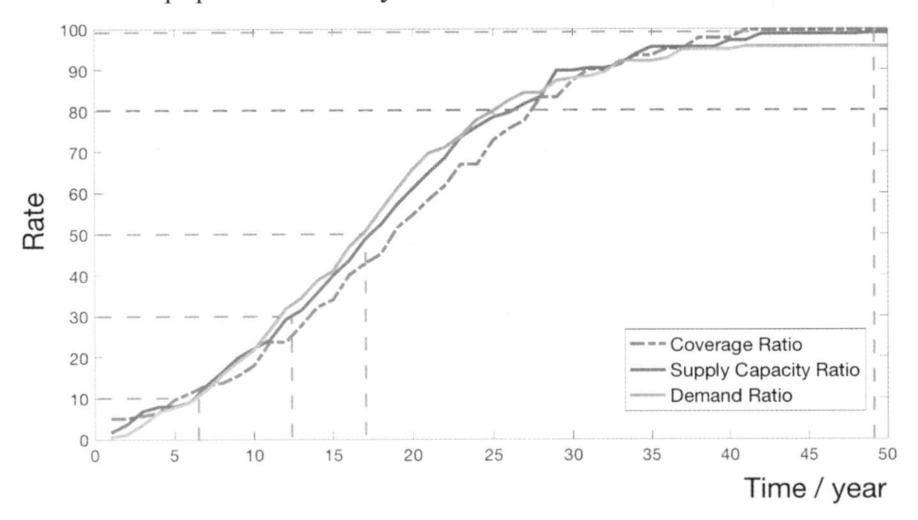

Fig. 7 Change of network during transition

Task 3: Model applied to countries with different conditions—take China as an example

In this section, we analyzed the differences between different countries. We have created a simple classification system that extended the applications of the model. Finally, we mainly analyzed the growth plan of the charging station in detail for China.

Creating a classify system

We searched the different **geographies, distribution of population and personal income** in China, Singapore, the United States, and South Korea. We found that there are two main differences:

• **Different rates of economic growth:** China is a developing country. Singapore, the United States, and South Korea are all belong to the developed countries. After investigation, GDP growth in developed countries generally does not increase over time, but the annual GDP and car ownership in developing countries is increasing year by year.

• **Geographic complexity varies:** China has more complicated geography than Singapore. But for our model, we only need to change the parameters of the nodes in different locations to adapt to different geographies. Geographical differences have little effect on our model.

Therefore, we can create a classification system. The **key factors** of the classification system are **economic development** depending on **population** and **personal income**. We divided different countries in the world into developed and developing countries. It is assumed that the model of developed countries has not changed and the developing countries' GDP and car ownership will grow linearly over the next 50 years. Also, the values of p and q in the demand forecasting model vary from different countries. Therefore, there are some amendments to the model of developing countries.

Model of Developing Countries - Taking China as an Example

According to the information, we keep the growth rate of China's GDP at 7%. Then, we fit the values of p, q by the data. In contrast to developed countries, p is smaller and q is larger in developing countries. This shows that the increase of EVs in developing countries comes more from imitation. Let $\mu=1.2$, we can get the coverage trend of China, as is shown in **Fig. 8**.

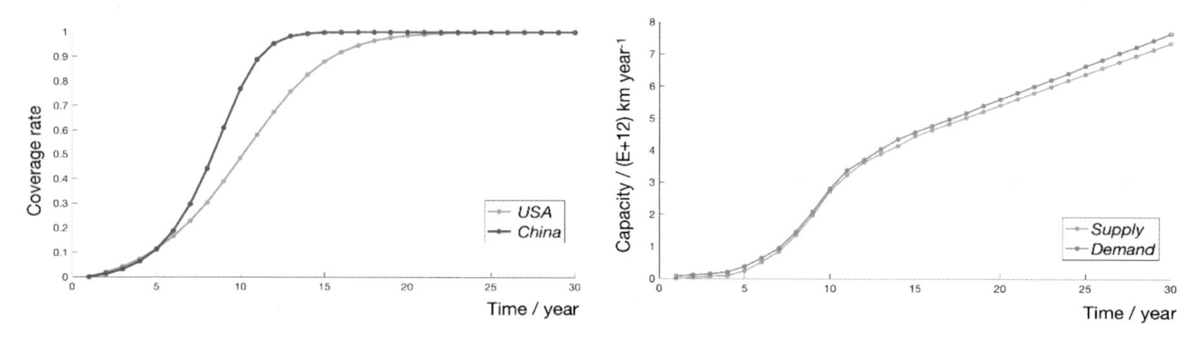

Fig. 8 Left: Change of China and US coverage rate over time. Right: China's supply and demand change over time.

The left shows that the trend of coverage rate in China is similar with US. However, some different details can be found in the comparison of the two countries. First, time cost of China to achieve 80 coverage and full coverage is lower than in the United States. In China, it takes only **10 years to achieve 80%** coverage and **15 years to achieve full coverage**. In the initial 5 years, the development of EVs is relatively slow compared to developed countries, so the supply of charging stations should not be too much. After 5 years, the development of EVs is faster than developed countries. The reason is that China is a country with **large population** but **low personal income**, compared with the developed countries. Therefore, the number of EVs is small and per capita vehicle ownership is also relatively low at the beginning, leading to the faster evolution of charging station.

The right figure shows that the demand and supply in China will keep growing, after the coverage rate has reached 100%. The reason is that China's vehicle has not yet reached its full capacity and its personal income increase constantly. Therefore, the number of Chinese EVs will continue to increase after 15 years, the supply and demand will increase until the pace of economic development slowed down and eventually reach saturated.

As a developing country like China, the economic development and electric car evolution happen at the same time. Even if a developing country migrate to all-electric cars, the demand for charging stations will increase as the economy grows. Compared to the developed countries, the supply of charging stations in developing countries will last for a long period of time.

China cannot represent all the developing country. Many of developing countries have much lower personal income than China. So, these countries cannot faster than other developed countries.

Task 4: The impact of technology

With the development of science and technology, new types of vehicles will have far-reaching effects on the popularity of EVs. We analyzed their impact on EVs through the changes of model parameters.

Sharing-car: reduces the total number of light car by improving the car's utilization. Rational planning of shared cars helps to improve the efficiency of charging stations. It is likely to promote the popularity of EVs.

Self-driving car: is similar with EVs. Electric self-driving is the ideal future vehicle which make our life more convenient and green. The development of self-driving will stimulate consumer interest in cars. Clean energy vehicles will be more consumers' choices.

Rapid battery-swap stations: promote the convenience of EVs and service capacity of charging stations. Specific to the network, it will shorten the transition period by improving the overall growth rate of supply capacity.

Task 5: Write a Handout (after References)

V Sensitivity Analysis and S&W

5.1 Sensitivity Analysis

Coefficients *p* and *q* in Demand Forecasting Model

The calculation of EVs ownership in the Demand Forecasting Model was recalled as Equation (1):

$$\frac{f(t+1)}{1-F(t)} = p \cdot R_{SD} + q \cdot \frac{N(t)}{m}$$

where,

• p is the coefficient of innovation (varies between 0.01 and 0.03). It captures the rate of EVs diffusion which only considers consumers will purchase the new product due to external influences.

• q is the coefficient of imitation (varies between 0. 3and 0.7). It captures the rate of EVs diffusion which only considers word-of-mouth consumers. And their buying decision depends on both the supply of the deployed charging network as well as the customer satisfaction of innovators.

We set p=0.015, and q=0.35 in United states.

In fact, the countries at different levels of development should not have same parameter, and the country in different stage also has changes in parameters. In countries with higher levels of development, the demand of their consumers to buy new products spontaneously is larger. So that p value should be much bigger. As for imitation coefficient, it relates more with the product features. Though developed countries have greater q value for their strong purchasing power, the change is not significant.

We test different combinations of p and q. The results are shown in **Fig. 9**. We verified the analysis described above. Countries with larger p-values have a faster promotion of charging stations. Further, we see only a tiny change when the q value varies with fixed p.

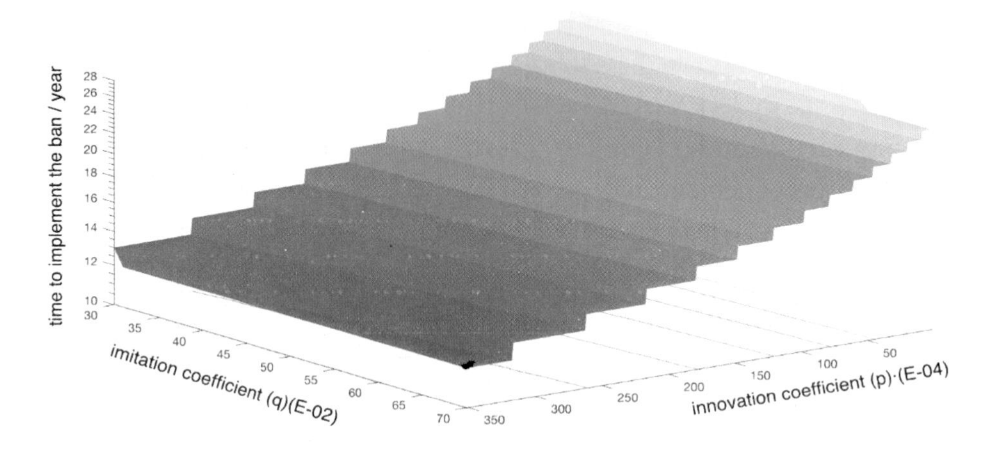

Fig. 9 Sensitivity of demand forecasting related with parameters of p and q

Optimization Model

In the optimal distribution model, we assumed that growth rate μ is constant. In fact, with the development of technology, the ability of different countries or enterprises to provide charging stations is also growing, such as the improvement of power stations, the emergence of automatic vehicles. Of course, economics and politics are also important factors affecting the increase of charging stations.

In general, μ should wavelike rise. But as shown in **Fig. 5**, it does not have big impact on the time to complete the transition to charging vehicles.

Error of commercial vehicles

We only focus on personal passenger vehicles in our model. However, there must be commercial vehicles in the future. Adding a large car to the model will have impact on our charging station plan.

In the Demand Forecasting model, we make the average annual mileage is 15,000km. average annual mileage will increase after adding commercial vehicles and charging demand will increase. Finally, unchanged supply will lead to the evolution of the charging station slows down.

5.2 Strengths and Weaknesses

Advantages

- **Extensibility & portability:** For instance, cities in the same state could be used as nodes and we can use this model to explore the charging station distribution plan in a state. Furthermore, we can also set the country's optimal distribution network by using different states as nodes.
- **Enormous flexibility:** Our demand forecasting model improved the Bass diffusion model. It includes some important factors into a single, robust framework, and can be applied to different countries.
- **Strong reliability:** We consider not only the objective factors, wealth distribution and population density, but the subjective factors, such as convenience and imitation of consumer behavior were also added to the model, making it more realistic.

Weaknesses

• Based on a large amount of data, our model predicts the distribution of charging needs. However, it is difficult for us to find accurate data for rural areas. As a result, our model cannot get the exact location of the charging station.

• The error keeps growing over time in prediction model and it may cause mistakes in the charging station plan.

VI Conclusion

We provided a detailed analysis of charging network distribution plan to achieve full adoption of all-EVs. We started with analysis of optimal distribution without transition time, then discussed the specific optimization process in multi-stage evolution. The model takes geographies, population and wealth distributions into consideration and create a classification system. Four sub-models, Demand Forecast model, Supply Capacity Model, Supply-Demand Network Model, and Optimal Distribution Model, comprise the integrated model. We conclude our exploration as follows,

• We proposed the optimal solution for charging station allocation. With and without considering the transition time respectively. We explored the key factors that affect the distribution and development of charging stations, and estimated the timeline of the full evolution to EV.

• We analyzed each stage of the EV's development of South Korea and the United States. It will cost approximately 20 years for the full evolution to EV in the case of optimal distribution. we suggest that the charging stations should be allocated in **city-based** regions in the early stages of the development of EVs. In the mid-term, there tends to be **a mixture of suburbs and cities**.

• We have established a classification system based on economic development of different countries. In the study of China's charging station distribute plan, we found that China can complete the conversion of an all-electric vehicle in just 15 years. However, with the development of economic after 15th year, China government would also need to add more charging stations every year to supply the new generation of EVs.

HANDOUT

distribution plan of charging station

Dear President,

Energy is the material basis for human survival and economic development. However, with the sustained and rapid development of the world economy, energy shortage, environmental pollution and ecological deterioration have gradually deepened, and the contradiction between energy supply and demands has become increasingly prominent. At current energy consumption estimates, energy sources such as fossil fuels can last less than a century. As a representative of new energy sources, there is a great chance that electricity will become a substitute for fossil fuels. The energy revolution will have a tremendous impact on the transportation industry, and the car will be the first be affected. In the past 10 years, the electric car industry is developing rapidly. More and more consumers are also beginning to buy electric cars. Electric vehicles are very likely to become substitutes for gasoline cars. Therefore, whether for economic or environmental reasons, a reasonable impetus to the transformation of electric vehicles is very necessary.

At present, our team is studying the optimal allocation of charging stations for electric vehicles, the purpose of our research is to develop a reasonable distribution plan to meet the needs in different periods of the growing demand for charging cars. We have developed a model to optimize the multi-stage network distribution of charging stations, and have found the key factors affecting the distribution plan for different countries.

We simulated and forecasted the development of electric vehicles in three representative countries- the United States, South Korea, and China. By analyzing the distribution of people's needs at different stages, a reasonable charging station allocation strategy was formulated. Based on the analysis, we have got several interesting conclusions that we hope will help you in formulating the policy:

- If you want to realize the transformation of electric vehicles as soon as possible, on the basis that the annual distribution plan can meet the demand growth of in the year (that is, the optimal distribution), then your country will need 10 to 20 years to make a transition, whether your country is developed countries or developing countries.
- The optimal distribution method can be roughly summarized as follows: If your country has not yet been assigned a charging station or charging station coverage is less than 30%, then your country is in the early stages of the development of electric vehicles, so, we are more inclined to suggest that the cities should be seen as the center distribution stations. As the city has a much large population density and relatively high economic capacity, the demand growth is faster than the suburbs and the countryside. When the charging station coverage is between 30% to 60%, indicating that your country has developed to intermediate stage, this time, you can consider the suburbs mixed model to set charging stations. When the charging station coverage capacity is about 80%, according to the needs of citizens, you can enact the ban on petrol cars to achieve the transformation of electric vehicles.
- If your country is a developed country, then I suggest that you provide more power stations each year than the amount of demand growth. The higher the supply, the faster the transformation speed is. It is worth emphasizing that, in the middle of the transition, demand will grow much faster than other periods, which means that you need to provide more charging stations each year to meet the demand growth.
- If your country is in development. The number of power stations that can be supplied each year may be constrained by the national economy. We recommend that you try your best to ensure that the number of power stations supplied each year is higher than the demand increase in the current year without affecting the country's development. If you can guarantee this every year, your country may transition faster than developed countries.

Expect the era of green energy to come earlier!

Sincerely yours,

Team#76639

Appendix

Personal consumption capacity and population by state/ city/ province In United States, Korea, and China									
PCE by Bureau of Economic Analysis (millions of dollars) Population by the U.S. Census Bureau the United States				*Personal Income: 2015 data at current price Korea			China		Population(ten thousand)
Fips	Area	PCE 2016	Population (2017)	Area	Personal Income	Population	Area	per capita disposable income	Population
00000	United States	12816386	325719178	Seoul	196.9	10349312	Hebei	20543.4	7470
01000	Alabama	152397	4874747	Busan	58.5	3678555	Shanxi	20411.7	3682
02000	Alaska	36758	739795	Incheon	46.9	2628000	Liaoning	23222.7	4378
04000	Arizona	239680	7016270	Daegu	40.8	2566540	Jilin	20208	2733
05000	Arkansas	92984	3004279	Daejeon	26	1475221	Jiangsu	29677	7999
06000	California	1641724	39536653	Gwangju	24.1	1416938	Zhejiang	34550.3	5590
08000	Colorado	236296	5607154	Suwon	22.0008436	1242724	Anhui	21024.2	6196
09000	Connecticut	173447	3588184	Goyang-si	18.7095366	1073069	Fujian	28055.2	3874
10000	Delaware	40813	961939	Seongnam-si	17.911537	1031935	Jiangxi	19860.4	4592
11000	District of Columbia	38720	693972	Ulsan	16.571579	962865	Shandong	25755.2	9947
12000	Florida	796537	20984400	Bucheon-si	14.3961794	850731	Henan	20442.6	9532
13000	Georgia	357332	10429379	Jeonju	11.6936236	711424	Hubei	20839.6	5885
15000	Hawaii	64460	1427538	Ansan-si	10.5161212	650728	Hunan	21318.8	6822
16000	Idaho	56642	1716943	Cheongju-si	10.2031604	634596	Guangdong	30226.7	10999
17000	Illinois	534750	12802023	Anyang-si	10.1987178	634367	Hainan	20917.7	917
18000	Indiana	236353	6666818	Changwon	8.561998	550000	Sichuan	20307	8262
19000	Iowa	115999	3145711	Pohang	7.591998	500000	Guizhou	18700.5	3555
20000	Kansas	104615	2913123	Uijeongbu-si	7.1873334	479141	Yunnan	21074.5	4771
21000	Kentucky	147697	4454189	Hwaseong-si	7.1321598	476297	Shaanxi	20733.9	3813
22000	Louisiana	160025	4684333				Gansu	17156.9	2610
23000	Maine	57281	1335907				Qinghai	17566.3	593
24000	Maryland	269223	6052177				Heilongjiang	17759.8	3799
25000	Massachusett	354085	6859819				Beijing	21644.9	1362.86

	s		
26000	Michigan	390497	9962311
27000	Minnesota	242489	5576606
28000	Mississippi	90261	2984100
29000	Missouri	228648	6113532
30000	Montana	42956	1050493
31000	Nebraska	74320	1920076
32000	Nevada	106364	2998039
33000	New Hampshire	65151	1342795
34000	New Jersey	438030	9005644
35000	New Mexico	74290	2088070
36000	New York	926168	19849399
37000	North Carolina	342753	10273419
38000	North Dakota	36552	755393
39000	Ohio	434951	11658609
40000	Oklahoma	129390	3930864
41000	Oregon	162684	4142776
42000	Pennsylvania	522697	12805537
44000	Rhode Island	44850	1059639
45000	South Carolina	165036	5024369
46000	South Dakota	34164	869666
47000	Tennessee	227596	6715984
48000	Texas	1018027	28304596
49000	Utah	107142	3101833
50000	Vermont	29761	623657
51000	Virginia	350198	8470020
53000	Washington	312817	7405743
54000	West Virginia	63022	1815857
55000	Wisconsin	224143	5795483
56000	Wyoming	23611	579315

Tianjin	7055.4	1044.4
Shanghai	19506.7	1450
Chongqing	8631.6	3392.11

Data source:

https://www.census.gov/

http://kostat.go.kr/portal/eng/pressReleases/1/index.board?bmode=read&aSeq=360213

http://worldpopulationreview.com/

www.stats.gov.cn/

http://data.stats.gov.cn/easyquery.htm?cn=E0103

References

Bass, Frank M., Trichy V. Krishnan, Dipak C. Jain. "Why the Bass Model Fits without Decision Variables". Marketing Science, 13.3(1994):203-223.

_____. Comments on "A New Product Growth for Model Consumer Durables The Bass Model". Marketing Science, 50.12 S.(2004):1833-18480.

Becker, Thomas A. ,Ikhlaq Sidhu (PI), BurghardtTenderich. *Electric Vehicles in the United States: A New Model with Forecasts to 2030.* Center for Entrepreneurship & Technology (CET). University of California, Berkeley, 2009.

Ceylan, H., Baskan O., Ozan C., Gulhan G. "Determining On-Street Parking Places in Urban Road Networks Using Meta-Heuristic Harmony Search Algorithm". In: de Sousa J., Rossi R. (eds) *Computer-based Modelling and Optimization in Transportation.* Advances in Intelligent Systems and Computing, Springer, Cham, 262(2014).

Hussein, E., Hafez AI, Hassanien AE, Fahmy AA. "Nature inspired algorithms for solving the community detection problem". LOGIC JOURNAL OF THE IGPL, 25. 6 (2017): 902-914.

IEA.GlobalEVOutlook2017. http://www.iea.org/publications/freepublications/publication/GlobalEVOutlook2017.pdf

Joana, Cavadas., Goncalo Correia., Joao Gouveia. "Electric vehicles charging network planning". Computer-based modelling and optimization in transportation. Advances in Intelligent Systems and Computing, 262(2014): 85－100

Li, Min, et al. "Demand Forecasting-Based Layout Planning of Electric Vehicle Charging Station Locations." 4(2016). Team #76639 Page 21 of 21

Sung, HoonChung, Changhyun Kwon. "Multi-period planning for electric car charging station locations: A case of Korean Expressways".European Journal of Operational Research, 242.2(2015): 677-687.

2.The virtual escape of the Louvre

2019
MCM/ICM
Summary Sheet

Abstract

Life is like a box of chocolates, you never know what the next piece is. Currently, terrorist attacks have occurred frequently near the Louvre in France. Our team discussed this task and researched into a plan of developing the current Louvre evacuation.

First, we divide the influencing factors affecting the evacuation process of tourists into two aspects: internal and external factors. Internal factors refer to individual characteristics of tourists; external factors refer to objective environmental characteristics. Our evacuation process for tourists is divided into three stages: event occurrence, command coordination and orderly evacuation. Then we summarized the characteristics of the three stages.

Then, we start from these two factors to build a model. We first establish a single-layer tourist evacuation model to determine the speed of tourists' escape By introducing the principle of Cellular Automata[?], we establish the rules of visitor movement and conflict avoidance rules to determine the best escape distance. From this, we can determine the shortest evacuation time for single-story visitors. Then, due to the distinction of each layer's layout and the choice between stairs and elevators, we put forward the coefficient of passing difficulty and the probability of selection of the way to quantify the impact of the distinction of each layer's layout on the time of escaping and transferring between the floors. Thereby, we establish a multi-layer tourist evacuation model.

Finally, we solve the model by opening only one entrance, adding three exits, hijacking scenarios and arson scenarios; simulate the evacuation process and get the best evacuation time under different conditions.. The time spent by visitors in all four cases is $736.5s$, $522.03s$, $444.78s$ and $563.52s$ respectively. The full opening of the entrance can shorten the evacuation time by 29.12%, which indicates that when a terrorist attack occurs, the export should be increased as much as possible to improve the efficiency of escape. The hijacking scenario saved $77.25s$ compared to the normal opening of the four exits, indicating that the current internal layout of the Louvre is unreasonable and needs further optimization. When we conduct sensitivity analysis, we find that the information transmission rate is a positive sensitivity index that affects the evacuation time. We summarize the advantages and disadvantages of the model, and put forward suggestions for improving the Louvre escape evacuation plan.

作　　者：王一名，李辉，樊帆

指导教师：罗虎明

奖　　项：2019年美国大学生数学建模竞赛F奖

1 Introduction

In recent years, terrorist attacks have occurred frequently, which makes people longing for a peace life. The conflicts of religious beliefs, the intensification of social conflicts and the intensification of the refugee crisis lead to the fact that Europe has been a major disaster area for terrorist attacks once upon a time. Correspondingly, the anti-terrorist struggle in Europe and even countries around the world in response to the terrorist threat will also be a "constant war" of "normalization". On February 3, 2017, a knife attack took place near the Louvre museum in Paris, France. A man armed with a knife tried to force his way into the Louvre museum in the area. Subsequently, the number of visitors to the Louvre plummeted, causing serious damage to the museum's economy.

The Louvre is one of the world's largest art museums both in scale and the flow of visitors, with an area of about 72,735 square meters and a wing length of 480 meters. It received more than 8.1 million visitors in 2017.once a terrorist attack occurs, how to safely evacuate the tourists inside the Louvre will become a very challenging problem. Our job is to explore and design an optimal evacuation plan for the Louvre in the event of an emergency,which needs to ensure that the evacuation process is both efficient and safe.

There are two main research perspectives on the evacuation of personnel under major accidents. On the micro perspective, we mainly consider the micro characteristics of pedestrians, emphasize the influence of psychological factors and external environment on the movement behavior, and thus can simulate the individual behavior of evacuation personnel. But this Angle is more computationally intensive, especially in complex cases, the computational efficiency is lower. The other is to discretize the evacuation area and simulate the situation of personnel evacuation according to certain evacuation rules. The calculation is simple and the operation speed is fast, but it is difficult to capture individual behavior. The typical model is Cellular Automata.The Cellular Automata model has been widely used in evacuation simulation thanks to its simple, flexible and efficient rules. Therefore, we decided to establish a model combining the advantages of two methods, taking human characteristics and external environmental factors into consideration, and conduct simulation and solution through Cellular Automata. According to different types of potential threats, a variety of possible scenarios are set, and certain modifications are made to the model in specific scenarios, so as to achieve a high degree of security while ensuring the efficiency of evacuation.

Therefore, we decide to combine the advantages of the two methods, consider the characteristics of human beings and external environmental factors, establish a model, and simulate and solve it by cellular automata. With regard to the potential threats that may arise in the evacuation process, we mainly consider arson with a high probability of occurrence and terrorists carrying guns, and modify the model in these two scenarios, respectively. In order to ensure the high efficiency of escape evacuation, a high degree of security is achieved.

2 General Assumption

- Incident caused by human factors —— the terrorist attacks, not including natural disasters, such as earthquakes, fires, and tornadoes.
- During the evacuation of tourists, the internal structure of the building will not suffer great damage.
- Each exit can only pass 5 tourists at a time. Due to the lack of information on secret exits, we do not consider such exits in the plan.
- Tourists are randomly distributed on every floor, and in the evacuation, tourists always aim to find the nearest exit.
- We do not consider the difference in tourists' response to emergencies.
- The data used in the model are authentic and will not cause great error to the results.
- Considering the potential threats in the evacuation process of tourists, we hypothesized the corresponding scenarios according to the threats in different situations.

3 Notations

As shown in Table 1.

4 The Model

4.1 Problem analysis

Before designing the model, we first analyze the basic situation of the Louvre Museum, the evacuation process of tourists and the composition of tourists during the emergency.

- **Louvre's exit setting**

  The Louvre has four main entrances, of which the glass pyramid entrance is the museum's main and most common public entrance, located on the second floor of the museum's underground. Three other entrances usually reserved for groups and individuals with museum memberships: the Passage Richelieu entrance, the Carrousel du Louvre entrance, and the Portes Des Lions entrance.[1] These three entrances are located on the ground floor. There are other entrances and passageways at the Louvre that

Table 1: Conforms to the meaning explanation

Abbreviation	Conception	Descripition
N	Total number of visitors	The total number of visitors stranded in the museum during emergencies
S	Building size	The open interior area on a layer
N_t	The number of stranded tourists	The number of visitors who did not leave the museum at time t
ρ_t	Visitor density	Number of tourists per unit area at time t
V_t	Flow velocity	Velocity of evacuation at time t
$D(i,j)$	Escape distance	Distance between A and the same floor's exit
T_1	The time taken in the first phase of evacuation process	time when the emergency personnel are fully in place
T_2	The time spent in the second/third phase of evacuation process	the time taken in the evacuation of all tourists with the help of the emergency personnel
φ	Information dissemination rate	ratio of the number of tourist using "Affluences" to the total number of tourist
$D(i,j)$	Escape distance of tourists	the straight distance between tourist coordinates and exit coordinates
Θ	Coefficient of passing difficulty	measure the difficulty of passing through the floor

are also known only to museum security staff, and the information about these entrances is not public.

The interior of the Louvre is intricate. Therefore, in the process of personnel evacuation, emergency personnel need to coordinate and direct tourists to quickly evacuate the scene, and emergency personnel play an extremely important role in this process.[?] The Louvre's elevators and staircases are randomly located, with partial staircases often connecting only two floors. Even on the same floor, visitors still need to enter different exhibition halls through various small stairs.

- **Analysis of tourists' evacuation process**

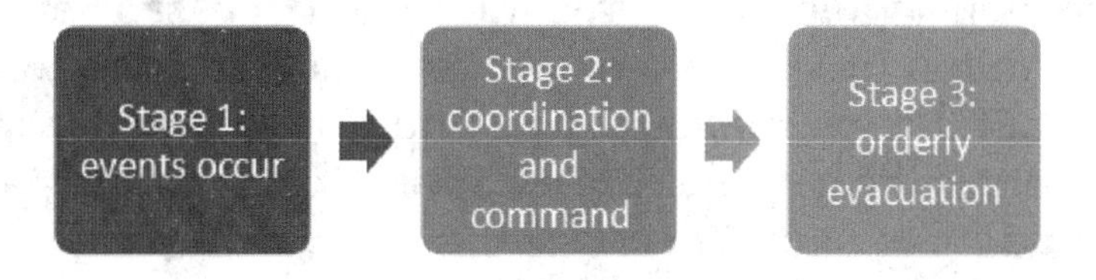

Figure 1: Flow chart of tourist evacuation

When an emergency occurred. Most visitors in the museum wanted to leave the hall, which caused the scene to be confused. At the same time, emergency personnel quickly enter each floor. Therefore, the time required for the event to occur is consistent with the time required for all emergency personnel to be in place.[2] Then, the emergency personnel command and coordinate the tourists to evacuate, but it may take a certain period of time to change the evacuation status from chaos to stability because of the diversity of tourists, language communication barriers and growth background.[3] The process of this transformation is defined as coordinated command's period.[4] Under the guidance of emergency personnel, the emergency personnel's instructions are quickly evacuated from the building and arrived at the safe area.[**?**]

- **The diversity of visitors**

  Visitors are made up of people from different countries, which leads to language barriers and customs in different countries. Secondly, visitors can personally visit the self-help, or they can form a group of no more than 25 people to visit. There are also some disabled people and pregnant women in the tourists. There are also elevators and special stairs for the disabled to provide convenience for visiting various pavilions.

- **Factors affecting the evacuation process of tourists**

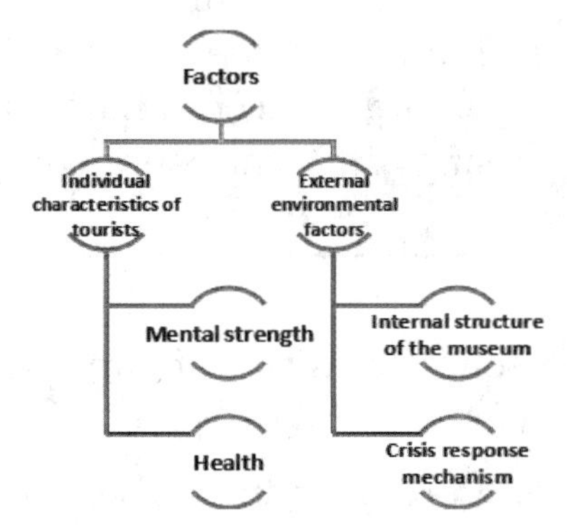

Figure 2: Factors affecting the evacuation process of tourists

The factors affecting the evacuation of tourists can be divided into Personal feature of tourists and external environmental factors. Personal feature of tourists refer to their own psychological factors and physical factors. Psychological factors refer to the psychological characteristics and bearing capacity of tourists in the face of emergencies.Physical factors refer to their own health conditions, which will affect the speed of escape and the tolerance of obstacles in the process of escape. External environmental factors refer to the fact that the internal structure and crisis response mechanism of the museum determine whether the escape process of tourists is smooth and smooth, including the number and setting of exits, stairs and elevators, as well as the number and overall comprehensive quality of emergency personnel.

4.2 Single-floor tourist evacuation model

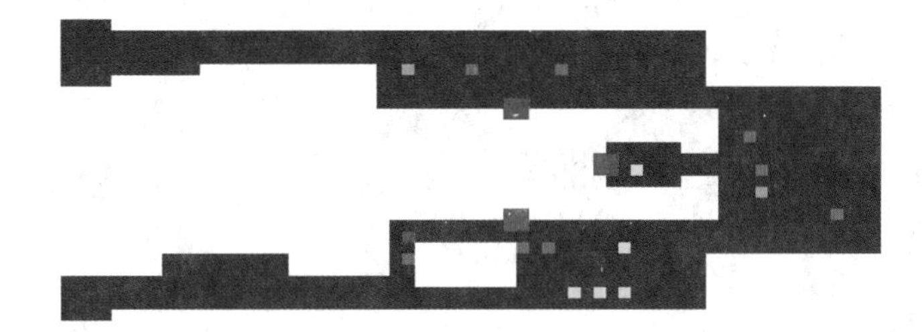

Figure 3: Structure of the Louvre's underground floor

Note: The red square indicates the stairs, the green square indicates the disabled elevator, the blue square indicates the up elevator, and the yellow square indicates the down and two-way elevator.

We regard the Louvre's underground floor as the research object, starting from the most basic model. As shown in figure 3, the underground floor has three exits that are marked by red dots in the figure3. Stairs are divided into one-way stair and two-way stair (We only consider the stairs connected different floors, not the stairs connected different exhibition halls). The number of them is m_1 and m_2 respectively[4], and the number of elevators is n. When the evacuation process reaches the first stage, the total number of tourists is N. Emergency personnel use staircases and elevators to enter the museum. The average time for each floor by stair is T_u, and elevator for T_d. When the elevator reaches the designated floor, the automatic closing time is T_w,and the number of emergency personnel is N_p, so emergency personnel in place of all time for T1. And there are

$$T_1 \in [0, 4T_u + 4T_d + 3T_d] \tag{1}$$

We can't give the exact formula for T_1, but we can determine its range of values. There are two extreme states in which emergency personnel arrive at each floor. Among them, the ideal state is that the emergency personnel are distributed on each floor, and the arrival time can reach 0. In the worst state, the emergency personnel are all located on the second or top floor, then the time required for all personnel to reach each floor is $4T_u+4T_d+3T_d$. Therefore, the total time for emergency personnel to be in place varies between these two values.

After all the emergency personnel were in place, the personnel evacuated into the second stage, and with the help of emergency personnel, the tourists evacuated in an orderly manner. The second and third phases of evacuation cannot be accurately divided in time. Therefore, we combine the time of the two phases with T_2. The process of solving T_2 needs to start from two aspects of speed and distance. The specific analysis is as follows:

- **Escape speed**

  In the study of Predrech-enskii, it is mentioned that the movement speed V_t of tourist evacuation has a logarithmic relationship with the density of tourists ρ_t, and its calculation formula is as follows:

$$\rho_t = \frac{N_t}{S} \tag{2}$$

$$V_t = \alpha A + \beta B + \gamma \tag{3}$$

$$A = 1.32 - 0.82 ln \rho_t$$

$$B = 3 - 0.76 \rho_t$$

Among them, N_t is the number of tourists stranded at time t, and S is the floor space of the building. There are a number of certain values for the coefficients in the subsequent studies by Ando, Hankin et al, but in this paper, we still use the results of Predrech-enskii, that is $\alpha = 0.24, \beta = 0.02, \gamma = 0.26$.

At the same time, the individual characteristics of tourists will have a certain impact on the speed of their progress. In terms of psychological factors, there is an asymmetry between the information of tourists, staff and terrorists, which will aggravate the negative emotions of tourists who are excessively worried about their safety in the event of terrorist attacks.[?] In the absence of information, such negative emotions are manifested as tourists aimlessly looking for an exit and eventually getting lost. On the contrary, when tourists have enough escape information, negative emotions caused by worry will increase their escape speed. In addition, the physical condition of tourists will also affect the speed of escape. We classify tourists into the disabled and the normal.

At the same time, the individual characteristics of visitors will have a certain impact on the speed . In terms of psychological factors, because of the asymmetry between information among tourists, staff and terrorists, this will aggravate the negative sentiment of tourists who are too worried about their own safety in the event of a terrorist attack. In the absence of information, this negative sentiment is manifested in the fact that tourists are aimlessly looking for exports and eventually lost their way. Instead, when tourists have enough information on escape, fears that negative emotions will increase the speed of escape. In addition, the physical condition of the visitors will also affect the speed of escape. We classify tourists as disabled and normal. Visitors' health status will affect their escape speed to varying degrees. The better the physical condition, the faster the speed of the student's escape process.Therefore, we define the health impact factor of visitors as ε. The greater the value of ε, the better the physical fitness of the visitors. In view of the impact factors proposed by WANG Ping (2005) that affect the personal physical characteristics of the high-rise building escape process, we will also use the same method for the value of ε according to the distribution of normal adults, the elderly or children, and the disabled. Considering the personal characteristics of tourists, namely psychological factors and health conditions, in the model, we correct the movement speed during the evacuation process of tourists to V_{ti}:

$$V_{ti} = (1 + P_i)v_{ti} + \varepsilon_i \tag{4}$$

$$P_i = \begin{cases} 0, & \text{If the tourist remains calm} \\ 0.1, & \text{If the tourist is nervous} \\ 0.2, & \text{If the tourist is too nervous} \end{cases}$$

$$\varepsilon_i = \begin{cases} 0.24, & \text{If the tourist is a healthy adult} \\ 0.2, & \text{If the tourist is an old man or a child} \\ 0.15, & \text{If the tourist is a disabled person} \end{cases}$$

Among them, P_i is the intensity of the psychological state of the tourists, and its value is positively related to the degree of tension of the tourists. Here we simplify the P_i to a certain extent, and take the values according to the three mental states of calmness, tension and over-stress.

We mentioned in the above analysis that information asymmetry will affect the speed of tourists' escape. When tourists have a lack of escape information, their fears will lead to a lack of rational thinking. They blindly follow a large number of people moving, often deviating from the correct route. This situation indicates that the lack of escape information will increase the interval of the first phase of the evacuation process to a certain extent. Conversely, when visitors have sufficient escape information, their anxiety will also increase their speed. But at this time, they have a clearer awareness and a greater chance of finding the right escape route. Therefore, in solving the problem of information asymmetry, the technological factors are even more critical.

Visitors come from different countries. Before the emergency personnel are in place, it is not good to rely on public broadcasting to guide tourists to escape correctly. Because of the existence of language barriers, it is impossible for all visitors to absorb the information expressed by public broadcasts. A software application called "Affluences" from the Louvre provides real-time updates of the estimated waiting time for each portal to help visitors enter the museum. Then, in the event of a terrorist attack, we can also publish escape routes and evacuation tips on this platform, so that visitors can get more information. At the same time, providing a multi-language version of the text description can largely overcome language communication barriers. The more visitors use this App, the more helpful it is to improve the escape speed of visitors. Therefore, we define the ratio of the number of people using "Affluences" to the total number of visitors as the information transmission rate φ, then its formula is:

$$\varphi = \frac{\text{The number of this APP's users}}{N}$$

$$V_{ti}' = (1 + P_i + \varphi)v_{ti} + \varepsilon_i \tag{5}$$

In the formula, V_{ti}' is the moving speed of the evacuation process of tourists after considering the information technology factor into the model.

- **Escape distance**

When a terrorist attack occurs, the tourist takes the route closest to the exit as the best escape route. Here we define two concepts. The escape distance refers to the total length of the escape route chosen by the visitor. The average escape distance is the average of the escape distances of all visitors from the initial location to the exit. Assuming that the initial position of the visitor i is at point $A(i, j)$ and the position at the exit is at point $B_m(x_m, y_m)$, the escape distance of the visitor can be expressed as:

$$D\left(i,\ \ j\right)=\sqrt{(i-x_m)^2+(j-y_m)^2} \tag{6}$$

But we know that the error caused by this kind of processing is very large. There are regular and irregularly shaped obstacles inside the building, which makes it impossible for visitors to reach the exit in a straight line when they escape. The escape distance calculated in this way is much smaller than the actual value. Therefore, we further optimize the model according to the principle of Cellular Automata to improve the accuracy of the model.

First, let's introduce the principle of the Cellular Automaton[**?**]. Cellular Automaton is a model that simulates the evacuation process of a person. It consists of a cell space and a transformation function defined in the function. which is

$$A=\left(w,\ \ S,\ \ n,\ \ F\right)$$

This formula represents a Cellular Automaton system. w represents the dimension of the system; S is the finite discrete set of states of the cell; n represents the combination of cells in the spatial neighbourhood, contains the space vector of each different cell state. n is equal to $(S_1, S_2, ..., S_m)$. m is the number of cells in the neighbourhood; F is the variation rule, and S_m is a local conversion function mapped to S. The cellular automaton has two models, Von Neumann and Moore. We use the latter. As shown in Figure 4, each cell can be moved to an unoccupied cell in eight adjacent directions.

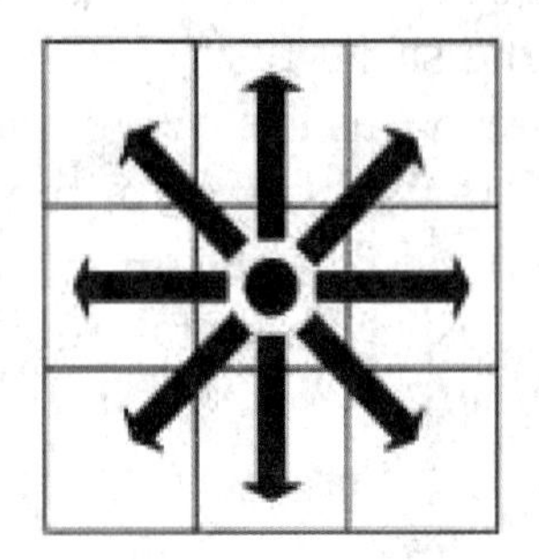

Figure 4: Cell movement direction

(i-1, j+1)	(i, j+1)	(i+1, j+1)
(i-1, j)	(i, j)	(i+1, j)
(i-1, j-1)	(i, j-1)	(i+1, j-1)

Figure 5: Cell location map

For a single floor, we set the layer space in a two-dimensional space, and divide it evenly according to the matrix form. Each grid is a cell, and all cells together form a cell space. Considering that the Louvre covers a large area, we allow each cell space to accommodate only 25 people (forming a $5*5$ square matrix), and divide 25 tourists into a tourist group. Divide groups and individuals to make the model more streamlined. There are three states in the cell: one is that the cell is empty, the other is that the cell is occupied by the tourist group, and the third is that the building occupies the cell, we can use a two-dimensional array to represent:

$$H=(i,\ \ j)=\begin{cases}1, & \text{If the cell is occupied}\\ 0, & \text{If the cell is idle}\\ 99, & \text{If obstacles occupy the cell}\end{cases}$$

It is assumed that the vertical projection of each visitor can be simplified to a square of $0.4m*0.4m$, so that the length and width of a cell are $2m$. The xOy coordinate plane is established with the left-most lower point of the two-dimensional space plane as the origin, and the grid is arranged in an orderly manner in the direction of rows and columns, so the central coordinate of any one cell (i,j) can be determined as $(2i-1,2j-1)$. If there are multiple exits and the position coordinate of exit I is (x_i,y_i), then the nearest distance between the cell and the exit can be expressed as:

$$minD\left(i,\ \ j\right)=min\left\{\sqrt{(2i-1-x_i)^2+(2j-1-y_i)2}\right\} \tag{7}$$

The nearest distance from the exit of the cell in the eight directions that the cell (i,j) is about to move is:

$$minD\left(i-1,\ \ j-1\right)=min\left\{\sqrt{(2i-3-x_i)^2+(2j-3-y_i)2}\right\}$$

$$minD\left(i,\ \ j-1\right)=min\left\{\sqrt{(2i-1-x_i)^2+(2j-3-y_i)2}\right\}$$

$$minD\left(i+1,\ \ j-1\right)=min\left\{\sqrt{(2i+1-x_i)^2+(2j-3-y_i)2}\right\}$$

$$minD\left(i-1,\ \ j\right)=min\left\{\sqrt{(2i-3-x_i)^2+(2j-1-y_i)2}\right\}$$

$$minD\left(i+1,\ \ j\right)=min\left\{\sqrt{(2i+1-x_i)^2+(2j-1-y_i)2}\right\}$$

$$minD\left(i-1,\ \ j+1\right)=min\left\{\sqrt{(2i-3-x_i)^2+(2j+1-y_i)2}\right\}$$

$$minD\left(i,\ \ j+1\right)=min\left\{\sqrt{(2i-1-x_i)^2+(2j+1-y_i)2}\right\}$$

$$minD\left(i+1,\quad j+1\right)=min\left\{\sqrt{(2i+1-x_i)^2+(2j+1-y_i)2}\right\}$$

The specific process of using cellular automata:

Step 1: Initializing the space array and setting the two-dimensional cell state value, the visitor group is randomly distributed in the two-dimensional cell space;

Step 2: After the terrorist attack occurs, calculate the number of tourist groups and convert them into the current number of stranded people;

Step 3: The tourist group moves and judges the direction that the cell can move by comparing the two-dimensional state value $H(i, j)$ of adjacent cells;

Step 4: Calculate and compare the shortest distance of each moving direction of the cell, select the direction corresponding to the minimum value and move one step along the direction, until all cells with the state value of "1" find their unique target grid;

Step 5: Calculate the number of people currently detained in the space and make a judgement. If the current number of stranded is not equal to 0, repeat Step 2 and Step 3; If the current number of stranded people is 0, then stop the program and get the number of steps K of the group of tourists finally moving through the exit.

However, we also need to take into account the conflicts that may occur in the escape process of the visitor group. The tourists group mainly consists of two types of conflicts: (1) The group of tourists faces the choice of the next moving direction with the shortest distances; the two groups of tourists compete for the same moving position at the same time. In this regard, we have separately formulated the following conflict avoidance rules.

- **Grid attraction value**

  To avoid the first type of conflict, we need to adopt a new rule: calculate the attractive value of each grid when Step1 is performed. When the first type of conflict occurs, we then compare the size of the attractive value, select the direction with the least attractive value for the next move, and if the attractive values are equal, randomly select a direction to move.

  We define the attractiveness value as the direction value that is derived from the exit and reaches each grid. We specify a direction value of 1 at the entrance. As shown in Fig6, since the moving direction in the four directions is relatively shorter than the four directions of the skew, we move the visitor group from the entrance to the forward four directions, and the direction value is cumulatively increased by one. After moving one step from the entrance to the four directions of deflection, the direction value is cumulatively increased by two. If there are multiple exits, the direction value determined by the exit point of each grid with the smallest straight line distance from each exit is determined.

Att represents the attractive value of a grid, and the center point of the exit with the shortest distance from the grid (i, j) is marked as (x_0, y_0). Then

$$Att\left(i,\quad j\right) = 1 + \left|\frac{x_0 + 1}{2} - i\right| + \left|\frac{y_0 + 1}{2} - j\right| \tag{8}$$

$$min \quad Att\left(i,\quad j\right)$$

- **Competitiveness of the tourist group**

If multiple visitor groups select the same grid as the moving target at the same time, then we need to develop a rule to resolve this type of conflict. In fact, tourists who are in good health have a higher escape speed. These tourists will flee first, and will not create obstacles for the tourists. Therefore, we define competitiveness π as an indicator of the overall health of visitors within the tourist group.**[?]** Then the competitiveness of the visitor group of the cell (i, j) is:

$$\pi_{ij} = \frac{1}{25}\sum_{i=1}^{25} \varepsilon_i \tag{9}$$

$$max \quad \pi_{ij}$$

In case of the second type of conflict, the principle of avoiding is that the most competitive group of tourists has priority, while other groups choose the second-best direction.

The cell moves in the positive four directions with a step length of $2m$, and the cell moves in the skewed four directions with a step length of $2\sqrt{2}m$. According to the principle of Cellular Automata, we can get the distance of the last group of tourists passing through the exit under the shortest path:

$$L = 2K_1 + 2\sqrt{2}K_2 \tag{10}$$

$$K = K_1 + K_2$$

Where, K_1 represents the number of steps that the tourist group moves forward; K_2 represents the number of oblique moves of the tourist group, and K is the total number of steps.

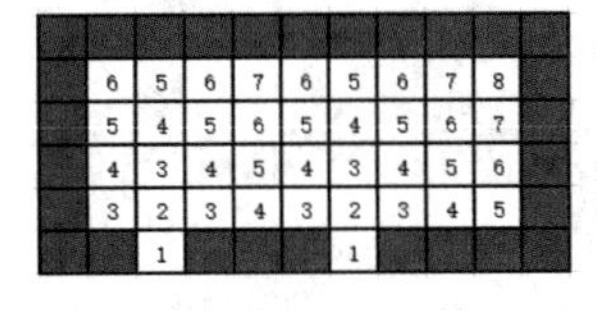

Figure 6: Schematic diagram of attraction value

In conclusion, we can conclude that the total evacuation time T of tourists is:

$$T_2 = \frac{L}{\overline{V}}$$
$$\overline{V} = \frac{\sum V_t}{t}$$
$$V_t = \frac{\sum_{t=1}^{25} V'_{ti}}{25}$$
$$T = T_1 + T_2$$

So we can get that

$$T = T_1 + \frac{t(2K_1 + 2\sqrt{2}K_2)}{\sum\sum_{t=1}^{25}[(1 + P_i + \varphi) * (-0.1968 ln\rho_t - 0.015\rho_t + \gamma_i + 0.3768) + \varepsilon_i]} \tag{11}$$

4.3 Multi-layer tourist evacuation model

From a single-storey building to a multi-storey building, we mainly consider it from two perspectives. On the one hand, during the evacuation process, visitors can use stairs or elevators between the floors. Obviously, the way to take the elevator is faster than walking the stairs, but the disadvantage is that the number and space of the elevator is limited. If a single ride is full, then other tourists will spend more time waiting for the elevator to arrive again. The stairs don't have to consider waiting time. But if the flow of people is large, the speed of movement will be slow or even stagnant. In general, the convenience of taking the elevator is higher than taking the stairs. On the other hand, the layout of the elevators and stairs on each floor and the setting of the exits are also very different. Reasonable layout of elevators and stairs can be easily found by tourists, saving escape time, dispersing people's traffic and speeding up evacuation. However, the wrong layout can not only reduce the escape time, but also cause waste of resources, and even endanger the safety of tourists in the escape process.

When considering the single layer tourist evacuation process, the distribution of different floors is not accurate enough, and the time delay caused by the tourists' choice of export is not taken into account. Therefore, from the single-layer transition to the multi-layer tourist evacuation model, we should start from these two perspectives.

- **Coefficient of difficulty**

Note: The red square indicates the stairs, the green square indicates the disabled elevator, the blue square indicates the up elevator, and the yellow square indicates the down and two-way elevator.

Carefully compare the distribution of elevators and stairs on the underground floor, floor 0, level 1 and floor 2, we found that the stairs and

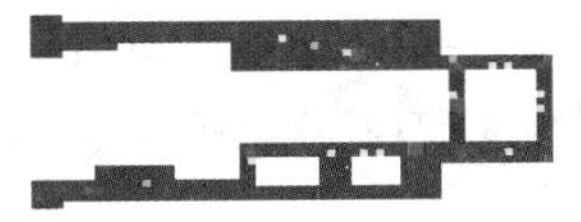

Figure 7: Louvre's 0th floor structure diagram

Figure 8: Louvre first floor structure diagram

Figure 9: Louvre second floor structure diagram

elevators are mainly distributed according to the exhibition hall. On the basement level, the number of elevators is upwards, the exhibition halls are evenly distributed in the south, the north and the east; on the 0th floor, the number of two-way elevators and stairs is 60%, and the exhibition hall is distributed relative to the basement. More dispersed, visitors can move more space; the first floor and the 0th floor are roughly the same, the elevator position and quantity are the same, the number of stairs is the same, but the 0th floor has the upstairs but no down stairs, and the first The case of the layer is the opposite. Therefore, emergency personnel can easily reach the first floor from the 0th floor, visitors can reach the 0th floor from the 1st floor faster; the second floor exhibition hall is concentrated in the north and east, and the number of elevators for the disabled is the largest, but concentrated in In the east, the disabled in the north during the escape will spend too much time on the move due to inconvenience.

Therefore, the difference in the layout of the escape route on each floor will affect the speed at which visitors can escape. The pass difficulty factor Θ_a is the intensity of the safety exit of a certain floor, the number of exits and the convenience of the exit affect the ease of the student's escape process on the floor. The difficulty of the safe exit is smaller, and the location is more concentrated. The greater the value of this coefficient, the longer it takes for the visitor to evacuate from floor a.

$$\Theta_a = \frac{u_a}{f_a} + \frac{\sum_{i=1}^{o} SP_i}{o} \tag{12}$$

Where f_a is the number of down and two-way elevators on floor a (excluding elevators for disabled people), u_i is the number of stairs on floor a (except for small stairs on the same floor), o is the number of all safety exits on this floor, SP_i It is the number of safe exits i and other safe exits not exceeding 50 meters. The ratio of u_a to fa reflects the convenience of the safe passage on the floor, and the mean value of SP_i reflects the channel density of the floor.

Table 2: Statistics of stairs and elevators on each floor

floor		-1	0	1	2
stairs	upside	1	3	0	0
	downside	3	0	5	3
	two-way	0	4	2	0
	total	4	7	7	3
elevator	upside	8	3	1	0
	downside	4	4	4	5
	two-way	1	8	5	0
	total	13	15	10	5

Calculated according to the data in Table 2, the travel difficulty factor from the underground floor to the second floor is $\theta_{-1} = 3.18, \quad \theta_0 = 4.26, \quad \theta_1 = 4.22, \quad \theta_2 = 6.2875$. From this we can see that the underground floor is more convenient, the layout of the 0th floor and the first floor are not much different, the traffic is difficult, and the second floor is too concentrated, so the traffic is not convenient compared with other floors. The results are consistent with the information reflected in the floor plan of each floor, and have certain authenticity. If you perform a single-layer modelling solution for each floor, you can make the results more accurate, but it takes a lot of time. Therefore, we can complete the evacuation time of other floors according to the time required for evacuation according to the single layer of the underground floor, combined with the difficulty coefficient of passage:

$$T_a = \frac{\Theta_a}{\Theta\_1}(T_{-1,1} + T_{-1,2}) \tag{13}$$

$$a = 0, 1, 2$$

In the formula, $T_{-1,1}$ is the time taken for the first stage of the evacuation and evacuation of the underground layer, and $T_{-1,2}$ is the total time used for the second and third stages.

- **Choice of mode of passage**

We divide the safety exit into two types: elevator and stairs. The elevator can be further divided into an ascending elevator, a descending elevator and a two-way elevator. In the process of escape, the type of safety exit will first make a choice according to its preference. The general tourist choice is based on the length of the escape time of different safety exits. Therefore, the number of tourists who choose elevators at the beginning is more. However, when the number of tourists who choose the elevator exceeds a certain threshold, the latecomer considers that the time waiting in line for the elevator is much longer than the time saved by the elevator escape, and the stairs are selected. Obviously, if the number of people choosing the elevator exceeds a certain level, it will have a negative feedback effect on the elevators selected by the tourists.Suppose the possibility that the visitor j chooses the elevator i to escape is φ_{ij}.

$$\varphi_{ij} = P_j - \frac{lnM_i}{100} \tag{14}$$

P_j represents the preference coefficient of the visitor j for selecting the elevator i, and takes a value between 0 and 1; Mi represents the number of visitors currently waiting for the elevator i. If the maximum number of passengers in the elevator is MC, then the time taken by the visitor to transfer between the two floors during the escape process is the greater of the time taken to take the elevator and the time taken by the last group of tourists to pass the stairs.

$$T_{ca} = max\left\{\frac{\varphi_{ij}N_{0a}}{f_a MC}, \frac{2(1-\varphi_{ij})N_{0a}}{u_a \overline{V}}\right\} \tag{15}$$

In the formula, N_{0a} is the number of tourists whose floor a stays at the time of the event, and T_{ca} is the time required for the visitor to descend from floor $a+1$ to floor a.

Therefore, based on the above analysis, we can conclude that if the evacuation of the Louvre, the time required is:

$$T' = \sum_{a=0}^{2}(T_a + T_{ca})$$

$$T' = \sum_{a=0}^{2}(\frac{\Theta_a}{\Theta\_1}(T_{-1,1} + T_{-1,2}) + max\left\{\frac{\varphi_{ij}N_{0a}}{f_a MC}, \frac{2(1-\varphi_{ij})N_{0a}}{u_a \overline{V}}\right\}) \tag{16}$$

4.4 Model solution

Before we solve, we first simplify the model. Because it is difficult to obtain private information such as the psychological status of tourists, we assume that visitors can calmly deal with emergencies in the event of a crisis., that is, Pi=0. Second, we are unable to get the number of user usage for the Affluences app. But users can make online bookings through the "Affluences" app, and we use the number of people who buy e-tickets instead of the value. We assume that the preference coefficient pi of the passengers for the way of elevator access is 0.8, that is, regardless of the number of waiting people, people are more willing to take the elevator. Since the specifications of the Louvre elevator are not known, easy for calculate, we assume that the maximum load capacity of the Louvre elevator is the same as that of the common elevator, which is 13 people. The Louvre has a human flow of more than 8.1 million people in 2017. The Louvre is closed on Tuesdays, with an average daily flow of 262,900 people. Moreover, terrorist attacks generally occur during peak hours of tourist visiting. so we assume that the number of visitors is 24,008 when the terrorist attack. We divide the model solution into four cases. The first case and the second are escape evacuation under normal conditions refers to, and the latter two are escape evacuation after potential threats happen. We mainly use Pathfinder software to solve the model. The software is based on the principle of cellular automata, which can realize the simulation evacuation process. Its advantages are simple to operate. avoiding a lot of complicated and error-prone programming, and the effect is also excellent.

- **Escape evacuation solution under normal conditions**
  Normally, the terrorist attack occurred near the Louvre Museum. In order to ensure the safety of tourists in the museum and to avoid the potential risks of terrorists entering the museum, emergency personnel need to evacuate the visitors in the museum to a safer place in time. We should also pay attention to the opening of the entrance during the escape process. Increasing the number of open entrances can increase the efficiency of evacuation and save time. However, increasing the number of opening entrances has also provided more ways for terrorists to take the opportunity to enter the Louvre, reducing the personal safety of tourists.[1]

a. Open only one main entrance: consider to open the entrance of the Pyramid, the other three entrances are close while visitors evacuating.

b. Add three open entrances: besides the entrance of the Pyramid, consider opening the Passage Richelieu entrances, the Carrousel du Louvre entrances, and the Portes Des Lions entrances at the same time.

Using Pathfinder software, the relevant data parameters are brought into the model, and simulations of both a and b cases are performed. It was found that in the case of opening only one main entrance, it takes $736.5s$ for all visitors to evacuate the museum; and by increasing the opening of the other three auxiliary entrances, the evacuation time of visitors is $522.03s$, which shortens the evacuation time by 29.12%. As shown in Fig. 10, the horizontal axis represents the evacuation time, the unit is second, the vertical axis represents the number of stranded people, red curve represents the scatter plot of the first case, and blue curve represents the scatter plot of the second case.[3] Intuitively, the red curve is far from the origin, while the blue curve is close to the origin, It indicating that after adding the opened exit, the evacuation speed of the tourists increases, and the efficiency of evacuation is greatly improved. At the same time, after increasing the opening exit, the number of stranded tourists is much smaller than the case which only open one exit. As shown in fig10.

We continue to analyze the flow of visitors at the evacuation in the four northern, southern and eastern regions of the four floors, and we further exploring the impact of increased exports on mitigating the flow of people in all areas of each floor (Figures 12–23 in the appendix). By comparing Figure 12-14 and Figure 15-17, we find that the flow of people on the underground floor drops sharply after open the new export, while the flow of people on the 0th floor increases sharply, and the flow of people on the first and second floor has slight fluctuation and there is little change. This is because the newly added three exits are located on the 0th floor. When the visitors are evacuated, the visitors from the 0th to the 2nd floor can choose to leave the hall directly from the 0th floor. The three newly opened exit shared most of the tourists who had to escape through the underground floor.[4] The escape congestion at the underground floor is alleviated, reducing the risk of stampede accidents during evacuation. Therefore, after the terrorist attacks, our proposal is to increase the opening of the other three entrances to reduce the flow of people and increase the speed of escape movement, thus shortening the evacuation time.

- **Escape evacuation solution after potential threats happen**

After the terrorist attacks, we believe that the potential threats to tourists in the process of escape include terrorists holding guns to control exits, and the evacuation process of tourists may cause fires or terrorist will arson and tourists may be trampled to death or hurt on the stairs. These potential threats will reduce the safety of the evacuating process, with arson and hijacking occurring most likely, which is the type of two potential threats we will focus on next. At the same time, applying the results of the above analysis, in the event of a terrorist attack, the emergency personnel will immediately open the four main exit after they arrive at the post.

a. Hijacking situation: The terrorists forcibly broke through the security check entrance on the second floor of the underground, entered the underground floor, and controlled the three elevators along the southmost side, causing the three elevators cannot be used when evacuating; b. Arson situation: During the evacuation process, the terrorists deliberately set fire at the northeast of the first floor, resulting one staircase and five elevators on this route being unusable.

The occurrence of two potential threats has caused some of the safe passages cannot be used when evacuating. In order to ensure the personal safety of tourists, emergency personnel should immediately inform the tourists to avoid the situations after discovering this unexpected situation. We use the Pathfinder software to set the impassable entry as an obstacle, continue to run the program, and simulate both scenarios. In the end, we concluded that in the event of a hijacking situation, all visitors evacuated the building in 444.78 s; in the event of an arson, all visitors required 563.52 s to evacuate. As shown in Figure 11, the number of tourists stranded in the three situations is roughly the same. It is incredible that after the hijacking scenario, the evacuation time of the tourists is less than the normal opening of the four entrances. After 100s, the Number of stranded generally smaller than the second case. This also indicates that the three elevators in the southernmost side of the underground floor play a minor role in the evacuation process, and the layout of the museum's safe passages needs further optimization.

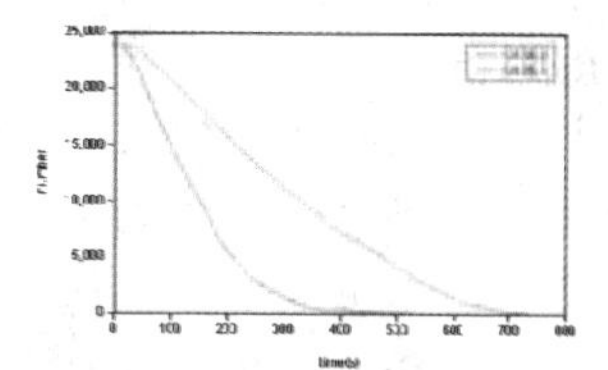

Figure 10: Change in the number of stranded persons in two cases

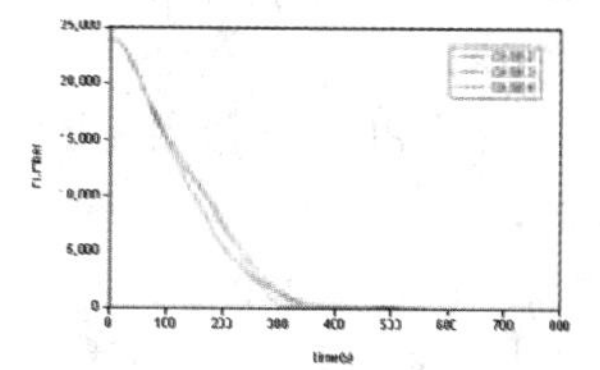

Figure 11: Comparison of Case2,3,4

5 Sensitivity Analysis

In our model, some inputs are not precise enough for the lack of actual data and some parameters are difficult to obtain directly. Those inputs or parameters may influence the result of our calculation, so we implement a sensitive analysis to test the robustness of our model.

- **Sensitivity Analysis for N**

Table 3: Sensitivity Analysis for N

	-10%	**-5%**	**0%**	**5%**	**10%**
T	7.023%	3.978%	0%	4.079%	7.318%

All the percentage presented on the table represents the maximum fluctuation among N. Therefore,N is not sensitive to the value of T.

- **Sensitivity Analysis for θ_α**

Table 4: Sensitivity Analysis for θ_α

	-10%	**-5%**	**0%**	**5%**	**10%**
T	9.854%	4.937%	0%	5.174%	11.315%

All the percentage presented on the table represents the maximum fluctuation among θ_α.Therefore, θ_α is not sensitive to the value of T.

- **Sensitivity Analysis for φ**

All the percentage presented on the table represents the maximum fluctuation among φ.It can be seen that φ is sensitive to the value of T,

Table 5: Sensitivity Analysis for φ

	-10%	-5%	0%	5%	10%
T	20.852%	11.106%	0%	12.897%	23.476%

namely the rate of information transmission lead to the great changes of T , so the rate of information transmission should be taken into account greatly to reduce evacuation time .

6 Conclusion

6.1 Conclusion

- **Increase the number of exits under the premise of ensuring safety**

  By comparing the difference in escape time between opening only one exit and opening four exits at the same time, we found that increasing the number of exit can greatly improve the efficiency of evacuation and shorten the escape time. The reduction in escape time can also circumvent potential risks. Therefore, after a sudden terrorist attack, the museum should increase the number of exits as much as possible, and at the same time establish a sound security prevention system at the exit, reducing the risk of providing more opportunities for terrorists due to more open exits.

- **Reasonable planning of safe channel layout**

  In the simulated hijacking situation, we found that the three elevators in the south of the underground floor were not available, but the evacuation time was shortened by 77.25s. It's indicating that the Louvre should rethink the layout and planning of safe access facilities. The internal structure of the building is an important factor affecting the safe evacuation of tourists. The safety passage should be appropriately dispersed to improve the utilization efficiency.

- **Attach importance to information technology factors**

  In the fifth part, we find that the information transmission rate is a sensitive indicator that affects the speed of tourists' escape. This shows that the acquisition of information can eliminate the information asymmetry of tourists, and help to improve the psychological state of tourists in the event of a terrorist attack, so that visitors can choose the escape route faster and improve the evacuation efficiency. The Louvre should increase the construction of the network platform and use network information technology to ensure the efficiency and safety of the tourists' escape process when guiding tourists to evacuate safely.

- **Consider building a "bridge" to connect the southern and northern region**

  In the analysis of the results of simulating the effects of two potential threats on evacuation time, we found that there is no direct connection between southern and northern region. The transfer of tourists from the eastern region to the south and north led to a surge in traffic during the first stage of the evacuation process. Excessive traffic will increase security risks and reduce the safety of the evacuation process. Therefore, we suggest that the Louvre can consider building a "bridge" that can connect the north and the south to make passenger flow circulates in evacuation process, and maintain a dynamic balance overall.

6.2 Strengths

i We consider the individual characteristics of the tourists and the external environmental factors in the model at the same time, which makes up for the shortcomings of the existing literature on the evacuation of the personnel from the micro and macro perspectives. Our consideration of the influencing factors is more comprehensive.

ii The model we built is very flexible. The model is simple, intuitive, easy to modify, and can be adapted to different needs and tasks to facilitate porting in different types of building development.

iii We take the information factor into account. Through in-depth analysis, we find that information asymmetry has a great impact on personnel evacuation, and gives relevant proposals in the text, which is a major innovation of this paper.

6.3 Weaknesses

i Although the model already has a certain degree of autonomy, the individual characteristic factors that still have subtle influences are not considered, such as the psychological state of the person. It can't achieve the effect of fully simulating a real person.

ii The model doesn't consider the impact of weather factors on the evacuation of tourists, and doesn't explore the situation of natural disasters that cause tourists to escape. We still need further research.

Appendices

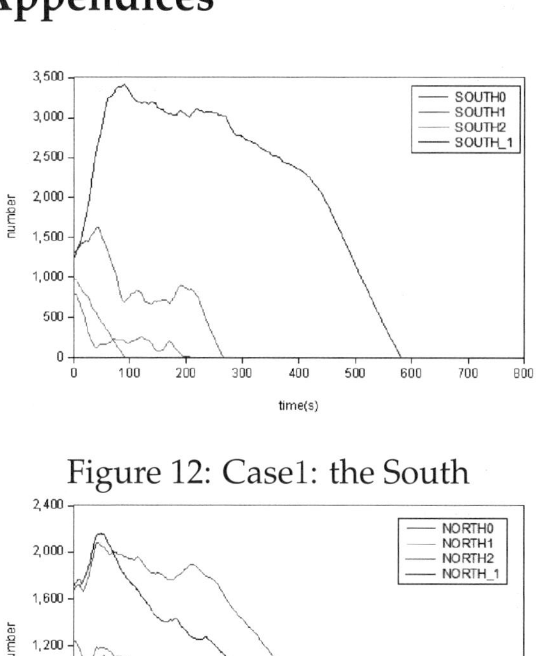

Figure 12: Case1: the South

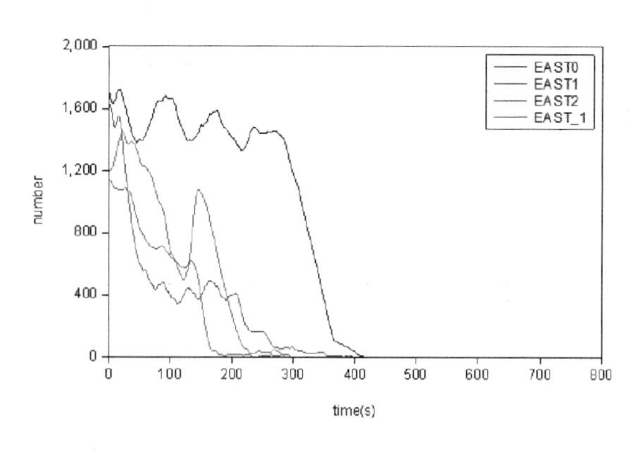

Figure 13: Case1:the East

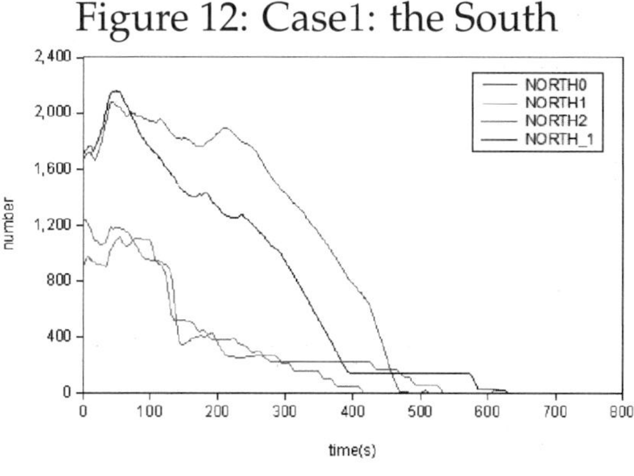

Figure 14: Case1:the North

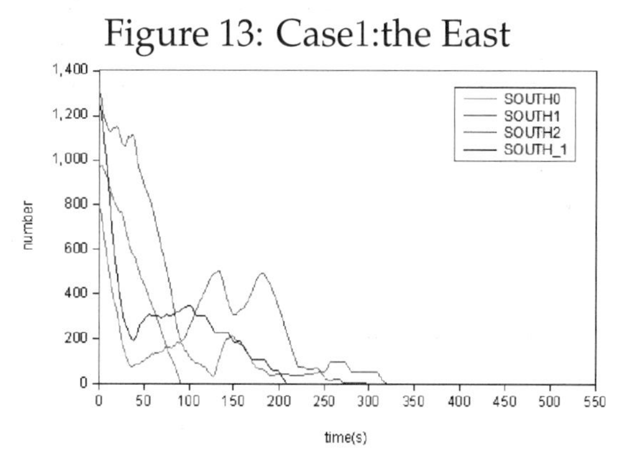

Figure 15: Case2: the South

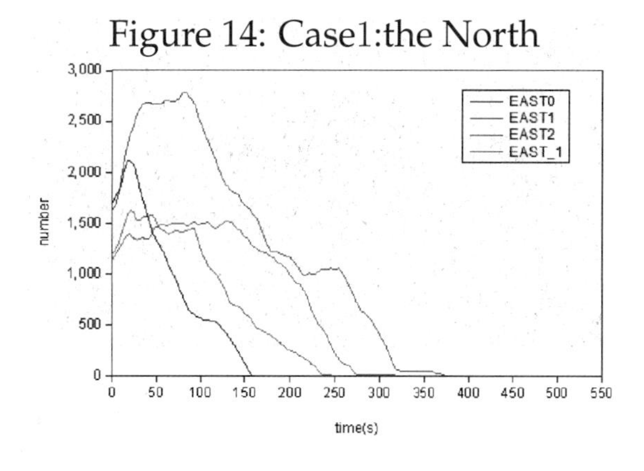

Figure 16: Case2:the East

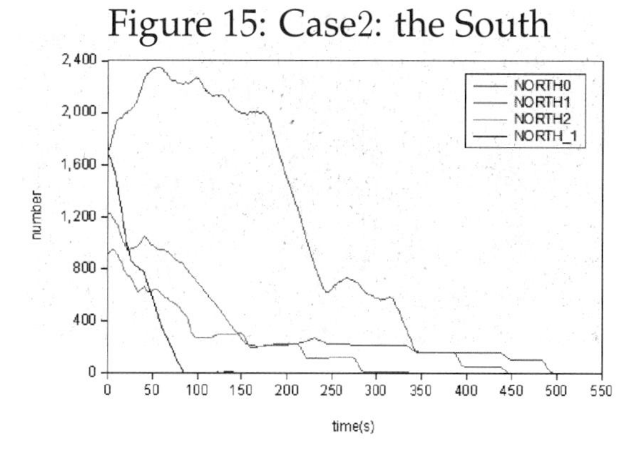

Figure 17: Case2:the North

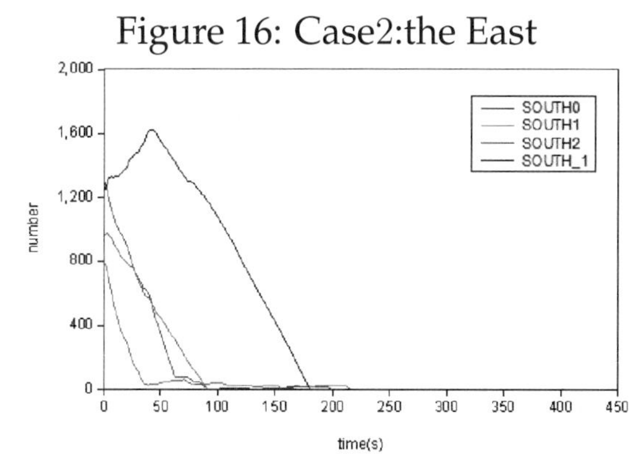

Figure 18: Case3: the South

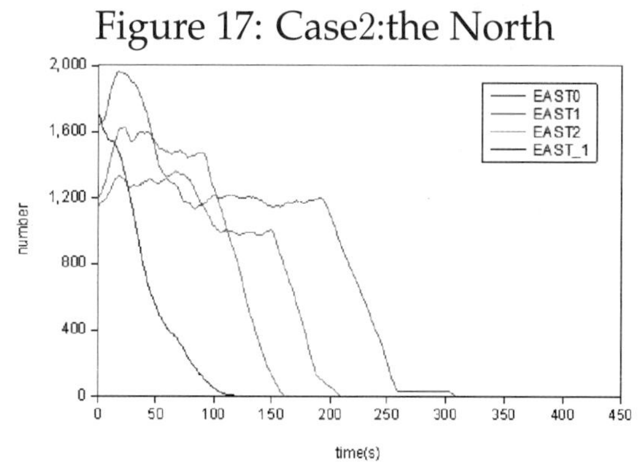

Figure 19: Case3:the East

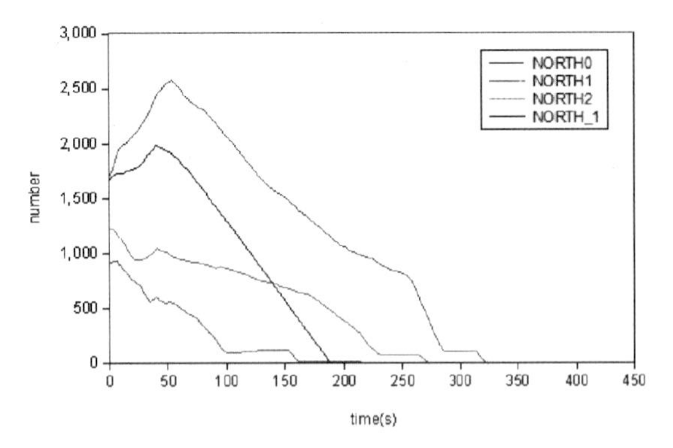

Figure 20: Case3:the North

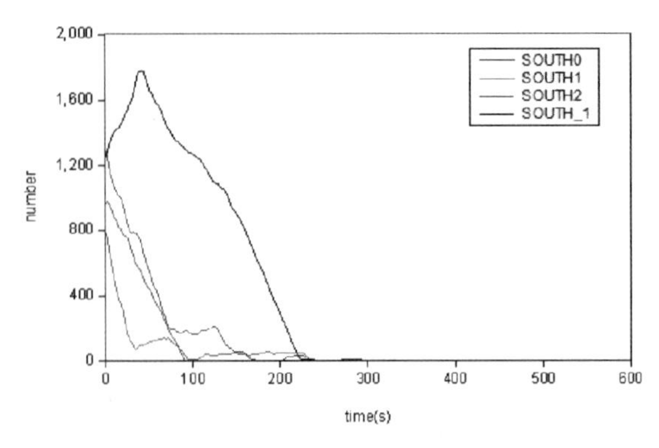

Figure 21: Case4: the South

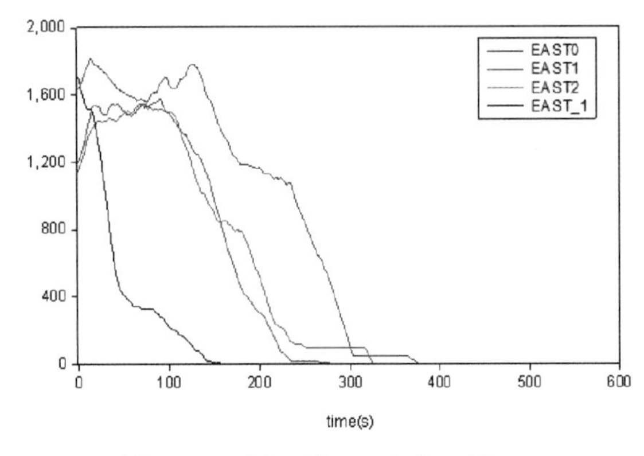

Figure 22: Case4:the East

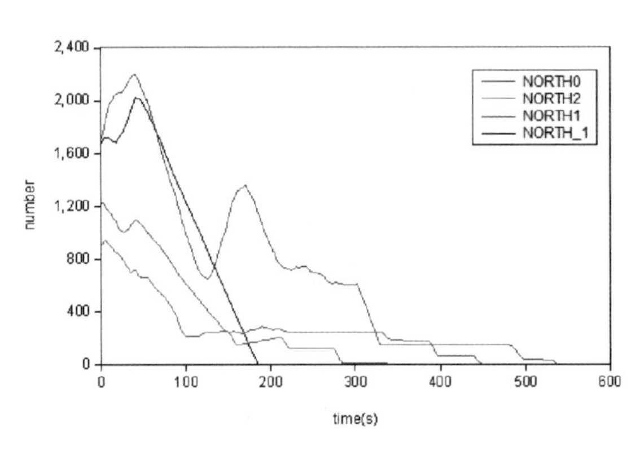

Figure 23: Case4:the North

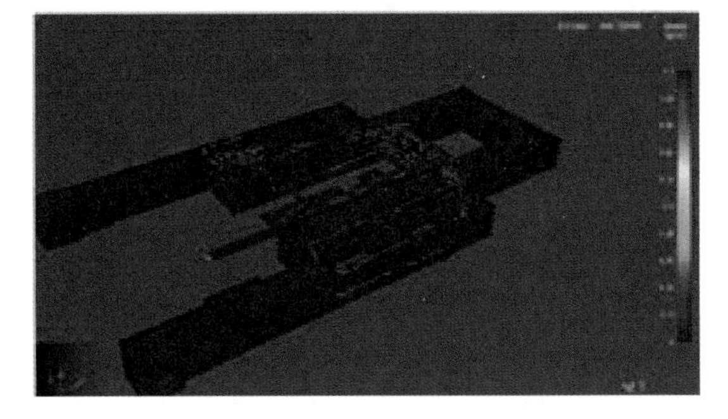

Figure 24: Speed heat map

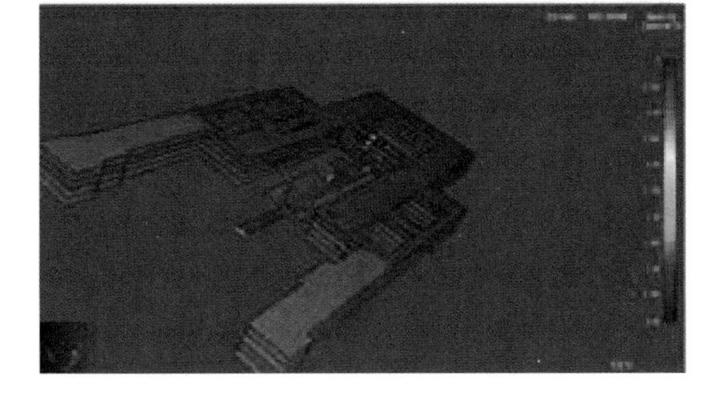

Figure 25: Density heat map

References

[1] M. Pidd, F. N. De Silva, and R. W. Eglese. A simulation model for emergency evacuation. *European Journal of Operational Research*, 90(3):413–419, 1996.

[2] Gunnar G. Løvs. Models of wayfinding in emergency evacuations. *European Journal of Operational Research*, 105(3):371–389, 1998.

[3] Xiaoshan Pan, Charles S. Han, Ken Dauber, and Kincho H. Law. A multi-agent based framework for the simulation of human and social behaviors during emergency evacuations. *Ai Society*, 22(2):113–132, 2007.

[4] David L. Bakuli and Mac Gregor Smith. Resource allocation in state-dependent emergency evacuation networks. *European Journal of Operational Research*, 89(3):543–555, 1996.

3.The mystery of data

Summary

Everyone has experience reviewing on e-commerce platforms. Reviews can directly reflect customer satisfaction with the product, and can even see the performance evaluation of the product. Therefore, based on the given data set, this article conducts a comprehensive analysis of consumer text evaluation. In particular, mining and analyzing the text information in the reviews. We can propose corresponding improve programs and marketing strategies for three products.

First, we perform descriptive statistics on the given data set and extract the high-frequency words in the reviews to reflect the main factors of customer concern. We can provide companies with reference ideas for product improvement. Then, we establish a sentiment analysis model based on the new rules to calculate the customer review index and establish a customer satisfaction evaluation system on three levels. Combine the review index and the star rating to calculate the total customer satisfaction index of the product. The results show that customer satisfaction with microwaves, hair dryers and pacifiers is at a high level. Fitting the curve of the total customer satisfaction index over time. We find that the total customer satisfaction index of the three products decreased year by year. Using the data obtained from the analysis, we can establish a GM(1,1) Gray Prediction model to predict the reputation trend of the three products in the next five years. This product has a risk of a declining reputation in the future. Then, using the review index as an intermediary, we study the relationship between specific quality descriptors of text-based reviews and star ratings and find that the two have a high correlation.

Secondly, we combine the customer satisfaction evaluation system, product sales and product popularity, and established a PCA (Principal Component Analysis) evaluation model to obtain the combination of reviews and star ratings that can best indicate the potential success of the product. We find that microwaves, hair dryers, and pacifiers all had the highest potential product success index in January 2015, and their corresponding star index and review index combinations were 87.50 and 76.93, 85.79 and 74.92, 87.61, and 75.75.

Third, we establish a Naive Bayesian classification model including BOW (bag-of - words) model. The BOW model converts text into word vectors that can be processed by a computer. In addition, we select 1-, 2-, and 3-star reviews within a specific time period. Training a Naive Bayesian classifier for reviews. Then classifying 20 1-star reviews over a period of time, and finding that only 11 are marked as 1-star. So, when a customer is reviewing a product, a particular star will influence the review to some extent.

Keywords: Sentiment Analysis; Customer Satisfaction System;PCA; Naïve Bayes Model

---

作　　者：徐萌，王展鹏，聂啸林

指导教师：魏宁

奖　　项：2020年美国大学生数学建模竞赛F奖

1 Introduction

1.1 Restatement of the Problem

Do you have a rating and review experience on Amazon Shopping? If so, is your review positive or negative? With the continuous development of the Internet and electronic technology, e-commerce and online marketing have gradually become a boom[1]. Amazon is currently the largest online e-commerce company in the United States, providing customers with a wide range of products online.

In addition, in order to investigate product customer satisfaction, improve customer loyalty, and attract consumers, Amazon encourages consumers to comment on products. Customers rate and write related reviews based on their shopping experience and perceived value of the product, providing potential consumers with decision-making assistance in purchasing. Ratings and reviews can also serve as important market feedback. Marketers can analyze customer satisfaction through ratings and reviews, identify product strengths and weaknesses, improve product characteristics, optimize marketing strategies, and increase product appeal and competitive advantage.

1.2 Our Work

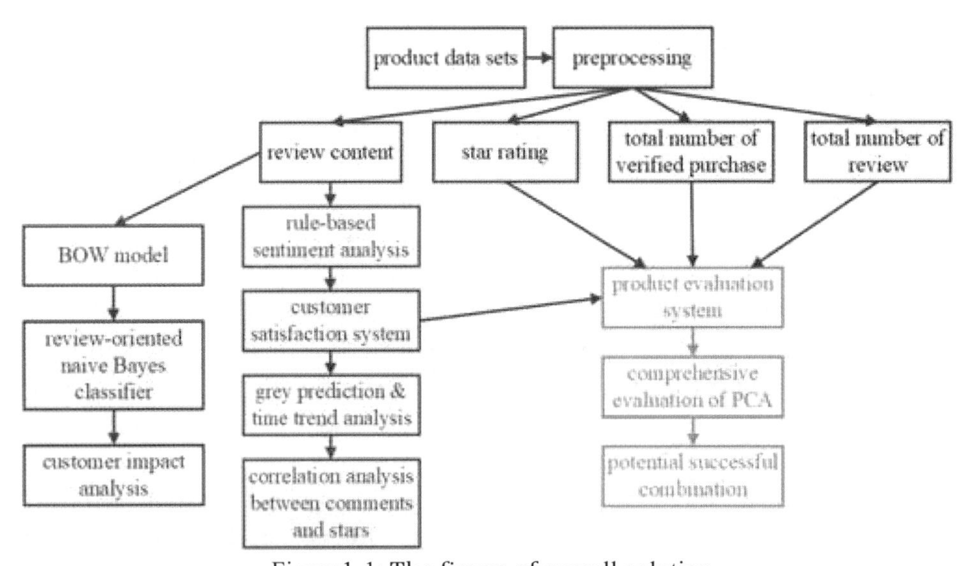

Figure1-1: The figure of overall solution

As shown above, our model is mainly divided into three parts. First, we perform a comprehensive descriptive statistics of the three products based on the data set provided in the attachment. The text reviews are then quantified. Based on the rated and quantified text reviews, we establish a three-level customer satisfaction evaluation system model, from a multi-dimensional perspective to describe the degree of customer satisfaction with the product, capture the defects of the product. Second, analyze the evolution trend of customer satisfaction evaluation indicators over time, use gray prediction models to predict the rising or falling trend of product reputation in the short-term future, and establish a comprehensive evaluation model of Principal Component Analysis to find the combination of rating indicators that can best reflect potential success or failure. Third, we establish a Naive Bayesian classification model based on the bag-of-words model to analyze the impact of specific star ratings of others on subsequent reviews. Finally, analyze the relevance between the keywords in each review and their own ratings. Based on the established model, try our best to put forward effective suggestions for improvement, and provide our model and analysis results to the marketing director.

2 Assumptions and Nomenclature

2.1 Assumptions

To simplify our problems, we make the following basic assumptions:

- Assume that the data after screening has certain reliability and usefulness, and can reflect the specific situation. For example, there is no malicious evaluation.
- Assume that reviews after purchasing a product are valid reviews.
- Assume that a short review is not very valuable for customers to buy products.
- Assume that each review is only related to the performance of the product itself and the impact of other reviews.
- Assume that there is a certain proportional relationship between product sales and the number of reviews, and this proportional relationship remains unchanged.
- Assume that customers will comment shortly after purchasing a product.
- Assume that the literature and conclusions cited in the article are correct and reliable.

2.2 Nomenclature

Table 2-2: the Nomenclature in our paper

VARIABLES	DEFINITION
Mic(Hdy,Pac)-SR	Star rating for microwave (hair dryer, pacifier)
Mic(Hdy,Pac)-Helpful votes	Microwaves (hair dryers, pacifiers) useful numbers in reviews
Val_i	Lexical sentiment value
Senti'	Sentence sentiment value
Fs	Score based on star rating
F_p	Text-based quantified sentiment score
F_z	Customer satisfaction index, which is the weight of star reviews and text reviews
TF_z	Overall customer satisfaction index
ρ_{xy}	Pearson correlation coefficient of sentiment score based on text quantification and score based on star rating
S_i	Product Potential Success Index
F_{sup}	Number of all reviews
F_n	Number of customer reviews after buying a product on Amazon

3 Data Preprocessing and Descriptive Statistics

3.1 Date Preprocessing

First, we divide the data in the worksheet into three categories: redundant data, invalid data and regular data. Then for more convenient discussion, delete the redundant data and the invalid data, at the same time, sort the regular data by time and extract the available data. The work we do in data processing is as follows.

3.1.1 Delete Error and Redundant Data

There are wrong lines, garbled, blank and redundant data in all three data files. In order to ensure the reliability and accuracy of the data, delete errors and redundant data.

3.1.2 Delete invalid data

Our analysis find that there are many other irrelevant products in the product data of pacifiers, and there are also reviews that the text content is not related to the product. We define it as invalid data and delete the invalid data based on the "product_title" field in each dataset. According to the assumption, delete the data in the dataset "verified_purchase" fields are" N" and "review_body" fields with less than 4 words. This guarantees the reliability of the data.

3.2 Descriptive Statistics and Analysis of Available Data

3.2.1 Descriptive statistics

Descriptive statistics refer to activities that use tabulation and classification, graphics, and calculation of general data to describe the characteristics of data, and are the basis of data analysis. We performed descriptive statistics on the pre-processed data set. Table 3-1 lists the star ratings,Helping vote averages,variances, and extreme value. The star ratings and helping vote averages reflect the customer's satisfaction with the product and the quality of reviews to a certain extent, the variances reflect the stability, and the extreme value reflects the data range.

Table 3-1: Descriptive statistics

Variable	Obs	Mean	Std.Dev.	Min	Max
Mic-SR	1006	79.344	27.135	20	100
Mic-Helpful votes	1006	3.863	15.856	0	345
Hdy-SR	8948	82.776	25.309	20	100
Hdy-Helpful votes	8948	1.796	13.072	0	499
Pac-SR	14350	85.868	23.628	20	100
Pac-Helpful votes	14350	0.653	4.072	0	209

3.2.2 Time Trend of Number of Comments

In marketing activities, sales can directly reflect the popularity of a product in the market. The number of reviews for certified purchases can indirectly reflect the sales of a product. Figure 3-1 is a time trend chart of the number of reviews for three products. It can be seen that over time, the online sales of the three products have continued to increase. This increase is mainly due to the development of the Internet and product improvements.

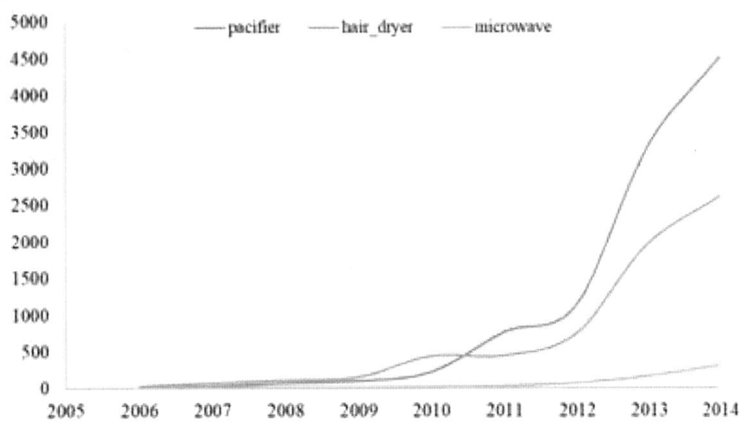

Figure3-1:The figure of the number of reviews

3.2.3 High-Frequency Vocabulary Analysis

High-frequency vocabulary can reflect consumers' general concerns about products. We extract nouns from online reviews of low and medium stars (including one and two stars). In order to visualize the factors that consumers are most concerned about, we use the data collected in the hair dryer data to generate a word cloud, as shown in figure 2 (For space reasons, we only show the word cloud of hair dryers. The word cloud of microwaves and pacifiers can be found in Appendix 1).According to the word cloud, consumers are most concerned about the wind power, temperature, speed of the hair dryer, and the use of the hair dryer buttons. etc. The taste and noise generated by the hair dryer are secondary concerns for consumers. Marketers can analyze the characteristics of products according to the major and minor factors of consumer concern, and improve products to enhance core competitiveness.

Figure 3-2:The figure of word cloud

4 Model

4.1 Sentiment Analysis Model for Online Shopping Reviews Based on New Rules

Sentiment analysis is a computational study of the opinions, attitudes, and emotions of an entity[2]. Entity objects include individuals, events or topics. At present, text mining technology based on sentiment analysis has been widely used in online reviews. Sentiment analysis of online reviews is to extract words containing positive, neutral, and negative terms in the text for analysis. By analyzing the data in the attachment, we find that there are many ways to express text comments. Based on traditional sentiment analysis models, we establish general rules that embody grammatical and syntactic conventions for expressing and emphasizing emotional intensity. Then, calculate the sentiment value of a review based on the rules and sentiment lexicon. The specific process is shown in Figure 4-1.

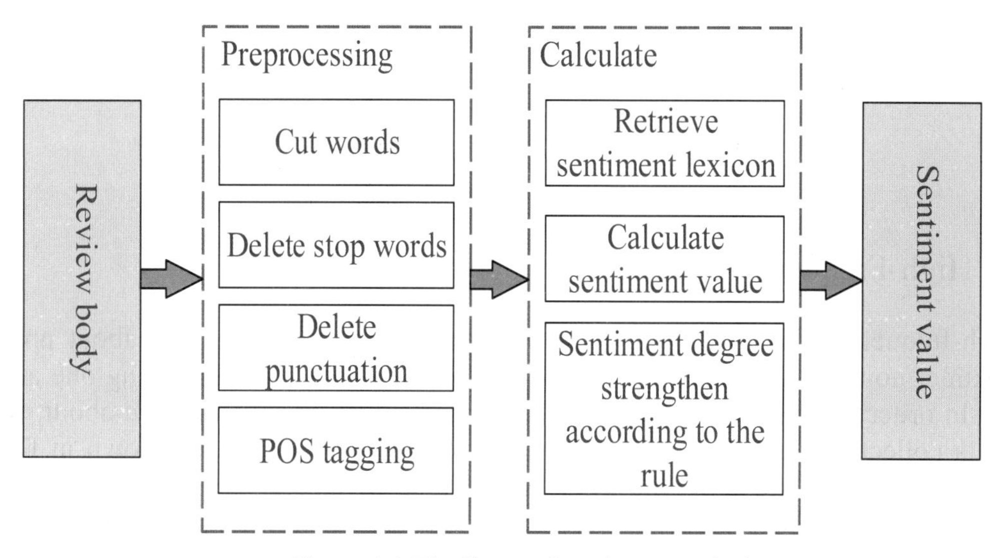

Figure 4-1:The figure of sentiment analysis

4.1.1 Word Segmentation and Emotion Words Recognition

Since computers cannot process sentences or paragraphs directly, they need to be segmented. We use the nltk library in Python for text word segmentation, removing punctuation and stop words (such as an, am, from). Part-of-speech(POS) tagging is then performed on each word to select emotional words that can express customer sentiment.

4.1.2 Establish New Rules

Traditional sentiment lexicons only contain sentiment values of words. When there is a certain sentence pattern or some form of emphasis, calculating only sentiment values will cause errors. In order to improve the accuracy of quantifying customer sentiment, the following four rules are designed to weight the sentiment value:

- An exclamation mark (!) increases emotional intensity without changing the semantic direction. For example, "This baby's pacifier is good !!!" is stronger than "This baby's pacifier is good." We set increase the emotional level by 30% when negative words appear.
- Degree modifiers (also called enhancers, auxiliary words, or degree adverbs) affect emotional intensity by increasing or decreasing intensity. For example, "This baby's pacifier is very good" expresses more intense emotions than "This baby's pacifier is good", while "This baby's pacifier is slightly better" reduces the intensity. We increase the emotional level by 20% in the former case and decrease the emotional level by 20% in the latter case.
- The contrastive conjunction "but" indicates a change in emotional polarity, and the emotions of the text following the conjunction dominate. For example, "microwaves heat up fast, but the operation is too complicated." People's moods are mixed, and the second half determines the overall score. When a sentence uses an contrasting conjunction, we reduce the emotional level of the previous sentence by 20%, and increase the emotional level of the following sentence by 30%.
- Negative words can reverse the polarity of the text. For example "This hair dryer is really not very good". We set the positive and negative emotion levels to change when negative words appear.

4.1.3 Sentiment Value Calculation

After identifying the emotional words, assign an emotional value to each emotional word according to sentiment lexicon, that is:

$$Val_i = (-1)^{OM} * Val' * \delta \tag{1}$$

Val' is sentiment value of sentiment feature words in sentiment lexicon, OM is the number of negative words in the sentence, δ is the weighting of emotional words according to the rules, Val_i is the sentiment value calculated according to rules.

After assigning sentiment values to words with different parts of speech, calculating the sentiment value of each text comment. The sentiment value can reflect the overall sentiment of the corresponding text review, that is:

$$Senti' = \frac{Val_1 + Val_2 + ... + Val_N}{N} \tag{2}$$

N is the number of sentimental features in a sentence, $Senti'$ is the sentence sentiment value.

4.2 Customer Satisfaction Evaluation System and Product Reputation Evaluation

The fundamental goal of marketing is to meet customer needs and improve customer satisfaction and loyalty of products. Online reviews can truly reflect the customer's feelings and needs for the product. Therefore, we establish a customer satisfaction rating system at different levels, reflecting customer satisfaction and demand for products at multiple levels.Figure 4-2 shows the hierarchy of the customer satisfaction rating system.

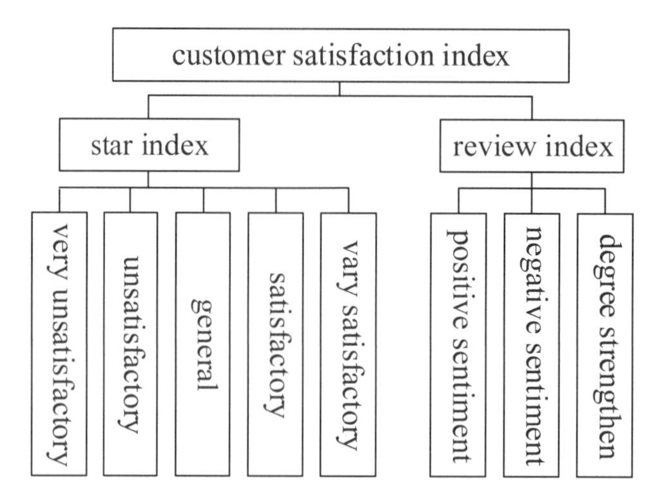

Figure 4-2:The figure of customer satisfaction system

4.2.1 Customer Satisfaction System Indicators

- Star rating index. In order to reflect the star rating more intuitively, we standardize the star rating and convert its value to 0-100, that is:

$$F_S = SR*20 \tag{3}$$

- Review index. According to the sentiment analysis model introduced above, convert its final score to 0-100, that is,

$$F_p = \frac{Senti - Senti_{\min}}{Senti_{\max} - Senti_{\min}}*100 \tag{4}$$

- Customer satisfaction index. The customer satisfaction index comprehensively reflects customer satisfaction with the product. Calculating the customer satisfaction index for each customer based on the star rating index and review index:

$$F_{Zj} = w_1 F_{Sj} + w_2 F_{Pj} \tag{5}$$

Reviews are more intense and specific than the star rating, so the weight of the star rating index should be differentiated from the weight of the review index. We set the weight of the star rating index to 0.4 and the weight of the review index to 0.6. We consider that vine's reviews are more authentic and authoritative, so we adjusted the star rating index weight to 0.3 and the reviewt index weight to 0.7.

According to the above model, we can find out the customer satisfaction index of each customer, and weighted average all the customer satisfaction indices to obtain the total customer satisfaction index. That is:

$$TF_Z = \frac{\sum_{j=1}^{n} F_{Zj}}{n} \tag{6}$$

Finally, the total customer satisfaction index of the hair dryer is 79.275, the total customer satisfaction index of the microwave is 75.619, and the total customer satisfaction index of the pacifier is 82.034.

According to the Likert scale, We divide customer satisfaction into five levels, namely very dissatisfied, dissatisfied, average, satisfied, very satisfied. We divide the total customer satisfaction index into 5 levels based on these five levels, that is, 0-20 points are very dissatisfied, 20-40 is not satisfied, 40-60 is average, 60-80 is satisfied, and 80-100 Very satisfied. We have calculated the total customer satisfaction index of these five levels, and the proportion of each level is shown in the Figure 4-3.

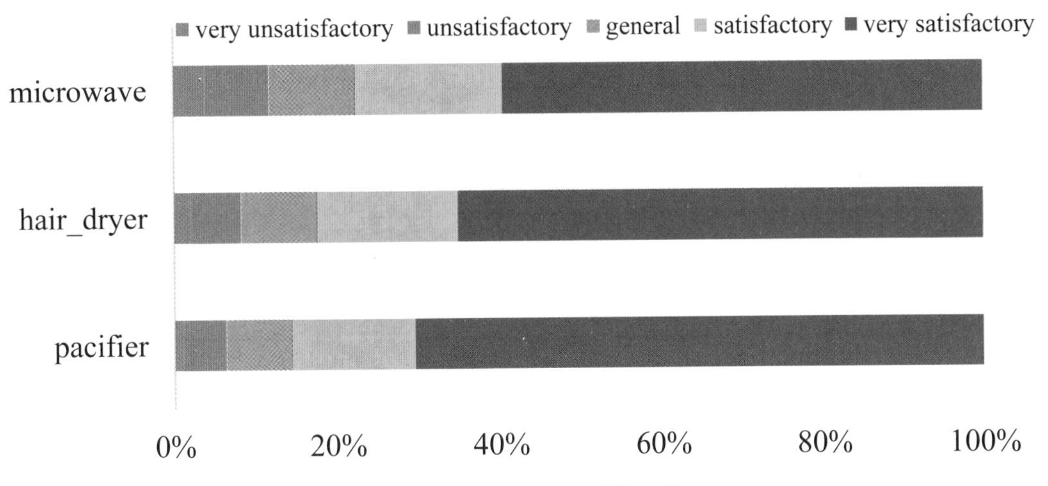

Figure 4-3: Percentage stacking diagram

From Figure 4-3 we can see the largest percentage of customer satisfaction among the three product reviews is 80-100. The total customer satisfaction index for pacifiers is the highest, indicating that pacifiers are more popular with consumers than the other two products. The total customer satisfaction index of microwave and hair dryers is at a satisfactory level, while the total customer satisfaction index of pacifiers is at a very satisfactory level. According to the longitudinal comparison, we find that the proportion of dissatisfied attitude, general attitude and satisfactory attitude is not much different for these three products. The difference in the total customer satisfaction index of the three products is mainly due to the difference between the very dissatisfied ratios and very satisfied ratios.

4.2.2 Product Reputation Evaluation and Prediction

The total customer satisfaction index directly reflects customer satisfaction with the product. In addition, customer satisfaction can indirectly reflect the reputation of the product in the target market. We take the year as the unit and use the total customer satisfaction index to make a time trend chart. In addition to analyzing the current status of product reputation, in order to enable more long-term analysis, a GM (1,1) Gray Prediction model is used to predict the trend of the total customer satisfaction index in the short term in the future so that companies can make effective adjustments.

First of all, according to the data we have compiled, we can get the initial sequence.That is:

$$x^{(0)} = (x^{(0)}(1), x^{(0)}(2), x^{(0)}(3), \ldots\ldots, x^{(0)}(9)) \tag{7}$$

Then we set up the grey differential equation and combined the least-squares method to find the prediction equation of the total customer satisfaction index, that is:

$$\hat{x}^{(1)}(k+1)=(x^{(0)}(1)-\frac{\hat{b}}{\hat{a}})e^{-\hat{a}k}+\frac{\hat{b}}{\hat{a}},k=0,1,.....,8 \tag{8}$$

Finally, we establish a differential equation to evaluate the reputation of the product.

$$\frac{dTF_Z}{dt}=a \tag{9}$$

If a> 0, then it indicates that the total customer satisfaction index of the product has continued to rise in a certain year, and the product reputation has increased. If a <0, then it indicates that the total customer satisfaction index of the product has a downward trend in a year, and the product's reputation is also falling.

Finally, the results of the Gray Prediction Models of the three products are shown in Table 4-1:

Table 4-1: Gray prediction evaluation model results

Product	Prediction model
Pacifier	$\hat{x}^{(1)}(k+1)=-9589.99e^{-0.009027553k}+9672.37$
Hair dryer	$\hat{x}^{(1)}(k+1)=-86392.28e^{-0.0434413O2k}+1989.05$
microwave	$\hat{x}^{(1)}(k+1)=-86392.28e^{-000881481k}+86475.67$

In order to ensure the accuracy of prediction, we use the residual test to test the Gray Prediction model. Let the residual error be ε (k), and the calculation formula is:

$$\varepsilon(k)=\frac{x^{(0)}(k)-\hat{x}^{(0)}(k)}{x^{(0)}(k)},k=1,2,...,9 \tag{10}$$

By calculation, the inspection residual value of microwave, hair dryer, and pacifier are all less than 0.2 (see the Appendix 2 for specific test results), so the gray prediction of these three products is highly accurate.

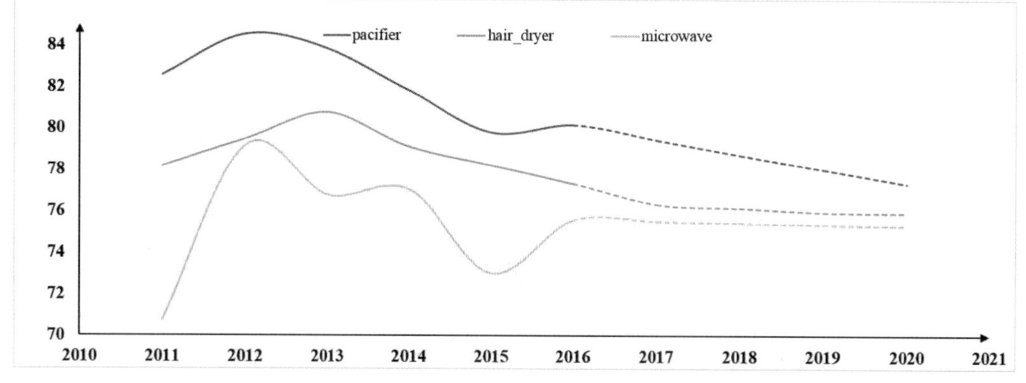

Figure 4-4:The figure of gray scale models predict trends

Figure 4-4 shows the time trend of the total customer satisfaction index for the three products: pacifiers, hair dryers, and microwaves. Overall, the trend chart reflects that between 2012 and 2015, the total customer satisfaction index of the three products continued to decline, and the reputation of the products continued to decline. Although the customer satisfaction index of microwave products increased in 2014, it is still lower than the total customer satisfaction index of 2012. According to the prediction model, from 2016 to 2020, the customer satisfaction of the three products will continue to decline, and the reputation of the products will continue to decline, but the decline is not significant. Therefore, in the future, three products can improve the customer satisfaction index by improving product characteristics or marketing strategies, and thus improve the reputation of the product.

4.2.3 Correlation Between Specific Quality Descriptors and Star Index

Because specific quality descriptors of text-based reviews have a direct relationship with the sentiment value of a review, if you explore the relationship between the descriptor and the star rating, you can perform a correlation analysis between the sentiment value of the review and the star rating. We use the Pearson correlation coefficient method to determine the correlation:

$$\rho_{F_s,F_p} = \frac{COV(F_s,F_p)}{\sigma_{F_s}\sigma_{F_p}} = \frac{E\left[(F_s-\mu_{F_s})(F_p-\mu_{F_p})\right]}{\sigma_{F_s}\sigma_{F_p}} \tag{11}$$

Through the calculation of the Pearson correlation coefficients of the three products the sentiment value of the review and star rating, the correlation coefficients of microwave, hair dryer, and pacifier are 0.733, 0.769, and 0.810 respectively. Therefore, there is a high correlation between the sentiment value of the review and the star ratings, so there is also a high correlation between quality descriptors and star ratings.

4.3 Best Potentially Successful Product Portfolio Model

This section establishes evaluation criteria for potentially successful products. Section 4.3.1 performs KMO inspection on the data. The results show that factor analysis can be performed. Section 4.3.2 establishes a Principal Component Analysis(PCA)[3] comprehensive evaluation model to calculate the product success index, finds the quarter with the highest potential product success index, and section 4.3.3 shows the calculation results.

4.3.1 Data Inspection

We use the KMO (Kaiser-Meyer-Olkin)[4] method to check whether the data is suitable for factor analysis. The KMO test statistic is an indicator used to compare the simple correlation coefficient and the partial correlation coefficient between variables. Its expression is:

$$KMO = \frac{\sum\sum_{i\neq j} r_{ij}^2}{\sum\sum_{i\neq j} r_{ij}^2 + \sum\sum_{i\neq j} r_{ij\bullet 1,2.....k}^2} \tag{12}$$

The KMO statistics are between 0-1. Generally, we think that 0.9 or above means that it is very suitable, 0.8 means that it is suitable, 0.7 means that it is normal, 0.6 means that it is not suitable, and 0.5 means it is extremely inappropriate. According to the test results, the KMO statistics of microwaves, hair dryers, and pacifiers were 0.799, 0.786, and 0.834, all-around 0.8. So The PCA model can be used to rank the products quarterly, as shown in Table 4-2.

Table 4-2: KMO test results

Product	Test Type	Test Form	Test Result
Microwave	KMO	KMO statistics	0.799
Hair dryer	KMO	KMO statistics	0.786
Pacifier	KMO	KMO statistics	0.834

4.3.2 The Principal Component Analysis Comprehensive Evaluation Model

Product success can be measured by customer satisfaction and product sales. In order to be able to more comprehensively evaluate the success of the product, the indicators of potential product success based on quarters are reflected in the following four aspects: star index and review index reflect customer satisfaction in a quarter, and the total number of reviews can reflect the consumer attention and product enthusiasm in a quarter, and reviews of certified purchases can reflect product sales in a quarter. Create the following regression equation:

$$S = \alpha_1 F_s + \alpha_1 F_p + \alpha_1 F_{sup} + \alpha_1 F_n \quad (13)$$

Because the weight of the four indicators cannot be determined, we use the PCA model to reduce the dimensions of the four indicators to calculate the weight. Then rank each quarter based on the calculated comprehensive evaluation value to get the quarter of the potential successful product. Proceed as follows:

First, the valid data in the three files of microwave, hair dryer and pacifier are standardized and transformed according to the Z-score method, that is:

$$Z_{ij} = (X_{ij} - \overline{X_j}) / S_j \quad (14)$$

$\overline{X_j}$ represents the average value of the j-th index, S_j represents the variance of the j-th index, X_{ij} represents the j-th index value of the i-th sample, and Z_{ij} represents the result of data standardization.

Then, a covariance matrix R is established based on the standardized data matrix. This index is a statistical index that reflects the closeness of the correlation between standardized data.

$$R = \begin{bmatrix} r_{11} & \cdot & \cdot & \cdot & r_{1p} \\ \cdot & & & & \cdot \\ \cdot & & & & \cdot \\ \cdot & & & & \cdot \\ r_{p1} & \cdot & \cdot & \cdot & r_{pp} \end{bmatrix} \quad (15)$$

Solving the characteristic root and the characteristic vector according to the correlation degree equation, and arrange the characteristic roots from large to small: $\lambda_1 > \lambda_2 > \lambda_3 > ... > \lambda_p$. Assuming Yi represents each principal component, and $M_1, M_2, M_3, ..., M_p$ are corresponding unit eigenvectors, the i-th principal component equation of x can be expressed as:

$$Y_i = M_{i1}x_1 + M_{i2}x_2 + M_{i3}x_3 + ... + M_{ip}x_p \quad (16)$$

The number of principal components is selected based on the angle of the variance contribution rate.That is:

$$CR = \lambda_1 / \sum_{i=1}^{p} \lambda_1 \quad (17)$$

$$CRA = \sum_{k=1}^{p} (\lambda_k / \sum_{i=1}^{p} \lambda_1) \quad (18)$$

According to the selected principal components, the variance contribution rate is weighted, and a small number of principal component factors are linearly combined to establish a relationship with the product's potential success coefficient.

Finally, we rank the products according to their potential success factors, and then derive the quarters of potentially successful products.

4.3.3 Final ranking

Using MATLAB to solve we can get the potential success index expression of microwave, hair dryer and pacifier.That is:

$$S_{mic} = 0.544Y_1 + 0.394Y_2 + 0.062Y_3$$

$$S_{hdy} = 0.546Y_1 + 0.329Y_2 + 0.125Y_3$$

$$S_{par} = 0.582Y_1 + 0.280Y_2 + 0.138Y_3$$

Then we can get the ranking of the product's potential success index. Table 4-3 shows the top five quarters of the three products and their potential success coefficients.

Table 4-3: PCA comprehensive analysis ranking results

microwave			**hair dryer**			**pacifier**		
Rank	**Quarter**	**S**	**Rank**	**Quarter**	**S**	**Rank**	**Quarter**	**S**
1	2015-1	3.62	1	2015-1	2.60	1	2015-1	3.44
2	2015-2	3.17	2	2002-3	2.24	2	2014-4	3.15
3	2015-3	2.97	3	2015-2	2.09	3	2015-2	3.08
4	2014-4	2.55	4	2014-4	2.06	4	2014-3	2.76
5	2014-3	2.11	5	2005-4	1.91	5	2015-3	2.01

According to the ranking of product potential success index, we conclude that the product potential success index of microwave ovens, hair dryers and baby pacifiers were the highest in the first quarter of 2015.

Finally, according to the previous statistics, the combination of star rating index and review index of the three products in the first quarter of 2015 can be obtained, as shown in Table 4-4:

Table 4-4: Potential successful product evaluation portfolio

Product	F_s	F_p	F_z
microwave	87.50	76.93	71.15832143
hair dryer	85.79	74.92	79.26824825
pacifier	87.61	75.75	80.49662017

It can be seen from the table that when the quarterly average star rating and review index of the microwave oven are 87.50 and 76.93, respectively, the microwave oven is in the potential success stage. When the quarterly average star rating index and review index of the hair dryer are 85.79 and 74.92, respectively, the hair dryer is in the potential success stage. When the pacifier's quarterly average star rating index and review index are 87.61 and 75.75, respectively, the pacifier is in the potential success stage. Due to space limitations, this article will not repeat the combination of star rating index and review index at the potential failure stage, the method is the same.

4.4 Sentence classification model based on bag of words model

When customers see a series of low-star reviews or high-star reviews when reviewing products, the reviews they write and the star ratings they receive will also be affected and become lower or higher. Therefore, it is necessary to design a Naive Bayes classification model for product reviews to prove this point. Section 5.1 describes the use of a bag-of-words model to convert text into word vectors that can be processed by a computer. Section 5.2 describes the construction and advantages of a feature-weighted Naive Bayes classification model. Section 5.3 design experiments prove the above points and analyze experimental results.

4.4.1 Bag of Words Model

Since text is unstructured data, computers cannot process unstructured data. Transform it with a bag-of-words model, and the quality of the conversion will directly affect the final classification result.

The bag-of-words model ignores the grammar, word order, and syntax of the text, and only treats the text as a collection of words. The appearance of each word in the text is independent, and does not depend on the presence of other words. For example, the following two sentences:

Sentence 1：“Useful, breathable, and basic pacifier.”

Sentence 2：“this pacifier is useful and lovely”

Create a dictionary of all the words in this sentence:

dictionary={1：“useful”，2：“breathable”，3：“basic”，4：“pacifier”，5：“lovely”}

The dictionary contains 4 words, each of which has a unique index. The order of the words in the dictionary is not related to the order in which they appear in the original sentence. Using the index number of the dictionary, we can get the vector form of the sentence:

Vector 1：[1，1，1，1，0]

Vector 2：[1，0，0，1，1]

Both vectors contain 5 elements, where the n-th element of vector 1 represents the number of times the n-th word in the dictionary appears in a sentence. The same applies to vector2. The above conversion process ignores the order of the keywords, but only uses the text as the probability set of some keywords. Each keyword is independent of each other. In this way, each document can be represented as a keyword occurrence Statistical collection of frequencies.

4.4.2 Feature-weighted Naive Bayes classification model

When customers see a series of low star reviews or high star reviews, the reviews they write and the star ratings they give will also be affected and will be low or high. Therefore, we design a naive Bayes classifier to prove this point. Naive Bayes text classifier is based on the prior probabilities and feature distributions of different classes, and uses Bayes theorem to perform a transformation to find the posterior probability that the object belongs to a certain class, and selects the class with the largest posterior probability to implement the object Text category[5]. The specific reasoning process is as follows:

Bayes' theorem can be described as conditional probability on random events A and B, as shown in formula (19):

$$P(A \mid B)=\frac{P(A)P(B \mid A)}{P(B)} \tag{19}$$

The probability that formula (19) is expressed as a text classification form is formula (20):

$$P(C \mid X_{F1},\ldots,X_{Fn})=\frac{P(C)P(X_{F1},\ldots,X_{Fn} \mid C)}{P(X_{F1},\ldots,X_{Fn})} \tag{20}$$

C is a certain category, $X_{F1},\ldots,X_{Fn}$ are characteristics of a certain review.

Naive Bayes classifier is finally expressed as formula (21)

$$\gamma=\arg\max_{c\in C} p(c)\prod p(X_{Fi} \mid c) \tag{21}$$

γ is the final calculated category, $p(X_{Fi} \mid c)$ is the frequency with which feature items of a sentence appear in a category, $p(c)$ is the prior probability of a category, the calculation method is as the formula(22):

$$p(c)=\frac{C_n}{C_N} \tag{22}$$

C_n is the number of reviews in a category, C_N is the number of total comments.

In order to get a more accurate classification, we use TF-IDF[6] algorithm to weight each $p(X_{Fi} \mid c)$ during calculation. Term Frequency(TF) is the Frequency of text features appearing in the document, and it is believed that the higher Frequency a text feature appears in a certain category, the more it can represent the document. The calculation method of TF is shown in formula (23) :

$$\mathrm{TF}_{i,j}=\frac{n_{i,j}}{\sum_k n_{k,j}} \tag{23}$$

$n_{i,j}$ is The number of times a text feature appears in a category, $\sum_k n_{k,j}$ is the total number of times the text feature appears in all categories

Inverse Document Frequency (IDF) is used to measure the general importance of a word. IDF believes that the higher the Frequency of a feature appears in all documents, the worse its ability to represent a certain class will be. The calculation method of IDF is shown in formula (24) :

$$IDF=\lg(\frac{N}{n_t}) \tag{24}$$

N is total number of categories, n_t is the number of categories that contain a certain text feature.

For each item $p(X_{Fi} \mid c)$ of TF-IDF method weight processing, formula (25) is used to calculate

$$p(X_{Fi} \mid c)=p(X_{Fi} \mid c)\times \lg(\frac{n+\pi^{-1}}{m}) \tag{25}$$

n is the total number of categories, m is the number of categories in which feature item X_{Fi} appears, π^{-1} is for smoothing

From this we can see that the Naive Bayesian classification algorithm has two major advantages: it does not have complex differentiation and matrix operations in its calculation process, and it is efficient. The Naive Bayesian model is derived from classical mathematical theory and has stable classification efficiency. The training process of the Naive Bayesian classifier is to calculate the prior probability and conditional probability of the classifier, and the calling process is to calculate the maximum posterior probability. The steps for classifying customer reviews are as follows:

Step 1: Perform word segmentation and remove stop words processing on 1-star reviews, 2-star reviews, and 3-star reviews by category, and build a bag of words model;

Step 2: Calculate the a priori probability of each category according to formula (22) to form the classifier parameters;

Step 3: Calculate the conditional probability of each category according to formula (25) to generate a classifier;

Step 4: Perform word segmentation and remove stop words processing on the classification items, and construct a bag of words model;

Step 5: Calculate the final classification result according to formula (21).

4.4.3 Experiment

In order to prove that customers will be affected by other reviews when reviewing a certain product, we select 1-star, 2-star, and 3-star reviews in pacifier.csv file from 2014/11/20 to 2014/12/5, and train them into a naive Bayesian classifier for reviews according to the steps in the previous section. Select 20 1-star reviews within 4 days from 2014/12/6 to 2014/12/9, and call the classifier to classify. Naive bayes classifier is a probabilistic classifier. The classification results are to be represented by the probability that the classification item belongs to a certain class, and the category with the maximum probability and greater than a certain threshold is usually taken as the final result. In this paper, 0.7 is taken as the threshold. The brief results are shown in Table 4-5 (the Detailed results in Appendix 3):

Table 4-5: Brief classification results

Tag（star rating）	original data	results
1-star	20	11
2-star	0	2
3-star	0	7

The chart or table shows that the selection of article 20 of the 1 star review, there are 7 points to three star, 2 has been assigned to 2 star, 11 have been assigned to 1 star, which means that the customer should give high star's reviews, due to the influence of low star comments, gives a low star's reviews. Our model shows that when customers review on a product, they are more likely to write some type of review after seeing a series of low star ratings.

5 Strengths and Weaknesses

5.1 Strengths

- We have added four new rules to the sentiment analysis model, taking into account the enhancement of emotional level by exclamation marks, degree adverbs, contrast conjunctions and negative words. This makes our mining of review text more accurate and reliable.
- We have established a three-level customer satisfaction index system to accurately measure customer satisfaction.
- When analyzing the trend of product reputation, the Gray Prediction method was used to successfully predict the future reputation of the product, enabling decision-makers to adjust and improve product marketing strategies through prediction.
- In calculating the potential success or failure of a product, in addition to star ratings and reviews, we also consider the product's popularity and product sales and establish a comprehensive evaluation model of principal component analysis, and accurately calculate the potential successful product portfolio.

5.2 Weaknesses

- When performing sentiment analysis on comments, expressions also increase the emotional intensity of sentences, but the rules we designed did not take into account.
- Since we use the prior and the data to determine the probability of the posterior to determine the classification, there is a certain error rate in the classification decision.
- The interpretation of the principal components is ambiguous. It is not as clear and precise as the meaning of the original variables. This is the price that has to be paid in the process of reducing the dimension of the variables.

6 Conclusion

In this article, firstly, we perform descriptive statistics and high-frequency vocabulary analysis on the data sets of microwave, hair dryer, and pacifier. Secondly, a sentiment analysis model based on the new rules was established to obtain the review score. We weighted the review score and product star rating to obtain the total customer satisfaction index of hair dryer, microwave, and pacifier, which were 79.275, 75.619, 82.034.Evaluate product reputation based on time, draw a time trend chart, and establish a gray prediction model. It is found that the total customer satisfaction index and reputation of the three products gradually decline and may continue to decline in the future. Third, the correlation analysis between the review index and the star index was found to have a high correlation between the review index and the star index. Fourth, establish a comprehensive evaluation model of principal component analysis, and find that microwaves, hair dryers, and pacifiers have the highest potential product success indexes in the first quarter of 2015. The corresponding star index and review index combinations are 87.50 and 76.93, 85.79, and 74.92, respectively, 87.61 and 75.75. Finally, we set up a Naive Bayesian classification model for product reviews to analyze the impact of specific star ratings on the ratings of others, and find that the existing star rating will affect the customer's evaluation of the product, that is, seeing many low-star reviews will lead to more low-star reviews. We provide the above results to the marketing director and make relevant marketing strategy recommendations.

7 Letter

From: MCM 2020 Team 2006037
To: Marketing Director of Sunshine Company
Date: 9, March 2020
Subject: Product analysis and recommendations

For your information, here we introduce our analysis of customer focus factors and sales of products, customer satisfaction, product evaluation optimization, and marketing strategies. The details are as below:

- **Customer attention factors and sales**
  - For microwave, the main factors of customer concern in the review in the review include the appearance of the microwave, the heating temperature, the convenience of installation and the operation of the buttons.
  - For hair dryer, the main factors of customer concern in the review include the wind power of the hair dryer, the heat of the hair blow, the speed of the hair blow, and the use of the hair dryer buttons.
  - For pacifier, the main factors of customer concern in the review include the appearance, convenience and safety of the pacifier.
  - From 2011 to 2015, reviews of these three products have continued to increase. This can reflect the continued growth in sales of these three products.
- **Customer satisfaction**
  - The total customer satisfaction index (out of 100) for microwaves, hair dryers, and pacifiers are 79.275, 75.619, 82.034. The overall customer perceived value of the three products is higher than the customer's expected value. Customers maintain a relatively satisfactory attitude towards these three products.
  - From 2012 to 2015, the total customer satisfaction index of these three products has gradually decreased, and the reputation of the products has also declined.
  - According to our prediction model, there is a risk that the reputation of the product will continue to decline in the next 5 years.
  - There is a high correlation between specific quality descriptors and star ratings in customer reviews.
- **Product evaluation optimization**
  - According to customer satisfaction, product sales, product popularity, microwaves , hair dryers, and pacifiers all had the highest potential product success index in the first quarter of 2015.
  - When microwaves, hair dryers and pacifiers are in the quarter with the highest success index, the combination of star rating index and review index are 87.50 and 76.93, 85.79 and 74.92, 87.61 and 75.75.

- **Marketing strategies and suggestions**
  - Microwaves can simplify operation buttons, reduce the complexity of microwaves appliances, and enhance its convenience.
  - Hair dryers can increase gears to increase the risk of high temperatures or strong winds, and increase the target population of hair dryer products.
  - Pacifiers are the most popular products among the three products. You can attract more consumer groups by beautifying their appearance.
  - The total customer satisfaction index of all three products showed a downward trend. In addition to improving the above performance, we believe that we can also adopt a differentiation strategy and a price strategy for durable products such as microwave and hair dryers to increase competitiveness. Pacifiers can use market segmentation strategies to attract consumers.

We hope you will find this report riveting and that it will be helpful for you and your staff as you analyze online reviews and develop marketing strategies. We wish you the best in your endeavors.

Respectfully,

Team 2006037

8 References

[1]. Yang, M. Research on Intellectual Property Rights of Electronic Commerce from the Perspective of Big Data. in Research on Intellectual Property Rights of Electronic Commerce from the Perspective of Big Data. 2018.

[2]. Huang, J., et al. Improving Sequential Recommendation with Knowledge-Enhanced Memory Networks. 2018: ACM.

[3]. Margaritis, A., et al., Identification of ageing state clusters of reclaimed asphalt binders using principal component analysis (PCA) and hierarchical cluster analysis (HCA) based on chemo-rheological parameters. Construction and Building Materials, 2020. 244.

[4]. Yılmaz, Ö. and T.N. Ş, Mechanism of KMO Catalyzed Hydroxylation Reaction: A Quantum Cluster Approach. The journal of physical chemistry. A, 2019.

[5]. Xu, S., Bayesian Naïve Bayes classifiers to text classification. Journal of Information Science, 2018. 44(1): p. 48-59.

[6]. Zhu, Z., et al., Hot Topic Detection Based on a Refined TF-IDF Algorithm. IEEE Access, 2019. 7: p. 26996-27007

Appendix

Appendix 1

Microwave and pacifier word cloud

Appendix 2

Test results of gray prediction model for pacifier

year	Original value	Predictive value	Residual
2007	82.38711111	86.18450682	0
2008	88.16446269	85.40997299	0.022457528
2009	82.59223469	84.64239984	-0.034116262
2010	84.83452727	83.88172481	0.002264732
2011	82.57969554	83.12788592	-0.015766942
2012	84.5391696	82.38082171	0.016693844
2013	83.83722269	81.64047132	0.01737177
2014	81.81924484	80.90677441	0.002184981
2015	79.81223749	80.17967117	-0.013713898

Test results of gray prediction model for hair dryer

year	Original value	Predictive value	Residual
2007	81.53026667	81.53026667	0
2008	81.29882569	81.0910842	0.022457528
2009	81.48114286	80.11241245	-0.034116262
2010	78.77364384	80.01231514	0.002264732
2011	78.19278378	79.13652436	-0.015766942
2012	79.50808835	79.34645765	0.016693844

2013	80.77002733	79.75634534	0.01737177
2014	79.12343054	78.65652342	0.002184981
2015	78.23422579	77.12435435	-0.013713898

Test results of gray prediction model for hair dryer Microwave

year	Original value	Predictive value	Residual
2007	83.39222222	83.39222222	0
2008	84.093625	76.11959766	0.094823208
2009	56.90066667	63.65189817	-0.118649427
2010	89.28341667	75.98551993	0.148940276
2011	70.71653846	75.91856965	-0.073561734
2012	79.2085303	75.85167836	0.04237993
2013	76.79019136	75.784846	0.013092106
2014	77.06148065	75.71807254	0.017432939
2015	73.02776485	75.6513579	-0.035925967

Appendix 3

Naive Bayes detailed classification results

review_id	1-star rating	2-star rating	3-star rating
r1yx38yubq3clp	0.68743	0.52698	0.74959
r1k59pfltxv2up	0.38491	0.46328	0.74745
r4ffm22fp25xw	0.70152	0.68247	0.65316
r15swcx12w37s	0.84546	0.46728	0.34617
r27pk6xnmw6dix	0.77632	0.55641	0.50239
r4pt40w1981dm	0.72333	0.70421	0.32637
r1qdi66zpsp78c	0.80012	0.45972	0.68493
R97QVNM3UB3WY	0.73481	0.39465	0.59148
R26QA7NABQ4N5Q	0.67245	0.79462	0.77831
R29CTXNP7EGY8I	0.69143	0.56731	0.70462
rs955jgdu0enl	0.52146	0.52187	0.72331
r3edc276hxx4mk	0.33685	0.73251	0.34578
r19tfpdw9mrv4c	0.92156	0.70581	0.65349
ryobd9uzobwd	0.78427	0.33129	0.33542
rplhbp6u0hbro	0.64937	0.51943	0.72500
r2f9nvcezt9pbl	0.88495	0.71241	0.61549
r39mocao781wtl	0.74941	0.42157	0.49912
rljj7tdckzptk	0.71251	0.65936	0.81748
R215V1JVH2RK1G	0.52152	0.33468	0.70983
r30ltctmhvthc6	0.83428	0.66724	0.61493

4.Optimization Scheme of Global Disposable Plastic Production Reduction Based on Multi-angle Evaluation of each Region

summary

Since the invention of plastics at the beginning of last century, plastics have gradually become a necessity for human survival. The great convenience that plastics have brought to human beings promotes the large-scale survival of plastics. However, the overproduction of plastics has a huge negative impact on the earth's environment. In order to ensure human survival and the earth's environment, we urgently need to put forward plans that can slow down plastic production.

In this paper, the characteristics of plastic production in different regions are analyzed based on the types and regional evaluation of plastics. According to the current policies of different regions for plastics, the most popular methods to slow down plastic production are evaluated, and the optimization scheme and its implementation time are given.

First, according to the actual life, disposable plastics are divided into four categories: Medical usage, Daily use, Catering usage, Industrial & Agricultural usage. The harm degree of different categories of plastics to the environment is obtained by using Analytic Hierarchy Process, and rank them with the harm degree.

Based on three popular methods to deal with plastic waste, we use entropy weight method combined with evaluation method to determine the weight of the three methods, and pick out the optimal strategy with the characteristics of cost saving, high efficiency, sustainable development of resources and virtuous cycle. The optimal solution is obtained by Cross Matrix Reinforcement and the minimum limit given by Poisson Optimization Algorithm.

Next, according to the factors of economy, culture, science and technology, we divide the world into nine regions, including East Asia (EA), Southeast Asia (ESA), Europe (EUR), Africa (AF), Oceania (OA), North America (NA), Latin America (LAN), South America (SA), Middle East (ME) (ranking in no order) The comprehensive evaluation of plastic processing capacity in the above areas was conducted from three perspectives of policies, production and technology.

Combined with damage degree of different categories of plastic and comprehensive plastic treatment capacity of different regions, we set three expected goals, and use Goal Programming to determine the optimization of different region.

According to the calculated residual of annual plastic output in each region, we define the regional comprehensive quality, and get the regions with less grade fluctuation under the angle of comprehensive plastic processing capacity and Goal Programming, and use the annual net plastic production of these regions to estimate the global annual net plastic output. Calculate the time required to reach the expected target using the estimated global annual net plastic production.

To solve more practical problems, we extend our model.

Finally, we verify the established model to ensure that our model has practical significance.

Keywords: Analytic Hierarchy Process; Goal Programming; Multi-angle Evaluation; Entropy Weight Method; Cross Matrix Reinforcement Method; Poisson Noise Matrix Optimization Algorithm

作　　者：王维浩，许泽东，王帅

指导教师：任争争

奖　　项：2020年美国大学生数学建模竞赛F奖

MEMO
FROM Team 2008157,MCM/ICM 2020
TO: International Council of Plastic Waste Management (ICM)
DATE: February 18, 2020
SUBJECT: Recommendations for reducing disposable plastic products

---

Dear International Council of Plastic Waste Management (ICM), we are honored to be hired to research methods to address the escalating environmental crisis. And it is excited that we found the methods.

According to the real life, we divide the disposable plastic products into four categories: Medical usage, Daily use, Industrial & Agricultural usage, Catering usage, and formulate Environmental pollution， Necessity degree， Processing difficulty， Production consumption and other standards to evaluate the damage degree of plastic.

We have integrated economic, technological and cultural factors to divide the world into nine regions, namely East Asia (EA), Southeast Asia (ESA), Europe (EUR), Africa (AF), Oceania (OA), North America (NA), Latin America (LAN), South America (SA) and the Middle East (ME), and investigated the production and policy and processing technology related to plastics of the representative countries in these regions by 2018. The regions are ranked according to the size of production, policy strength, science and technology level.

In order to solve the serious environmental problems, we set the following three goals:

1. The total annual plastic production in each region shall not exceed the previous global plastic production ;
2. Priority should be given to reducing the production of Medical usage, followed by Daily use and Industrial & Agricultural usage, and finally Catering usage;
3. The total annual plastic production in each region shall not exceed the total known production in the previous year.

Based on these three objectives, the objective planning is carried out to obtain the global acceptable annual plastic output and the changes that each region should make in order to achieve the objectives.

In order to evaluate the impact of different levels of production, policy and science and technology on the comprehensive capacity of a region, we define the comprehensive capacity Q.

In order to propose solutions, we evaluated the effects of the three popular methods.

According to my analysis and evaluation, our team gives the following opinions;

For Medical usage, Daily use, Industrial & Agricultural usage, Catering usage, in order to minimize the harm of plastics to natural life and human life, according to the weight ranking, we give the following suggestions:

- Through today's science and technology, priority should be given to the researching of the medical plastic substitutes' development ;
- Reduce the use of living plastics in life, such as buying bulk goods, shopping bags, reducing the use of packaging plastics in express delivery;
- Attention should be paid to the recycling of plastics in industrial and agricultural production, and plastic products that can be used for a long time and repeatedly should be used as far as possible.

For the three characteristics of output, policy and science and technology level, the negative impact of output is the largest, the positive impact of policy is the largest, and the positive impact of science and technology level is general. It is suggested that the country or region should give priority to the introduction of policies and laws on disposable plastics.

For the plastic treatment method, the effect of reducing unnecessary disposable plastic is the best; the second is recycling compost plastic packaging; the effect of manufacturing 100% reusable packaging plastic is the worst. In the upgrade process, priority is given to reducing the use of unnecessary disposable plastics.

For the world's largest acceptable annual production of plastic, we calculated it to be 332.8 million tons / year, which is 7.56% lower than the annual production of 360 million tons / year in 2018.

If we adopt the above suggestions, the annual output of global plastics will decrease at the rate of 4.65 million tons / year. If there is no force majeure, the expected target will be achieved in 2025.

Our team sincerely hope that ICM can adopt the scheme we provide.

1. Introduction

1.1 Background

In recent years, the problem of plastic waste pollution has received increasing attention. According to research, only 9% of plastic waste produced by humans is recycled, 12% is incinerated, and 79% goes to landfills or the natural environment [1]. A large amount of plastic garbage enters the ocean through various channels, causing the marine environmental pollution problem to become more serious. At present, 85% of marine litter is plastic. The report of the First Global Comprehensive Ocean Assessment stated that marine plastic litter is a serious threat to marine health. We have found plastic components in most species such as fish, shellfish and turtles, and These animals are human food, and waste plastic entering the ocean poses a huge threat to the natural environment and animal and human health. At the same time, marine debris also affects the development of aquatic product trade, and has a negative impact on society economy.

In 2019, the European Parliament voted to pass a ban on disposable plastic products, and began to completely ban disposable plastic products from 2021 to control environmental pollution caused by plastic waste. It is estimated that curbing plastic pollution could reduce environmental damage costs by nearly $25 billion by 2030， while halving the ten most common wastes in the ocean [2]. Therefore, reducing or eliminating waste of plastic products is of great significance to human society as a whole.

1.2 Our Goal

In order to highlight that different factors will have different degrees of impact on the formulation of the plan, we have classified plastic categories, regions, comprehensive capabilities and solutions.

Based on our understanding of the problem and the data we found, we focus on the following issues in this paper:

- Calculate the damage degree of different kinds of disposable plastics, and give targeted suggestions;
- Calculate the difference of effect brought by different schemes, and give targeted suggestions;
- Define the impact of production, policy and technology level on the disposable plastic processing capacity of the region;
- Calculate the acceptable global annual plastic output value under the current situation, and estimate how long it will take to achieve the expected effect after using the optimization scheme provided by us.

2. Assumptions and Notations

2.1 Assumptions

Based on data and reality, in order to help us to model, we make the following assumptions as follows:

- Plastic products in this paper refer to single-use or disposable plastic products;
- Disposable plastic products are plastic materials or products that are not recyclable and become trash;

- Raw materials and the manufacturing process of different kind of plastic products are different;
- Plastic demand and consumption in different countries and regions are roughly the same (the general consumption of human life in different countries and regions is roughly the same);
- GDP in different regions can reflect the corresponding consumption of plastics.

2.2 Notations

Here are the notations and their meanings in this paper

Table 1

Notation	Meaning
R_P	Rank of Policies
R_T	Rank of Technique
R_{PR}	Rank of Production
R_{RA}	Rank of Ratio
R_O	Rank of quality
R	comprehensive Rank
GDP_{TI}	GDP of the tertiary industry
W_{MU}	Weight of Medical usage
W_{CU}	Weight of Catering usage
W_{DU}	Weight of Daily use
S	Total global plastic production of last year
N_C	Calculated value of plastic production
N_R	Ture value of plastic prodution
Q_{PR}	Quality coefficient of Production
Q_P	Quality coefficient of Pollicies
Q_T	Quality coefficient of Technique
Q	Comprehensive quality
CI	Consistency Indicators
RI	Average random consistency Index
CR	Consistency Ratio
e	Entropy
P_{ij}	Probability
R	Orijinal matrix
W_i	Weight
B	Combined weight
B_{ij}	Beneficial resault index
b_{ij}	Prime cost index
V	Ideal separate
S	Rate of progress
MID	Middle best fit separate
Z_i	Resultant vector
prox	Nearest neighbour operator
R_{mn}	Real
h	Gradient
A	Linear observation operator

3.Preparation

We classify disposable plastics and make evaluation standards to estimate plastic products' properties. The properties will be used in modeling.

3.1 Plastic Classification

According to function, we classify disposable plastics into four categories. The categories are given in Table 2.

Table 2

Category	Meaning
MU	Medical Usage
DU	Daily Use
CU	Catering Usage
IAU	Industrial & Agricultural Usage

3.2 Evaluation Standards

We set up four standards to estimate the properties of disposable plastic products. The indicators and definitions are listed below:

- Environmental pollution: this standard indicates the pollution degree of plastic products to the environment;
- Necessity degree: this standard indicates the demand degree of plastic products in real life;
- Processing difficulty: this standard reflects the resource consumption of plastic products' disposal;
- Production consumption: this standard reflects the resource consumption of plastic products in the production process.

3.3 Plastics Estimation

We use Analytic Hierarchy Process (AHP) to estimate the properties of the four categories. The evaluation standards are set as criteria, and the four categories are alternatives.

Figure 1

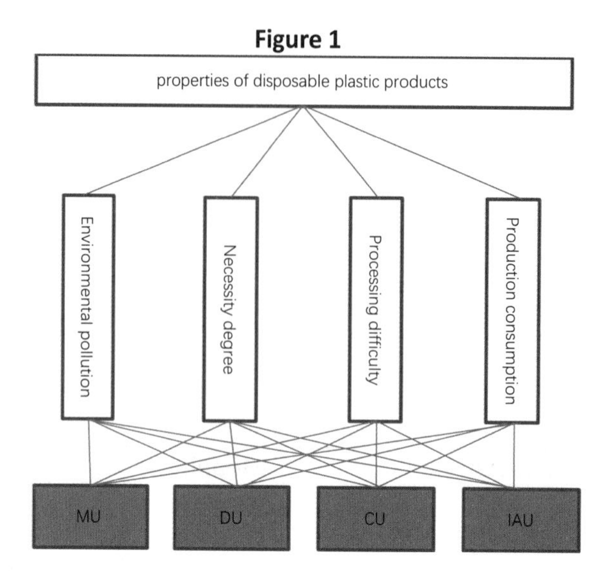

By looking up literature [1.3], we judged the priorities and get the judgement matrix of criteria (Table 3) and alternatives (Table 4).

Table 3

A	B1	B2	B3	B4
B1	1	1/5	1	6
B2	5	1	7	8
B3	1	1/7	1	2
B4	1/6	1/8	1/2	1

Table 4

B1	MU	DU	CU	IAU	B2	MU	DU	CU	IAU
MU	1	5	3	1/5	MU	1	1	5	7
DU	1/5	1	2	1/7	DU	1	1	4	6
CU	1/3	1/2	1	1/9	CU	1/5	1/4	1	3
IAU	5	7	9	1	IAU	1/7	1/6	1/3	1
B3	MU	DU	CU	IAU	B4	MU	DU	CU	IAU
MU	1	4	6	1/5	MU	1	1/2	1/4	1/8
DU	1/4	1	2	1/6	DU	2	1	1	1/6
CU	1/6	1/2	1	1/8	CU	4	1	1	1/7
IAU	5	6	8	1	IAU	8	6	7	1

The determinants we have listed are indicators that are harmful to human survival, and the larger the numbers are, the worse they are.

Through AHP calculation, we get the corresponding weights of each alternative and sort them. The sorting results are as follows.

Table 5

Rank	Alternatives	Weight
1	MU	0.3538
2	DU	0.2887
3	IAU	0.26
4	CU	0.0975

Since the determinants we have listed are indicators that are harmful to human survival, the larger the numbers are, the more harmful they are to human beings.

Through the analytic hierarchy process, we get that the Medical usage (MU) disposable plastic products is the worst, the corresponding weight is 0.3538, Daily use (DU) is the second, the corresponding weight is 0.2887, the next is Industrial & Agriculture usage (IAU), the corresponding weight is 0.2600, Catering usage (CU) is the best, the corresponding weight is 0.0975.

4. Model Construction

4.1 Evaluation Model for Treatment Methods

The recycling of plastics can provide great convenience for the environment on which humans depend, and encourage plastic production and processing enterprises to innovate recycling technologies to control environmental pollution caused by plastic waste. We must first solve the non-essential disposable plastic packaging. Starting from this point is the most cost-effective project at present.

p_i $(i = 1, 2, \ldots m)$ Here we use the entropy weight method. Information is a measure of the order of the system; entropy is a measure of the degree of disorder of the system. If the system may be in many different states, the probability of each state

$$e = -\sum_{n=1}^{i} p_i \ln p_i$$

Obviously when$p_i = \frac{1}{m}$ $(i = 1, 2, \ldots m)$, That is, when the states have the same probability, the entropy takes the maximum value, that is $e_{max} = \ln m$

Now we have m pieces of indexes to be evaluated, n pieces of indexes to be complain, Form the original evaluation matrix$R = (r_{ij})_{m\times n}$, Information entropy for a certain r_j:

$$e_j = -\sum_{i=1}^{m} p_{ij} \ln p_{ij}$$

$$p_{ij} = \frac{r_{ij}}{\sum_{i=1}^{m} r_{ij}}$$

It can be seen from the above description that the greater the degree of variation in the entropy value of an indicator, the more information it provides. In comprehensive evaluation, the greater the work done by the indicator, the greater its weight. Therefore, the entropy weight of each index can be calculated by using the entropy of each index value, and all indexes are weighted by the entropy weight of each index, so as to obtain an objective evaluation result, so the calculation process is : if we have m pieces of projects to be evaluated , and n pieces of indexes to evaluated $R = (r_{ij})_{m\times n}$:

$$R = \begin{pmatrix} r_{11} & \cdots & r_{1n} \\ \vdots & \ddots & \vdots \\ r_{m1} & \cdots & r_{mn} \end{pmatrix}_{m\times n}$$

(1) Calculate the proportion p_{ij} of the index value of the jth index and the item index

$$p_{ij} = \frac{r_{ij}}{\sum_{i=1}^{m} r_{ij}}$$

(2) Calculate the entropy value e_j of the jth index

$$e_j = -k\sum_{i=1}^{m} p_{ij}\ln p_{ij}，\ k = \frac{1}{\ln m}$$

（3）Calculate the entropy weight of the jth index $w_j = \frac{(1-e_j)}{\sum_{j=1}^{n}(1-e_j)}$

（4）Determine the comprehensive weight of the indicator β_j

Assuming that the evaluator weights the importance of the indicator α_j according to his own purpose and requirements，$j = 1,2,\dots n$

Combining the index's entropy weight w_j to get the comprehensive weight of j:

$$\beta_j = \frac{\alpha_i w_i}{\sum_{i=1}^{m}\alpha_i w_j}$$

From the above methods, our team believes that eliminating problematic or unnecessary disposable plastic packaging has the largest weight to achieve the plastic circular economy. The following analyzes other factors and the assessment and weight analysis of factors that my team believes: We are divided into the following three goals Compare:

- Decision 1: Reduce unnecessary disposable plastics

Table 6

	Benefit	Cost	Section
Entropy	0.9994	0.9995	0.9993
The entropy weight	0.1016	0.0952	0.1335

- Decision 2: Recycle compost plastic packaging

Table 7

	Benefit	Cost	Section
Entropy	0.9940	0.9603	0.9000
The entropy weight	0.0130	0.1227	0.1312

- Decision 3: Manufacture 100% reusable packaging plastics

Table 8

	Benefit	Cost	Section
Entropy	0.9014	0.4367	0.8972
The entropy weight	0.0213	0.8465	0.1212

The evaluation is as follows

① Constructing a weighted change matrix

$$V = \begin{bmatrix} 0.9994 & 0.9995 & 0.9993 \\ 0.9940 & 0.9603 & 0.9000 \\ 0.9014 & 0.4367 & 0.8972 \end{bmatrix}$$

Determine the ideal solution

$$V_+ = (0.0267,0.0266,0.0144,0.0255,0.0355,0.0370,0.0171,0.0327,0.0198)$$
$$V_- = (0.0239,0.0235,0.0133,0.0232,0.0316,0.0329,0.0290,0.0180,0.0270)$$

② Determining closeness to reasonable comprehensive income

Table 8

Decision	1	2	3
S+	0.0059	0.0036	0.0073
S-	0.0047	0.0075	0.0028
Close degree	0.5566	0.3243	0.7228

It can be seen that the closeness of decision 1 is 0.5566, which is the closest to a reasonable standard.

Weighted summation of indicators is $Z_i = \sum_{j=1}^{n} w_i b_{ij}$

Then the decision closeness is obtained as a decision matrix $B = \begin{bmatrix} 0.0059 & 0.0036 & 0.0073 \\ 0.0047 & 0.0075 & 0.0028 \\ 0.5566 & 0.3243 & 0.7228 \end{bmatrix}$

(1) Profitable Reasonable Index

$$B_{ij} = \begin{cases} 1 \\ \dfrac{\alpha_{ij} - S_i}{\vartheta_j - S_j}, S_j < \alpha_{ij} < \vartheta_j \\ \sigma_j \end{cases}$$

(2) Reasonable cost indicators

$$b_{ij} = \begin{cases} 0, & \alpha_{ij} \gg \vartheta_j \\ \dfrac{\vartheta j - \alpha_{ij}}{\vartheta j - Sj} & , S_j < \alpha_{ij} < \vartheta_j \\ 1, & \alpha_{ij} \ll \vartheta_j \end{cases}$$

(3) Interval indicator

$$\begin{cases} 0, \ \alpha_{ij} \gg \vartheta_j \\ \dfrac{\vartheta_j - \alpha_{ij}}{\vartheta_j - MID_j}, MID_j < \alpha_i < \vartheta_j \\ 1, \ \alpha_{ij} = MID_j \\ \dfrac{\alpha_{ij} - Sj}{MID_j - S_j}, S_i < \alpha_i < MID_j \\ 0, \alpha_{ij} \ll S_j \end{cases}$$

MID_j is Middle optimum

After the index is met, the decision matrix is returned. We first discuss the general algorithm.,

$$H_j = \frac{\sum_{i=1} f_{ij} \cdot \ln f_{ij}}{\ln m} (i = 1,2, \dots m; j = 1,2, \dots n)$$

H_j has the entropy of j indexes, the number of m schemes, and the number of n .

$$f_{ij} = \frac{1 + b_{ij}}{\sum_{i=1}^{n} (1 + b_{ij})}$$

To make $\ln(f_{ij})$ meaningful and consistent with the definition of entropy, $w_j = \frac{H_j}{n - \sum_{j=1}^{n} H_j}$

$$B = \begin{bmatrix} 0.857 & 0.144 & 0.064 & 0.009 & 0.089 & 1.000 & 0.353 & 0.600 & 0.700 & 0.700 & 0.571 \\ 1.000 & 0.086 & 0.116 & 0.011 & 0.100 & 0.556 & 0.800 & 0.700 & 0.286 & 0.053 & 0.071 \\ 0.927 & 0.014 & 0.111 & 0.400 & 1.000 & 0.333 & 0.770 & 0.900 & 0.800 & 0.900 & 0.385 \end{bmatrix}$$

Then the weight vector is $W_j = (0.097, 0.160, 0.095, 0.018, 0.058, 0.000, 0.176, 0.024, 0.016, 0.013)$

Integrated vector $Z_i = (0.322, 0.271, 0.119)$

Z_i is the biggest that our team considers to reduce unnecessary plastic products, which shows that this decision is the most practical, close to the ideal, saving manpower and resources, and practical.

In summary, the advantages of the entropy weight method over those of the subjective assignment method are high accuracy and strong objectivity, which can better explain the results obtained.

Adaptability: It can be used in any process that requires weights. It can be evaluated in combination with the method of evaluating weights inward, and the best is selected.

4.2 Optimization Model of Reducing Plastic Damage

There are many countries and regions in the world. According to the economic level and the similarity degree of the relevant policies of each country on plastics [4,5], we divide the world into the following nine regions: East Asia (EA), Southeast Asia (ESA), Europe (EUR), Africa (AF), Oceania (OA), North America (NA), Latin America (LAN), South America (SA), Middle East (ME). (ranking in no order) After the regional division, it can better reflect the impact of policies and conditions in different regions on the production of disposable plastic products.

We have developed three evaluation criteria for the comprehensive ability of a region to process plastics:

- Policies: it reflects the enthusiasm of the region for the management or treatment of disposable plastic products;
- Production: It reflects the production capacity of disposable plastic products in this area
- Technique: It reflects the process capacity of disposable plastic products in this area.

Based on the network data [5] and the policy analysis of disposable plastic products issued by various countries, we rank the policies and techniques of each region from 9 - to 1, 9 represents the strongest, 1 represents the weakest, and the ability decreases from 9 to 1; for production, we define the PPR as the percentage of the region's 2018 global plastic GDP × 45 The percentage of regional annual plastic production in global production. (in order to be obvious in the following three indicators, namely percentage (P_{PR})$\times\sum_{i=1}^{9} i$) R_P, R_T, R_{PR} represent the ranking of regions in Policies, Technique and Production. The comprehensive level data of each region are as follows.

Table 9

Region	Rank	
	Policies (R_P)	Technique (R_T)
EA	5	6
ESA	3	4
NA	8	9
LAN	4	3
SA	7	7
OA	6	5
EUR	9	8
AF	1	1
ME	2	2

Table 10

Region	Production (P_{PR}*45)	Rank (R_{PR})
EA	0.25*45	9
ESA	0.08*45	3
NA	0.18*45	8
LAN	0.09*45	5
SA	0.12*45	6
OA	0.1*45	4
EUR	0.15*45	7
AF	0.01*45	1
ME	0.02*45	2

Figure 2

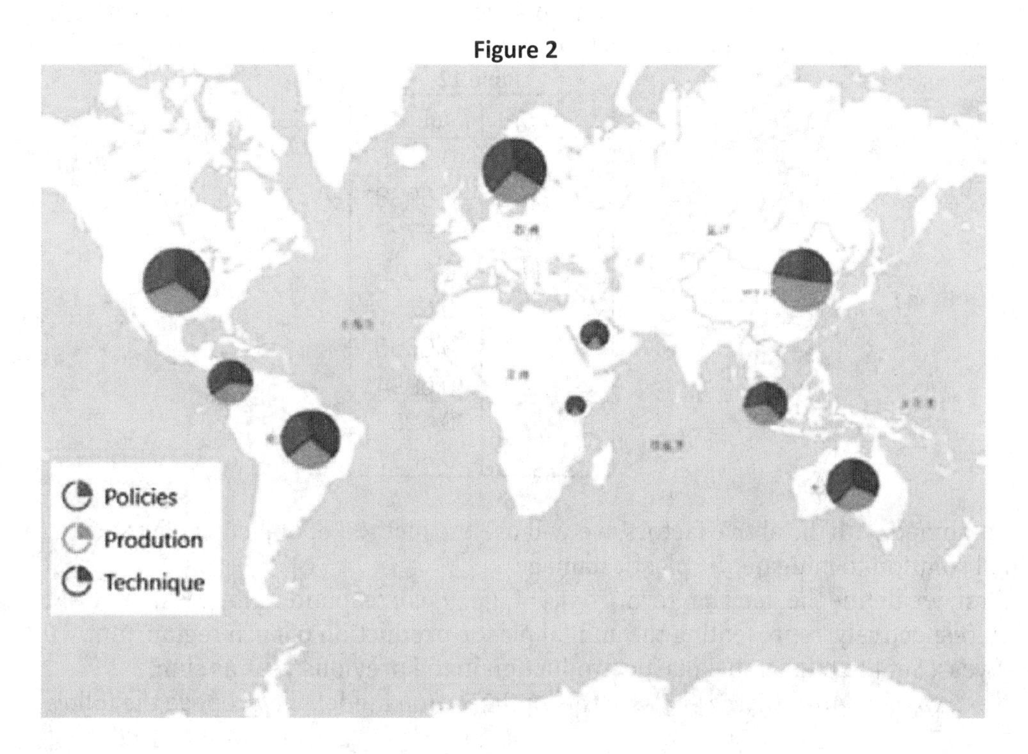

Figure 2 shows different regions condition by the standards.

We set R as the comprehensive ranking of regional processing plastic products, and the calculation formula is

$$R=\frac{R_P+R_T+R_{PR}}{3}$$

After calculation, the comprehensive ranking of each region is as follows

Table 11

Region	R
EA	7
ESA	3
NA	9
LAN	4
SA	7
OA	5
EUR	8
AF	1
ME	2

In the classification calculation, according to the data [7.8], we call Industrial & Agricultural usage (IAU) as the first and second industry plastics, and Medical usage (MU), Daily use (DU) and Catering usage (CA) as the tertiary industry plastics. The following table shows the percentage of tertiary industry plastics in total production in 2018 in each region.

Table 12

Region	GDP_{TI}
EA	0.5216
ESA	0.6938
NA	0.7737
LAN	0.6014
SA	0.6263
OA	0.6656
EUR	0.6994
AF	0.4267
ME	0.4837

Combined with the above factors, we will use the method of objective planning to find the optimal solution to mitigate the plastic damage.

First, we define the decision term. x_j=j(j=1,2,...,9) correspond to EA, ESA, NA, LAN, SA, OA, EUR, AF, ME respectively, representing the annual plastic production of each region, unit: 100 million tons / year. Set the total global plastic production in the previous year as sum.

Next, we need to clear the goals. In this optimization model, we propose the following objectives:

1. The total annual plastic production in each region shall not exceed the previous global plastic production (P_1);
2. Priority should be given to reducing the production of Medical usage, followed by Daily use and Industrial & Agricultural usage, and finally Catering usage (P_2);
3. The total annual plastic production in each region shall not exceed the total known production in the previous year (P_3).

The above three conditions are arranged according to the priority level, that is, goal (1) has the highest priority and needs to be satisfied first, followed by goal (2), and goal (3) has the lowest priority and could not be satisfied. Among them, we set the goal (1) as a rigid constraint, that means goal (1) must be satisfied, and the rest goals as flexible constraints.

Based on the above objectives, the corresponding goal programming model is

$$\min\ z = P_1 d_1^+ + P_2(d_2^- + d_2^+) + P_3(d_3^+ + d_4^+ + \cdots + d_{n+2}^+) \quad (1)$$

$$\begin{cases} \sum_{j=1}^{9} x_j + d_1^- - d_1^+ = SUM & (2) \\ \sum_{j=1}^{9} x_j \cdot GDP_{TIj} + d_2^- - d_2^+ = (W_{MU} + W_{DU} + W_{CU}) \cdot SUM & (3) \\ x_j + d_{j+1}^- - d_{j+1}^+ = W_{PR}S & (4) \\ x_j > 0\ j = 1,2, \ldots n \\ d_i^-, d_i^+ \geq 0\ i = 1.2, \ldots, n+2 \end{cases}$$

In the above model, Medical usage, Daily use and Catering usage are uniformly calculated as the tertiary industry plastics, in which GDP_{TIj}（j=1.2...9）refers to the percentage of the tertiary industry plastics output in the total GDP of each region in the previous year, W_{MU}, W_{DU}, W_{CU} refer to the total ranking weight calculated in the corresponding AHP model. In the case of the three goals set, formula (2) represents the global annual plastic output, formula (3) represents the plastic output of the tertiary industry, and formula (4) represents the annual plastic output of each region.

5. Extension of the Model

The above model is based on the damage degree of four categories of plastics obtained from AHP. When the model is extended, the following steps need to be carried out:

(1) Number of countries or regions to be investigated clearly ;

(2) Rank Policies, Technique and Production from 9 to 1 according to the actual situation; (note that 9 in production indicates the maximum output, 1 indicates the minimum output, 9 in policies and technology indicates the maximum policy support, the most advanced processing technology, 1 indicates the minimum policy support, and the most backward processing technology)

(3) Calculate the comprehensive ranking r according to the formula;

(4) Inquire the percentage of the annual output value of the tertiary industry plastics in the total output value of the country or region to be investigated;

(5) Query the annual output value of total plastics in the country or region to be investigated;

(6) Put the data into the model.

6. Answers

We found that the total output value of the world's plastics in 2018 is about 3.6*100 million tons, that is, sum = 3.6. In the above table (indicate the table), we listed the percentage of the plastic production of each region in the global total plastic output value in 2018 and the percentage of the plastic output value of the tertiary industry in the total plastic output value of the region.
x_j=j(j=1,2,...,9) respectively correspond to EA, ESA, NA, LAN, SA, OA, EUR, AF, ME, put the data into the Goal Programming model. We get the following results

Table 13

Comparation		N_C	N_R	Residual	Ratio	R_{RA}
Region	EA	0.88	0.9	-0.02	-2.22%	3
	ESA	0.263	0.288	-0.025	-8.68%	5
	NA	0.563	0.648	-0.085	-13.12%	8
	LAN	0.33	0.324	0.006	1.85%	2
	SA	0.378	0.432	-0.054	-12.50%	7
	OA	0.318	0.36	-0.042	-11.67%	6
	EUR	0.464	0.54	-0.076	-14.07%	9
	AF	0.062	0.036	0.026	72.22%	1
	ME	0.07	0.072	-0.002	-2.78%	4
Sum		3.328	3.6	-0.272	-7.56%	

According to the table, the annual production of the world's largest plastic acceptable under the constraint conditions is 3.328*100 million tons / year, which is 7.56% less than the annual production of global plastic in 2018.

Combined with Figure 2 and the Tables above, we define R_{RA} as the ranking of decline range, Q as the regional comprehensive quality, Q_{PR} as the quality coefficient corresponding to production in each region, Q_P as the quality coefficient corresponding to policies in each region, and Q_T as the quality coefficient corresponding to technology in each region. We define Q as the calculation formula

$$Q = Q_{PR} \cdot R_{PR} + Q_P \cdot R_P + Q_T \cdot R_T$$

$$|Q_{PR}| + |Q_P| + |Q_T| = 1$$

$$Q_{PR} < 0, 0 < Q_P, Q_T < 1, |Q_{PR}|, |Q_P|, |Q_T| < 1$$

In the formula, we define $Q_{PR} < 0$ to indicate that the production of the region has side effects on the comprehensive plastic processing capacity of the region, and policies and techniques of the region with $0 < Q_P, Q_T < 1$ have positive effects on the comprehensive plastic processing capacity of the region; and , $|Q_{PR}|, |Q_P|, |Q_T| < 1$ indicate that the three quality factors can not completely determine the comprehensive plastic processing capacity of a region alone.

We have known the R_P, R_T, R_{PR}, R_{RA} of each region in 2018. Let the R_{RA} of x_a region be higher than that of x_b region, we can calculate theQ_{PR}, Q_P, Q_Tthat meet the conditions through the following inequality group.

$$Q_{PRx_a} \cdot R_{PRx_a} + Q_{Px_a} \cdot R_{Px_a} + Q_{Tx_a} \cdot R_{Tx_a} \geq Q_{PRx_b} \cdot R_{PRx_b} + Q_{Px_b} \cdot R_{Px_b} + Q_{Tx_b} \cdot R_{Tx_b}$$
$$1 \leq b < a < 9$$

Figure 3

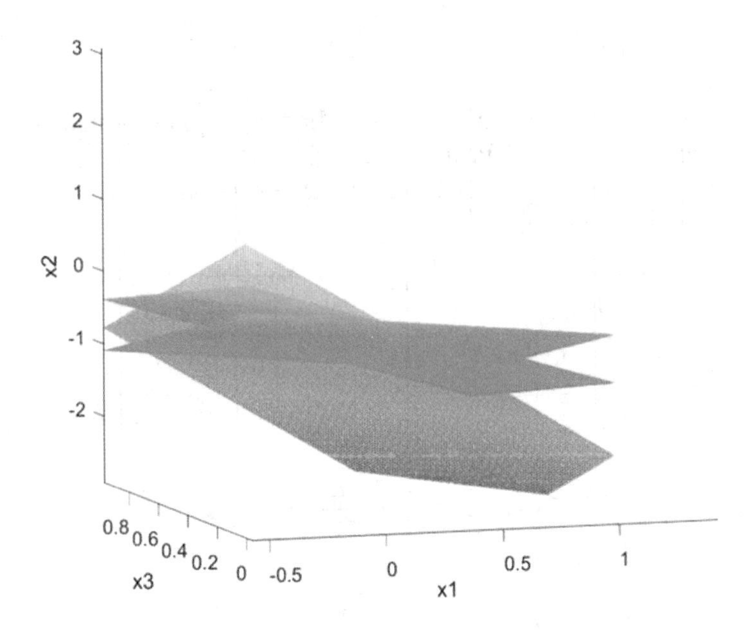

By calculation, in the feasible region, we can get

$$Q_{PR} = -0.35$$

$$Q_P = 0.20$$

$$Q_T = 0.45$$

These inequalities are satisfied. We calculate and rank W with the obtained W_{PR}, W_P, W_T and compare it with theR_{RA} calculated before, and get the following table

Table 14

Region	Q	R_W	R_{RA}
EA	0.55	3	3
ESA	1.35	5	5
NA	2.85	8	8
LAN	0.4	2	2
SA	2.45	7	7
OA	2.05	6	6
EUR	2.95	9	9
AF	0.3	1	1
ME	0.6	4	4

From the above table, we can see that the rank of W calculated by us is consistent with the rank of ratio calculated by using objective planning, that is, $R_{Qx_i} = R_{RAx_i}(i = 1,2 \dots,9)$. It is proved that the w calculated by us is valuable and can be used to evaluate the comprehensive ability of a certain area for plastics.

Combining the calculation formulas and results of $Q_{PR}, Q_P, Q_T, Q, R_P, R_T, R_{PR}, R_{RA}, R$

$$|Q_T| < |Q_{PR}| < |Q_P| < 1$$

$$-1 < Q_{PR} < 0 < Q_{PR} < Q_{PR} < 1$$

The results showed that Policies had the greatest impact on the comprehensive ability of plastic processing in a region, followed by Production and Technique.

Then we compare the difference between the comprehensive ranking R and R_Q The comparison results are shown in the table below

Table 15

Region	R_Q	R	$\|R_Q-R\|$
EA	3	7	4
ESA	5	3	2
NA	8	9	1
LAN	2	4	2
SA	7	7	0
OA	6	5	1
EUR	9	8	1
AF	1	1	0
ME	4	2	2

From the analysis of the above table, we can see that the fluctuation of R, R_Q is not obvious when it is large or small, which indicates that the countries and regions with strong and weak comprehensive processing capacity of plastics generally will not change greatly under any evaluation standard.

From the above table, we select the regions NA, SA, OA, AF with no obvious fluctuation to calculate the annual net plastic output of these four regions under the goal we set, that is, M = -0.085 + (-0.054) + (-0.042) + 0.026 = -0.155. Considering the influence of production, policies and technology on the annual net plastic output, we introduce v as the deceleration coefficient.

$$v = Q_{PR} + Q_T + Q_P$$

In the scenario we calculated, $v = 0.3$ so we get the annual net plastic output under the set target of M * V = -0.0465.

According to our calculation, the acceptable annual plastic production of the earth is $3.328*10^2$ million tons / year, so the time t required to reach the expected annual plastic production is

$$T = \frac{\Delta S}{M \cdot v}$$

In our calculation scenario, T = 5.849, that is to say, the plan will be implemented in 2019 and the target will be achieved in 2025.

7.Conclusions

- Under Environmental pollution, Necessity degree, Processing difficulty, Production consumption, the rank of the categories is Medical usage, Daily use, Industrial & Agricultural usage, Catering usage ;
- The most effective method among the three popular plastic treatment methods is to reduce unnecessary disposable plastic; the second is to recycle compost plastic packaging; the effect of manufacturing 100% reusable packaging plastic is the worst;

- The plastic output has the greatest negative impact on a country or region's comprehensive plastic processing capacity, and the policy has the greatest positive impact;
- No matter from which perspective, countries or regions with strong comprehensive capacity are always strong, and countries or regions with weak comprehensive capacity are always weak. Countries or regions in the middle of the ranking will fluctuate in different evaluation perspectives.

8.Model Analytics

8.1 Validation of evaluation model of treatment method

Amendment: We introduced Cross Matrix reinforcement method to improve and find the minimum.

Determine the index system for the evaluation purpose, establish an evaluation system for the evaluation problem, and record the decisions F1, F2, F3, and extend to F1, ... Fn ....
Determine the relationship between influencing directories. Factors have influence on factors. We construct a directed graph to reflect the relationship between factors.

Figure 4

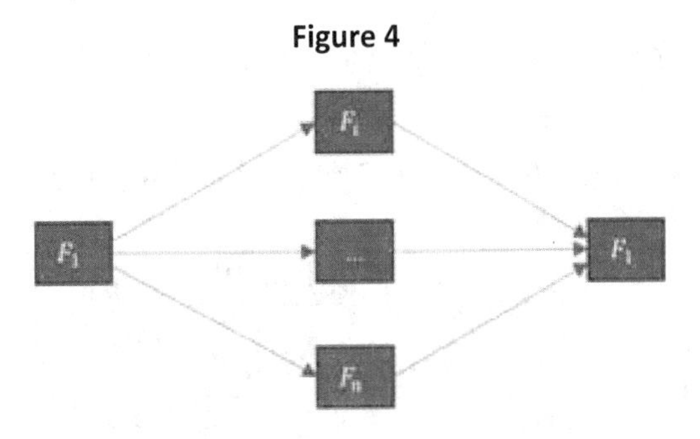

l_x is the initialization matrix, and the matrix is used to represent the direct interaction between various indices

Calculate the combined impact matrix between the indexes, $T = x + x^2 + \cdots + x^n$
Here, the comprehensive influence matrix reaction indexes are correlated with each other, and as the indexes, see cross reinforcement matrix.

$$w_i = \frac{\sum_{i=1}^{n} \beta_0 w_j}{\sum_{i=1}^{n} \sum_{j=1}^{n} \beta_{ij} w_j}$$

Calculate the new weight of the evaluation index by cost, benefit, interval weight

$$w = d_{w_i} + (1 - d) w_{ij}, 0 \ll d \ll 1$$

Calculate the comprehensive value weight
Calculate the comprehensive benefit weight

$$w = \frac{1}{2}\left(w_i + \frac{\sum_{i=1}^{n} \beta_{ij} w_j}{\sum_{i=1}^{n} \sum_{j=1}^{n} \beta_{ij} w_j}\right)$$

Table 16

Decision	1	2	3
Sum of cross reinforcement	0.3065	0.0963	0.1118
Effect weight	0.0998	0.0873	0.0764
The initial weights	0.0267	0.0064	0.1034
Normalize	0.1045	0.3199	0.1796

After we introduced the comprehensive impact matrix, in order to further find the minimum, we introduced a log -sum penalty algorithm and found the minimum. Formula introduction:

Matrix: Matrix$M = (m_{ij})_{m \times n} \in R^{m \times n}$, Frobenius Norm

$$\|M\|_F = \sqrt{\sum_{i=1}^{m}\sum_{j=1}^{n} m_{ij}^2} \; \|M\|_* = \sum_{i=1}^{n} \sigma_i(m),$$

The last one is the nuclear norm

Nearest neighbor operator h is $R^{m \times n}$ a real-valued convex function. To any $x \in R^{m \times n}$,

$$prox\hbar(x) = arg \min_m \{\frac{1}{2} \|m - x\|_F^2 + h(m): x \in R^{m \times n}\}$$

$prox\hbar$ is $R^{m \times n}$ deterministic nearest neighbor

A is Linear observation operator,m Is the information of the matrix to be recovered, Poisson (k)(Intensity is k)Poisson Noise, Build

$$p(y|m) = \prod_{i=1}^{N} \frac{[A(m)]Y_i^i}{y_{i!}} e^{-[A(m)]}$$

Posterior promotion :

$$p(m|y) = cp(y|m)p(m) = e \prod_{i=1}^{N} \frac{[A(m)]Y_i^i}{y_{i!}} e^{-[A(m)]}$$

Here is the noise reduction process based on the Poisson probability distribution, which is beneficial to protect the relative error and lost data in the process of data processing

$$p(m|y) = cp(y|m)p(m) = e \prod_{i=1}^{N} \frac{[A(m)]Y_i^i}{y_{i!}} e^{-[A(m)]}$$

First, the upper bound control function of the original objective function is constructed based on the current iteration point, then the next iteration point is obtained by minimizing the upper bound control function, and then an iterative sequence is obtained by repeating this process. It is easy to verify that the loss control function is smooth. As we all know, here is the step size corresponding to the gradient descent method, which can be solved by the threshold shrinkage method. The solution can be expressed by the nearest neighbor operator. The dynamic update of the step size can speed up the convergence speed of the algorithm.。

In summary, the table data on the previous page is optimized by one step:

decision ① k = 1,

decision ② k = 2,

decision ③ k = 3.

Take decision 1 as an example

(1) Initialize k = 1, initialize the step size parameter L through the BB criterion, and start with k = 1 first

(2) Gradient descent$x_k = M_k - \frac{1}{L_k^{-b}} \vartheta g(M_k)$

(3) Threshold contraction$M_{k+1} = prox\Omega\lambda/L_k(x_k)$

(4) Step parameter adjustment$L_k^{t+1} = rL_k^t$

(5) Check if step search criteria are met$|f(M_{k+1}) - f(M_k)| \ll rL_k^t$,

Meet the output Then the minimum is evaluated by the log-sum penalty algorithm as:

$$\frac{M_{k+1}=arg \min_{m}\{\frac{1}{2}\left\|M-(M_k-\frac{1}{L}\vartheta g(M_k)\right\|_F^2+\frac{\lambda}{L}\Omega(m))}{f(M_k)+\frac{ML_k^t}{2}\|M_{k+1}-M_k\|_F^\alpha} = 70\%$$

After Poisson matrix noise reduction processing and log-sum algorithm evaluation, we see that the minimum amount is 70%. If our world government uses decision 1 to reduce unnecessary plastics, it will have the highest efficiency. After the minimum reaches 70%, it will be possible. Contribute to the environment

8.2 Verification of Optimization Model of Reducing Plastic Damage

The utilization model of disposable plastic products was tested. The first is to test the AHP model. Here we introduce the maximum eigenvalue λ max of matrix A. Because the feature roots continuously depend on the scale of AHP, we need to test the consistency of the judgment matrix to verify the rationality of the model. The steps of one-time inspection are as follows:

(1) Calculate CI

$$\mathrm{CI} = \frac{\lambda\ \mathrm{max} - \mathrm{n}}{n - 1}$$

(2) Search RI

Table 17

n	1	2	3	4	5	6	7	8	9
RI	0	0	0.58	0.90	1.12	1.24	1.32	1.41	1.45

(3) Calculate CR

$$CR = \frac{CI}{RI}$$

When CR<0.10, the result can be accepted.

Table 18

Judgement matrix	CR
A	0.0804
B1	0.0839
B2	0.0265
B3	0.0821
B4	0.0531
Total	0.0444

CR<0.1000, which means the model is correct.

9. Strengths & Weaknesses

9. 1 Strengths

1. Use AHP. AHP is used to evaluate the hazard degree of four kinds of common disposable plastic products, which is beneficial to make targeted policies for specific types of plastic products;

2. Multi angle evaluation of different regions. From the perspectives of Production, Policies and Technique, this paper evaluates and ranks the comprehensive capabilities of different regions in the world, which is conducive to the adjustment of local policies;

3. Use goal planning. Use goal planning to calculate how to achieve the expected goal under the set goal. The result of goal planning is flexible, which allows the result to fluctuate on the premise of reality;

4. Popularization. It can be applied to all kinds of analysis process that need weight, and can also be used together with some analysis methods. The cross matrix analysis method introduces reinforcement matrix and Poisson optimization algorithm, so as to reduce the error,

9.2 Weakness

1. Hypothesis weakens the reality of the model. We set up five hypotheses before modeling, which will weaken the practical significance of modeling;

2. Subjectivity. Although the data is objective, there are inevitably subjective ideas when ranking the impact of AHP and Production, Policies, and Technique on a region;

Appendix

1. Appendix A Analytic Hierarchy Process Scales

Scale	Meaning
1	Indicates that the two factors are of the same importance
3	The former is slightly more important than the latter
5	The former is more important than the latter
7	The former is more important than the latter
9	The former is more important than the latter
2,4,6,8	Represents the intermediate value of the above adjacent judgments
Reciprocal	If the importance ratio of factor I to factor j is a_{ij}, then the importance ratio of factor j to factor I is $a_{ij}=1/a_{ji}$

2. Appendix B Analytic Hierarchy Process Indexes

```
txt3.t
1       1/5     1       6
5       1       7       8
1       1/7     1       2
1/6     1/8     1/2     1
1  5  3  1/5
1/5     1  2  1/7
1/3     1/2     1  1/9
5  7  9  1
1  1  5  7
1  1  4  6
1/5     1/4     1  3
1/7     1/6     1/3     1
1  4  6  1/5
1/4     1  2  1/6
1/6     1/2     1  1/8
5  6  8  1
1  1/2  1/4     1/8
2  1  1  1/6
4  1  1  1/7
8  6  7  1
```

3. Appendix C Analytic Hierarchy Process (Matlab)

```
clc,clear
```

```
fid=fopen('txt3.txt','r'); n1=4;n2=4;
a=[];
for i=1:n1
tmp=str2num(fgetl(fid));
a=[a;tmp];
end
for i=1:n1
str1=char(['b',int2str(i),'=[];']); str2=char(['b',int2str(i),'=[b',int2str(i),';tmp];']);
eval(str1);
for j=1:n2
tmp=str2num(fgetl(fid));
eval(str2);
end
end
ri=[0,0,0.58,0.90,1.12,1.24,1.32,1.41,1.45];
[x,y]=eig(a); lamda=max(diag(y));
num=find(diag(y)==lamda);
w0=x(:,num)/sum(x(:,num));
cr0=(lamda-n1)/(n1-1)/ri(n1)
for i=1:n1
[x,y]=eig(eval(char(['b',int2str(i)])));
lamda=max(diag(y));
num=find(diag(y)==lamda);
w1(:,i)=x(:,num)/sum(x(:,num));
cr1(i)=(lamda-n2)/(n2-1)/ri(n2);
end
cr1, ts=w1*w0, cr=cr1*w0
```

4. Appendix D Multi-objective Programming (Lingo)

```
model:
sets: level/1..3/:p,z,goal;
variable/1..9/:x;
s_con_num/1..11/:g,dplus,dminus;
s_con(s_con_num,variable):c;
obj(level,s_con_num)/1 1,2 2,3 3,3 4,3 5,3 6,3 7,3 8,3 9,3 10,3 11/:wplus,wminus;
endsets
data:
ctr=?;
goal=? ? 0;
g=3.600 2.664 0. 900  0. 288  0. 648  0. 324  0. 432  0. 360  0. 540  0. 036  0. 072;
c=1 1 1 1 1 1 1 1 1 0. 5216  0. 6938  0. 7737  0. 6014  0. 6263  0. 6656  0. 6994  0. 4267  0. 4837  1  0  0
0  0  0  0  0  0  0  1  0  0  0  0  0  0  0  0  0  1  0  0  0  0  0  0  0  0  0  1  0  0  0  0  0  0  0  0  0  1  0  0  0  0  0
0  0  0  0  1  0  0  0  0  0  0  0  0  0  1  0  0  0  0  0  0  0  0  0  1  0  0  0  0  0  0  0  0  0  1;
wplus=1 1 1 1 1 1 1 1 1 1 1;
wminus=0 1 0 0 0 0 0 0 0 0 0;
enddata
min=@sum(level:p*z);
p(ctr)=1;
```

```
@for(level(i)|i#ne#ctr:p(i)=0); @for(level(i):z(i)=@sum(obj(i,j):wplus(i,j)*dplus(j)+wminus(i,j)* d
minus(j))); @for(s_con_num(i):@sum(variable(j):c(i,j)*x(j))+dminus(i)-dplus(i )=g(i));
@for(level(i)|i #lt# @size(level):@bnd(0,z(i),goal));
end
```

References

[1] Wang Juying, Lin Xinzhen. Analysis of marine governance system to deal with plastic and microplastics pollution [J]. Pacific Journal, 2018, 26 (04): 79-87.
[2] The EU will completely ban disposable plastic products from 2021 [J]. Jiangxi Building Materials, 2019 (04): 18.
[3] Chen Weiqiang. Evolution of global plastic recycling system and China's response strategy [n]China environment daily, January 9, 2020 (003)
[4] Liao Qin, Wang Jinping, Wang Fang. International Single-use Plastic Pollution Control Policy and Its Implications for China [J]. Global outlook on science, technology and economy, 2018,33 (07): 63-71.
[5] Plastics policies of various countries in the world [J]. Plastics manufacturing, 2012 (08): 65
[6] https://www.breakfreefromplastic.org/
[7] https://data.worldbank.org.cn/indicator/SH.XPD.TOTL.ZS
[8] https://www.kylc.com/stats/global/yearly_overview/g_industry_value_added_in_gdp.html

5.基于遗传算法的智能RGV动态调度模型

摘　要

近年来，随着 RGV 在自动化生产中的广泛应用，如何构建和优化智能 RGV 的动态调度策略具有十分重要的现实意义。由于 RGV 的作业状况复杂多样、作业路径随机性大，本文综合采用了混合遗传算法、随机故障模拟以及降维的思想构建了四种不同生产条件下的 RGV 动态调度策略模型，分别求出了在一道工序无故障、一道工序有故障、两道工序无故障、两道工序有故障这四种条件下，连续作业 8 小时且完成成料最多时，RGV 的最优工作顺序及每个步骤的具体时间等相关数值。

本文利用混合遗传算法对智能 RGV 动态调度策略问题进行模拟，以提高 RGV 系统的作业效率；运用随机化算法对故障发生进行模拟以排查故障所用的时间；采用降维的方法对二维路径数据进行处理，以使在多重工序时仍可简便的使用遗传算法。

针对问题一，一道工序无故障调度策略问题，首先根据智能加工系统合理给出约束条件和目标函数，构建基于遗传算法的一道工序无故障模型，运用 MATLAB 软件遗传算法工具箱代入三组不同工作参数值分别进行求解验证，得出三组参数在 8 小时内分别最多加工完成 356、336、366 个成品。

针对问题二，一道工序含故障调度策略问题，首先运用随机化算法对故障进行模拟，再在问题一无故障模型的基础上构建一道工序含故障模型，运用 MATLAB 软件遗传算法工具箱代入三组不同工作参数值分别进行求解，得出三组参数在 8 小时内分别最多加工完成 346、326、353 个成品。

针对问题三，两道工序无故障调度策略问题，首先运用降维思想，将具有顺序性的工序转化为定向路径，构建了基于遗传算法的两道工序无故障模型，运用 MATLAB 软件遗传算法工具箱代入三组不同工作参数值分别进行求解，得出了三组参数在 8 小时内分别最多加工完成 66、55、57 个成品。

针对问题四，两道工序含故障调度策略问题，对问题二的随机化算法和问题三的降维思想进行综合利用，构建了基于遗传算法的两道工序含故障模型，运用了 MATLAB 软件遗传算法工具箱代入三组不同工作参数值分别进行求解，得出了三组参数在 8 小时内分别最多加工完成 62、53、55 个成品。

最后，对模型进行了合理性分析、对多模型的优缺点进行客观评价、对模型效率进行计算，并对包含故障模型中的随机量进行了灵敏度分析。同时，还对模型进行了优化和推广。

关键词：RGV 动态调度；遗传算法；随机化算法；MATLAB

作　　者：陈佳澄，宋宇格，王梅

指导教师：吴养会

奖　　项：2018年全国大学生数学建模竞赛国家二等奖

§1 问题的重述

一、 引言

RGV 是一种智能的有轨制导车辆。由于其低成本、高稳定性、高安全性以及灵活的轨道设计，在各种生产线中占据主导地位。而“如何调度 RGV 能在一定的时间带来最大的利润”也成为近期人们关注的热点。

由于小车的调度环境和任务存在着不可预测的扰动，因此本问利用遗传算法模型对智能 RGV 的动态调度策略问题提出了能够针对生产现场实际情况，给出具有可操作性的决策方案，并进行验证。

此模型解决了直线往复 RGV 的动态调度策略问题，并给出了在不同的生产工作要求下，系统作业最高效率的具体方案。提高了生产的效率。

二、 具体问题

1. 如果加工生产物料只需一道工序时，给每台 CNC 安装同样的刀具。要求给出一道工序无故障 RGV 的动态调度模型以及相应的求解算法。同时，利用题目给出的三组数据检验并评价模型，最后给出具体的求解结果，写入表格中；
2. 生产物料仍需一道工序，但每台 CNC 均会以 1%的概率发生故障，解决故障的时间为 10 到 20 分钟。要求给出一道工序含故障 RGV 的动态调度模型以及相应的求解算法。同时，利用题目给出的三组数据检验并评价模型，最后给出具体的求解结果，并填入表格中；
3. 生产物料需要两道工序，且两道工序有前后顺序。每台 CNC 中途不能换刀片，即规定其只能选择两个步骤中的一个进行加工。要求给出两道工序无故障 RGV 的动态调度模型以及相应的求解算法。同时，利用题目给出的三组数据检验并评价模型，最后给出具体的求解结果，并填入表格中；
4. 生产物料仍需两道工序，但每台 CNC 均会以 1%的概率发生故障，解决故障的时间为 10 到 20 分钟。要求给出两道工序含故障 RGV 的动态调度模型以及相应的求解算法。同时，利用题目给出的三组数据检验并评价模型，最后给出具体的求解结果，写入表格中。

§2 问题的分析

一、 研究现状综述

目前，国内许多专家从不同角度对智能 RGV 的动态调度策略问题进行了研究，下面简单介绍一些专家学者对调度策略问题的大致思路。

陈华、孙启元[1]针对直线轨道的往复两穿梭车系统可能存在的问题，采用在给定运送序列的情况下，应用禁忌搜索算法进行 2-RGV 的调度研究以达到最小化出库时间的目的，最后使用算例对模型进行测试，其不足之处是没有考虑 RGV 的运行时间及其变化性。

而聂峰、程珩[2]针对这个问题，利用顺序排队原则，更进一步的提出了就近算法，同时，其新颖的提出了使用策略设置方法来优化作业调度，同样达到了提高入库效率的目的。该方法容易实现和操作，且应用结果良好，但是相对来说该方法的考虑不够全面。

我们都知道目前有大量针对智能 RGV 的动态调度问题进行研究的文献，但现有文献或多或少都有不足之处，需要加以完善。

二、 对问题的分析和解题思路

本文主要研究智能 RGV 的动态调度问题，针对此问题，我们按照两种不同的物料加工作业方法进行研究，并根据有无故障将问题分成四个小问题，逐一进行研究并建立相应的模型。

首先，可以运用遗传算法建立单一工序下的 RGV 动态调度模型，该模型能够根据不同工作环境，提供具体的 RGV 调度方案。其次为了模拟 CNC 故障，考虑加入随机化算法，用以模拟 CNC 发生故障、排除故障时已工作的时间及检修故障时所花费的时间。这样建立模拟 CNC 故障的单一工作模式模型，此模型适用范围广，并且更接近实际情况。

但是该模型只能解决单一生产问题，因此为了增大其实用性，我们还需要建立两道工序的 RGV 动态调度模型，以及模拟 CNC 故障的多步骤 RGV 动态调度模型。这个可以采用降维的方法，将所有的步骤编号，再用遗传算法进行建模，并在更复杂的生产要求下，给出 RGV 的具体调度方案。

根据本文的研究思路，现做出整体的思路流程图，如图一所示。

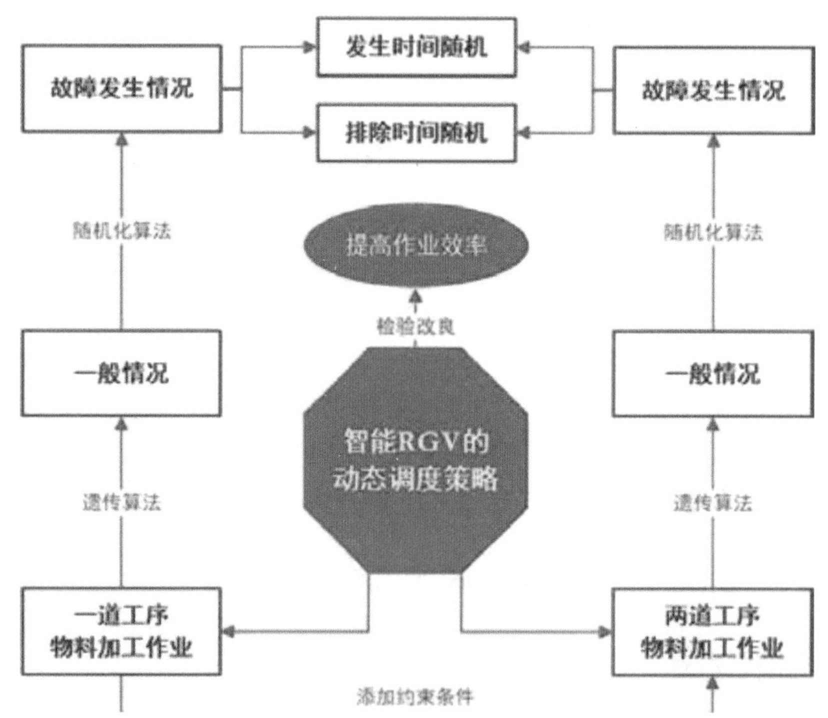

Figure 1 基于遗传算法的智能 RGV 动态调度模型整体思路图

§3 问题的假设

1. 假设 RGV 在接收到任务后，立刻行动，设备反应时间忽略不计；
2. 假设 RGV 从上传送带取物料的时间已经包含到了为 CNC 一次上下料所需时间中；
3. 假设 RGV 可以判断有无熟料，若无熟料，则不用进行清洗步骤；
4. 假设机械臂移动半成品时，RGV 转动所消耗的时间可忽略不计；

5. 在两道工序物料加工作业中，奇数编号的 CNC 完成第一道工序，偶数编号的 CNC 完成第二道工序；
6. 假设时间截至后，机器全部停止工作，仅完成整个清洗过程的物料算作成料；
7. 假设 RGV 的移动、上下料、清洗作业及 CNC 的加工作业所花费的时间没有误差；

§4 符号说明

序号	符号	符号说明
1	r	计算机数控机床（CNC）的编号，$r=1,2,\dots,8$
2	i	轨道式自动引导车（RGV）作业编号，$i=1,2,3,\dots,n$
3	J_i	RGV 第i次作业所对应的物料
4	x_{ri}	当物料J_i是 RGV 给编号为r的 CNC 上下料时，其值为 1，否则为 0
5	y_{rki}	当物料J_i是编号为r的 CNC 第k次加工的时，其值为 1，否则为 0
6	n_i	RGV 第i次作业所对应的 CNC 编号，$n_i=1,2,\dots,8$
7	T_{ri}	RGV 给编号为r的 CNC 运送第i个物料的出发时间
8	S_d	RGV 移动d个单位的所需时间，$d=0,1,2,3$
9	d_{ab}	RGV 从编号为a的 CNC 移动到编号为b的 CNC 的距离
10	W_j	RGV 为 CNC 一次上下料所需时间，$j=1,2$ W_1为为奇数号 CNC，W_2为为偶数号 CNC
11	C	RGV 完成一个物料的清洗作业所需时间
12	G	CNC 加工完成一个一道工序的物料所需时间
13	U_i	第i个物料的上料开始时间
14	F_i	第i个物料在 CNC 上完成加工作业的时间
15	D_i	第i个物料的下料开始时间
16	E_i	第i个物料完成清洗作业被送出系统的时间
17	I	完成清洗作业被送出系统的物料总数
18	B_i	第i个物料在 CNC 上加工时，若发生故障则值为 1，否则为 0
19	SUM_b	某一时间段内发生的故障总数
20	$rand$	一个在$(0,1)$之间的随机数
21	P_{work}	系统工作效率

§5 模型的构建

一、 问题一的分析与求解

1. 对问题的分析

问题一是只有一道工序的物料加工作业，所以每一台 CNC 均相同，不用单独考虑 CNC 的类别及加工顺序问题。该问题的目的是要通过对 RGV 和 CNC 进行合理的调度，使得在每班次连续作业的 8 小时中，这一套智能加工系统可以尽可能多的加工出成料。由于该问题的设置情况较为繁杂，使用精确搜索算法过于繁琐和费时，所以我们使用遗传算法对该题进行求解。

首先考虑单个物料的运送过程，除了刚开始 RGV 第一次为各 CNC 进行上下料作业时，因为之前并未有熟料产生所以不用进行清洗作业以外，之后的每一次作业都需要依次完成上下料和清洗作业才可获得成料，因此我们在分析和计算除第一次以外的作业时将上下料和清洗作业当作一个整体来处理。

其次是对于整个系统运作过程，根据该系统的构成和作业流程，我们要确保 RGV 每次仅为一台 CNC 进行上下料作业且每台 CNC 仅能加工一件物料。同时，因为 RGV 是在接收到某一 CNC 的需求信号后才对该 CNC 进行作业，所以还需保证每次 RGV 开始向某一发出信号的 CNC 移动时，该 CNC 已经完成了上一件物料的加工作业。

最后是对于目标的转换。由于直接使用遗传算法比较在每班次作业的 8 小时中完成的成料数较为麻烦，所以我们先假设极端情况算出一个较小的粗略值，使得该系统在 8 小时内加工的成料绝对不可能达到该值。再利用遗传算法得出加工该极限数量的用时最短的近似解，最后将这些解的用时缩短至 8 小时以内再进行比较得出最优解。

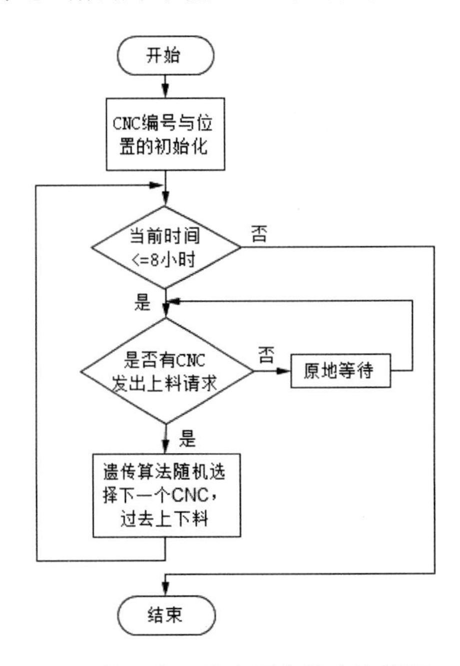

Figure 2 **一道工序无故障调度算法流程图**

2. 模型的建立

1）约束条件

由于 RGV 每次仅为一台 CNC 进行上下料作业且每台 CNC 仅能加工一件物料，所以首先对一次运送的物料数量进行约束。定义当物料J_i是 RGV 给编号为r的 CNC 上下料时，其值为 1，否则为 0，这样就可以通过累加值对运送数量进行约束，即

$$\sum_{r=1}^{8} x_{ri} = 1 \quad (1)$$

同时，由于只有在某台 CNC 的物料加工完成之后，才能进行上下料作业开始新的物料的加工，所以还需要保证 RGV 开始移动的时间不早于 CNC 完成物料加工的时

间。定义y_{rki}为当物料J_i是编号为r的 CNC 第k次加工的时，其值为 1，否则为 0，那么编号为r的 CNC 第$(k-1)$次加工的i值是$\sum_{i=1}^{n} iy_{r(k-1)i}$。而 RGV 从编号为$n_{i-1}$的 CNC 移动到编号为$n_i$的 CNC，其距离可表示为$d_{n_{i-1}n_i} = \left|\left\lfloor\frac{n_i+1}{2}\right\rfloor - \left\lfloor\frac{n_{i-1}+1}{2}\right\rfloor\right|$。则该约束为

$$T_{ri} \geq T_{r\sum_{i=1}^{n} iy_{r(k-1)i}} + S_{d_{n_{i-1}n_i}} + (W_1 + C)(r \bmod 2) + (W_2 + C)[(r+1) \bmod 2] + G \quad (2)$$

另外，为限制其运送顺序，还增添约束

$$T_{r(i-1)} \leq T_{ri} \quad (3)$$

2）目标函数

为了简化求解的过程，我们对目标进行了一个转化。首先，我们假设 RGV 一直处于工作状态，根据三组数据得出该智能加工系统的极端加工物料件数并进行适当取整，认为在现机器运作情况下其物料加工的上限为 500 件。在此基础上，我们以加工完成 500 件成料的用时最短为目标函数，建立基于遗传算法的 RGV 调度模型，并根据以该目标求出的最优解对一个班次 8 小时内完成的成料进行比较，最终输出 8 小时内完成成料数量最多的最优解。

首先，将所有的 RGV 运送情况分为两种，一种是接收物料的 CNC 为第一次接收物料，即该 CNC 上还没有已经加工好的熟料；另一种是接收物料的 CNC 不为第一次接收物料，即该 CNC 上已经有加工好的熟料。针对第一种运送，RGV 在进行上下料作业之后并不需要进行清洗作业，仅是在上一个物料上料开始之后完成上下料作业并移动至现在的位置，所以其第i个物料的上料开始时间为

$$U_i = U_{i-1} + W_j + S \quad (4)$$

该物料在上料开始之后完成了上下料作业和 CNC 的加工作业，所以该物料的加工完成时间，即其向 RGV 发出需求信号的时间为

$$F_i = U_i + W_j + G \quad (5)$$

而针对第二种运送，在计算时间时首先要判断同一 CNC 加工的上一件物料是否已经加工完成。通过比较该台 CNC 加工的上一件物料t的加工结束时间F_t与 RGV 运送的上一物料$(i-1)$的清洗作业结束时间$U_{i-1} + W_j + C$，若 RGV 运送的上一物料先完成清洗作业，则 RGV 需要原地等待一段时间直到该 CNC 加工完毕发出需求信号，所以这一物料的上料开始时间为

$$U_i = F_t + S \quad (6)$$

若该 CNC 先完成加工作业，则 RGV 在进行上下料作业之后还需要进行清洗作业，在上一个物料上料开始之后要先后完成上下料作业和清洗作业，再移动至当前位置才能进行又一次的上下料作业，故第i个物料的上料开始时间为

$$U_i = U_{i-1} + W_j + C + S \quad (7)$$

该种运送中物料的加工完成时间表达同第一种运送一样，即式(5)

至于某一物料i的下料开始时间D_i，可认为其等于该物料所在的 CNC 要加工的下一物料s的上料开始时间U_s，若该物料所在的 CNC 已经没有要加工的下一物料，则令其等于一个非常大的实数M，即

$$D_i = \begin{cases} U_S, & s \leq 500 \\ M, & s > 500 \end{cases} \quad (8)$$

最后，可据此列出物料i的清洗作业完成时间表达式，即物料称为成料被放到

下料传送带上送出系统的时间

$$E_i = D_i + W_j + C \tag{9}$$

所以，目标函数应为完成清洗作业被送出系统的时间小于 8 小时的物料个数I，寻求其最大值，即

$$\mathrm{f} = -\mathrm{I}, \quad E_i \leq 28800 \tag{10}$$

3. 模型的求解

完成目标函数的编写之后，利用 MATLAB 遗传算法工具箱，代入变量范围、整数指数、遗传种群大小、初始种群、停止准则等参数进行求解，并生成最佳适应度和最佳个体的图形。

其中，为了提高遗传算法求解的效率，我们为其人为设定了一个初始种群。一般主观认为 RGV 从近到远依次为 CNC 进行上下料作业，即按照编号为 1、2、3、4、5、6、7、8 的顺序进行作业，是一种较为省时的方法，所以可以认为这是一个较优解，并以此为初始种群产生新的子代种群进行优化。

针对题目中给出的智能加工系统作业的三组数据，代入建立的调度模型进行求解，分别给出一个班次 8 小时中近似最优的加工 CNC 编号、上料开始时间、加工完成时间、下料开始时间和清洗完成时间。题目所给的三组数据得出的近似最优解均为我们所假设的初始种群，即依次为编号为 1、2、3、4、5、6、7、8、1、2……的 CNC 进行上下料和清洗作业，能够在 8 小时中完成最多的成料，一、二、三组完成的成料数量分别 356、336 和 366。三组数据的具体运送步骤及各步骤时间见附录一。

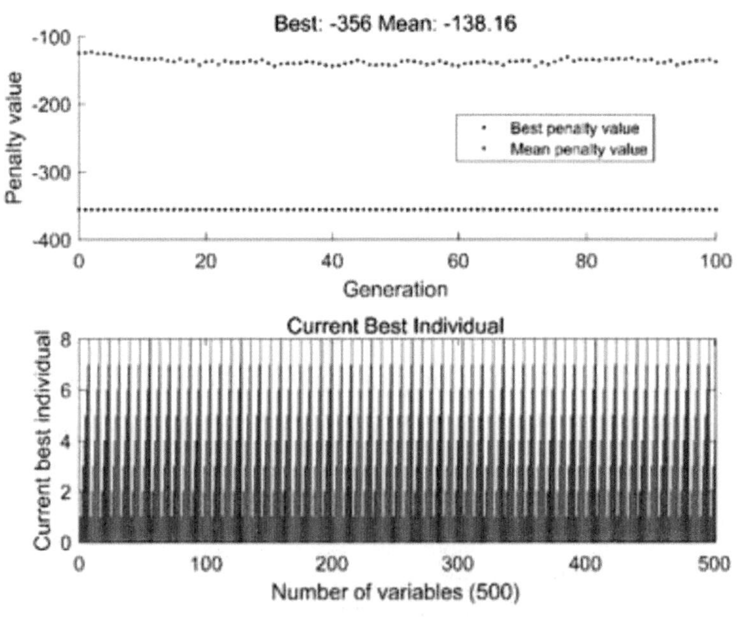

Figure 3 一道工序无故障问题第一组数据结果

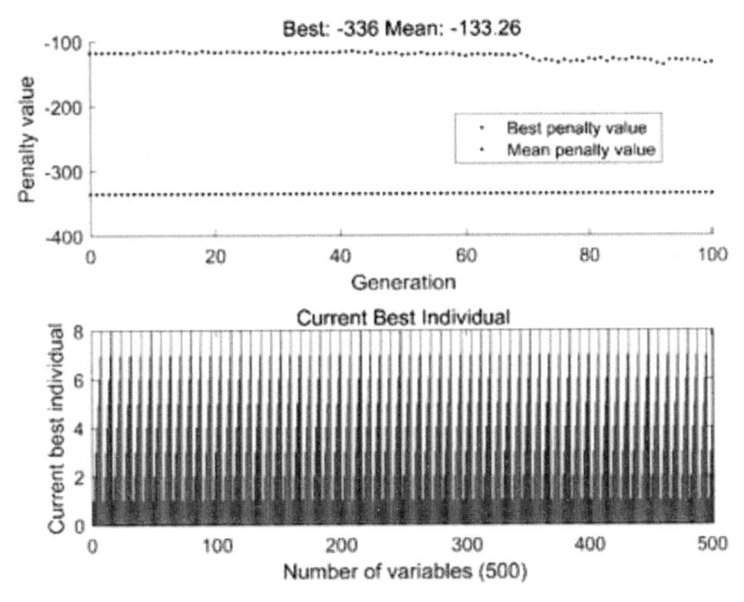

Figure 4 一道工序无故障问题第二组数据结果

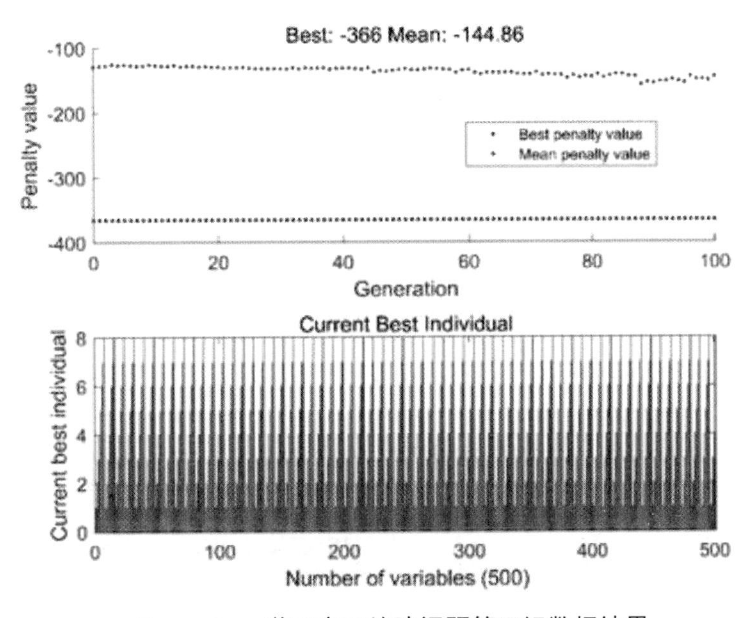

Figure 5 一道工序无故障问题第三组数据结果

二、 问题二的分析与求解

1. 对问题的分析

问题二是在问题一的基础上提出的，同样是一道工序的物料加工作业 RGV 调度，但其中 CNC 在加工过程中可能会发生故障并需要花费一定的时间去排除故障，从而降低整个智能加工系统的作业效率。据统计故障发生的概率约为 1%，因此可以通过生成随机数来对发生故障进行模拟，同样，每次故障发生的时间点和故障排除花费的时间都可通过随机数模拟。

这样，在 RGV 的安排调度过程中，如果某一 CNC 发生故障，那么该 CNC 上的物料必然报废，且在故障排除之后才能发出需求信号重新开始作业。故障排除的过程也可以看作物料加工的过程，但加工完成时间为故障排除完毕时间，且该次加工不产生熟料。所以问题二的目标函数一个班次 8 小时内完成成料的件数，即为在 8 小时内整个系统全部过程的轮转次数减去因故障而报废的物料数，需要寻找该数值的最大值。

2. 模型的建立

问题二的模型可以在问题一的模型上建立，在问题一模型的基础上增加一些关于故障的发生与排除的参数。首先，可以使用字母B_i来记录故障的发生，默认所有正常工作的 CNC 该B_i值为 0，当第i件物料在 CNC 上发生故障时其B_i值变为 1，则总共发生故障的次数或因故障报废的物料件数为

$$SUM_b = \sum_{i=1}^{n} B_i \quad (11)$$

再考虑故障发生的时间及其排查所用的时间，针对其发生的不确定性，可以使用随机化算法，用$rand$表示一个在(0,1)之间的随机数。用t_{br}记录 CNC 故障发生在一次加工中的时间，用t_{rp}表示故障排除花费的时间，那么

$$t_{br} = \mathrm{G} * \mathrm{rand} \quad (12)$$

$$t_{rp} = 600 * \mathrm{rand} + 600 \quad (13)$$

则发生故障时，故障的实际发生时间为$(U_i + W_j + t_{br})$，故障的结束时间即该 CNC 在此向 RGV 发出需求信号的时间，为$(U_i + W_j + t_{br} + t_{rp})$，而在此时刻之后 CNC 进入正常工作状态，可与问题一的基础模型一样，按照两种 RGV 运送情况进行探讨。

这样，其目标函数可表示成一班次 8 小时内的 RGV 运送总物料件数SUM_t减去因为 CNC 故障而报废的物料件数，并求该目标函数的最大值，即

$$\mathrm{f} = -\mathrm{I} = -(SUM_t - SUM_b) \quad (14)$$

3. 模型的求解

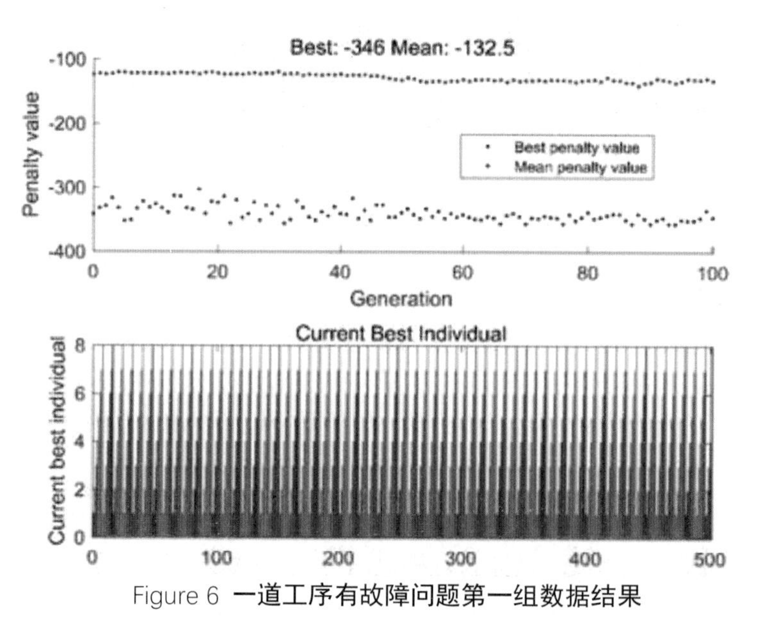

Figure 6 一道工序有故障问题第一组数据结果

首先编写目标函数。RGV 每次给 CNC 上下完物料时，程序运用随机化方法对是

否发生故障进行判断。若发生故障，则再次随机生成故障的发生时间及故障排除所需时间，且将故障排除完成的时间与 RGV 接收到信号的时间进行对比，仅当故障排除完成之后 PGV 才会接收信号并向其运送物料。若不发生故障，则与问题一的求解步骤相同。

与问题一设定同样的初始种群，在 MATLAB 软件的遗传算法工具箱中将编好的目标函数代入，并对变量进行上下限与整数约束，合理的选择种群大小和停止准则等参数，即可得到题目所给的三个组在一班次连续 8 小时的工作中最多分别完成 346、326、353 件成料。三组数据的具体运送步骤及各步骤时间见附录二。

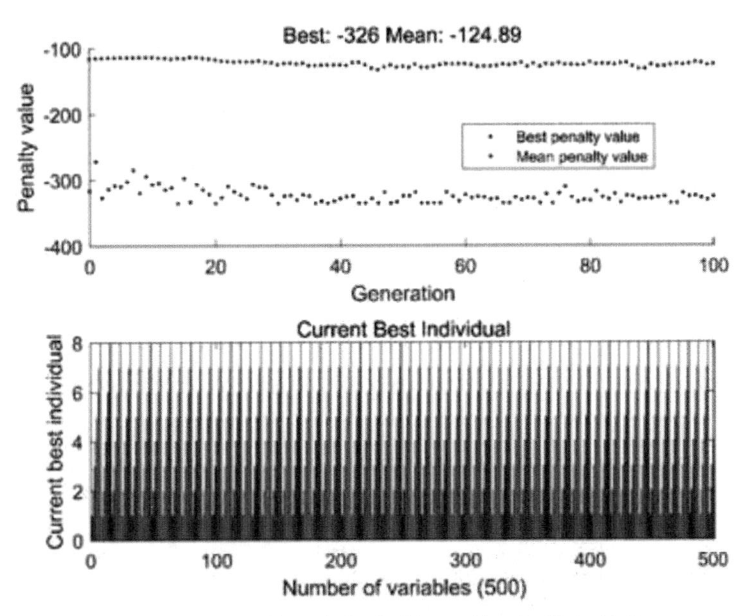

Figure 7 一道工序有故障问题第二组数据结果

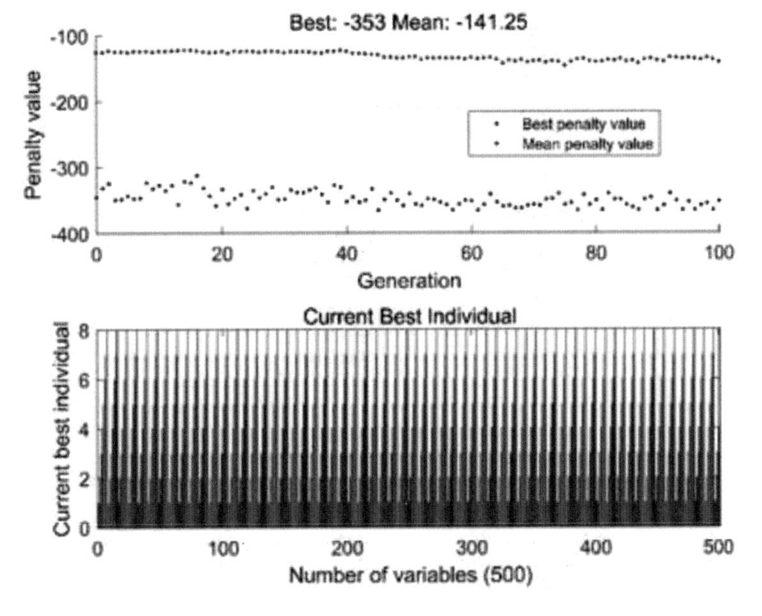

Figure 8 一道工序有故障问题第三组数据结果

三、 问题三的分析与求解

1. 对问题的分析

问题三是对问题一的拓展，将仅有一道工序的加工作业变成了有两道工序并且有先后顺序的加工作业。因此，本题必须考虑物料运送的次序问题。题目要求在每班次连续作业的 8 小时中，这套智能加工系统可以尽可能多的加工出成料。与一题模型相似，但是由于维度不同，我们必须对数据进行降维处理。算法流程如图所示。

首先，遗传算法模型需要提供一个数列来进行遗传变异。而本题由于需要两个步骤，无法直接使用遗传算法来实现，因此我们对数据进行处理。假设 CNC1#、3#、5#、7#加工完成第一道工序，CNC2#、4#、6#、8#加工完成第二道工序，且每一件物料必须先经过第一道工序加工才能去第二道工序。为了方便叙述，使用奇数侧与偶数侧作区分。完成一次成品的步骤为：因此，上料口直接放料到奇数侧，完成第一道工序后，经过 RGV 传送半成品到偶数侧，进行第二道工序的加工，最终清洗，将成品放回到下料口。

将由 1 到 2、1 到 4、1 到 6、……、7 到 8 的 16 种走法分别用序号表示。因此，RGV 所能走的全部路径都能由一个 1 到 16 的数组表示，并且这个数组仍可对应回 RGV 所走的具体位置。

其次是对整个运作过程的特殊情况进行分析。由于是分步完成的，所以生料必须先进入奇数组加工，再进入偶数组加工。因此，物料想要进偶数组加工，16 条路

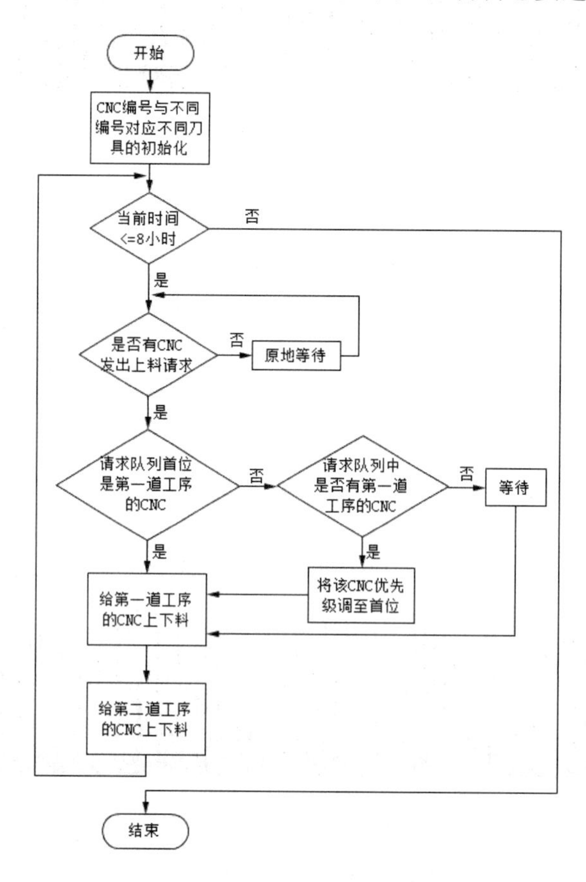

Figure 9 两道工序无故障调度算法流程图

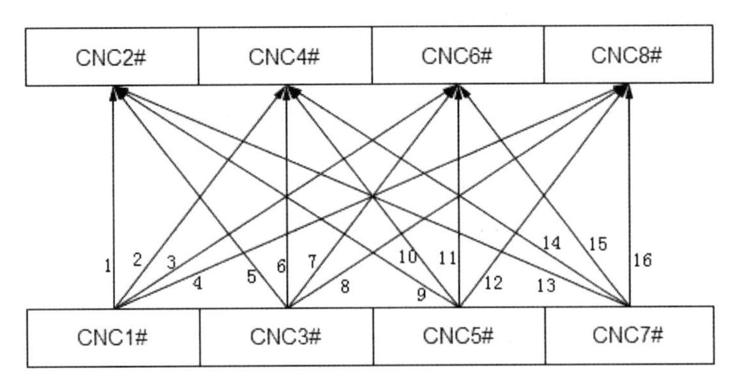

Figure 10 降维方法图示

径中至少应有一条必须满足其对应CNC的上一组有半成品加工完成并且要保证送往的偶数组CNC没有在工作。解决方案就是用最大值进行比较，选取较大的最为任务开始的条件。

最后，目标转换。由于遗传算法有自己的缺陷，无法在较短的时间内给出最优解，因此与问题一相似，先求出众多局部最优解，最后寻找在这些局部最优解之间的最优解。

2. 模型的建立

由于有先后顺序，所以如果是 CNC 奇数端初次上料时，物料不向偶数侧运输，此时 RGV 仍然处于初始位置。如果是给其第二次上料，小车运动到偶数侧，物料被运向偶数侧，不进行清洗操作。如果是第三次或第三次以上上料时，则完成整个流程。

并且，在给数组规定赋值规则时，需要注意，如果为第一次上料，则偶数侧的数据并不是简单的（加上一个上下料时间）、RGV的实时位置也不能直接由路线来判定，还要看是否满足到达偶数侧的条件，要经过处理。本文做出的处理为，给每次结果一个位置，是它上一路程结束的位置。

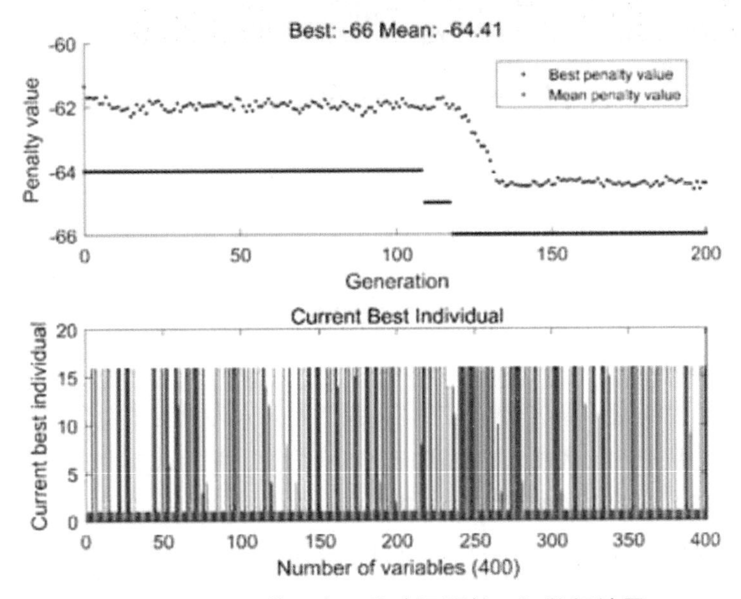

Figure 11 两道工序无故障问题第一组数据结果

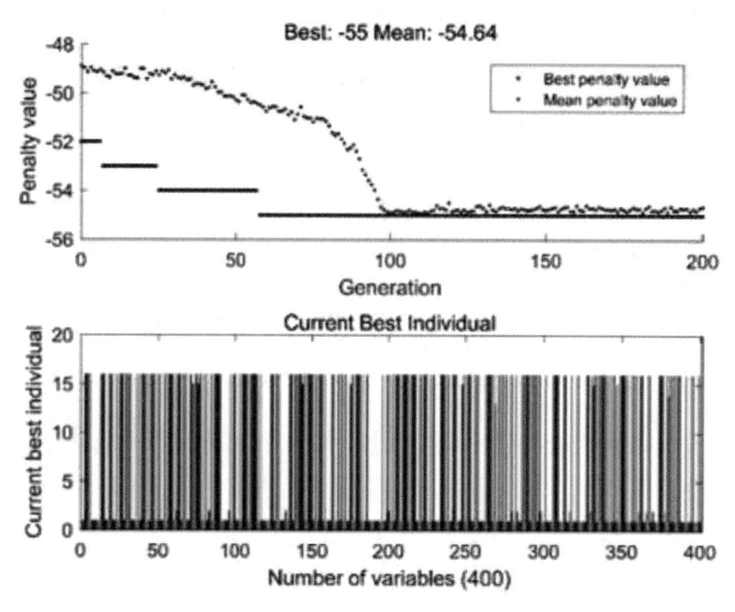

Figure 12 两道工序无故障问题第二组数据结果

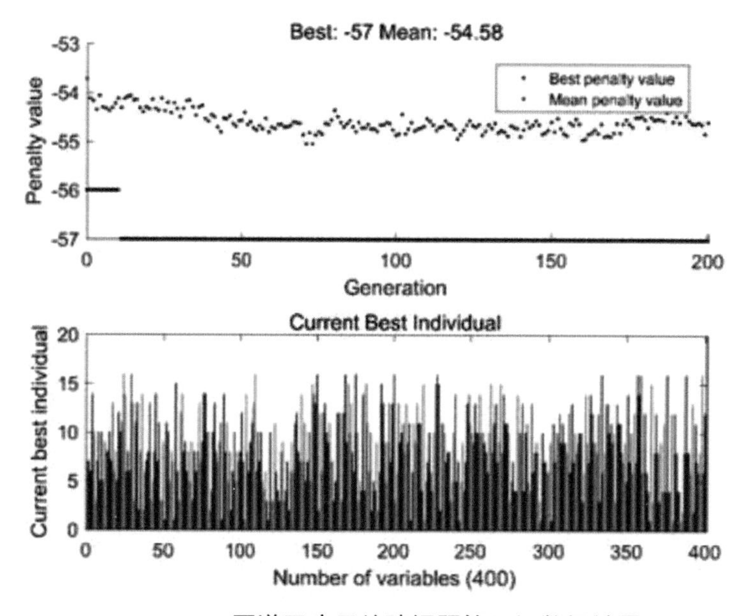

Figure 13 两道工序无故障问题第三组数据结果

3. 模型的求解

完成目标函数的编写之后，利用 MATLAB 遗传算法工具箱，代入变量范围、整数指数、遗传种群大小、初始种群、停止准则等参数进行求解，并生成最佳适应度和最佳个体的图形。

其中，为了提高遗传算法求解的效率，我们为其人为设定了一个初始种群。一般主观认为 RGV 从近到远依次为 CNC 进行上下料作业，并且认为由奇数侧的 CNC 直接将半成品料给与其正好相对的对应编号的偶数侧 CNC，编号即为：1、6、11、16。验证此解是一种较为省时的方法，所以可以认为这是一个较优解，并以此为初始种群产生新的子代种群进行优化。

针对题目中给出的加工系统作业的三组数据，代入建立的调度模型进行求解，分别给出一个班次 8 小时中近似最优的加工 CNC 编号、上料开始时间、加工完成时间、下料开始时间和清洗完成时间。在 MATLAB 遗传算法工具箱中求解，其 8 小时内完成的最大成料数量分别为 66、55、57。三组数据的具体运送步骤及各步骤时间见附录一。

四、 问题四的分析与求解

1. 对问题的分析

为解决两道工序并且包含故障的 RGV 动态调度策略问题，我们在两道工序无故障模型的基础上加入随机化算法。相比于无故障模型，此模型中加入了不确定定量，使模型可以用于故障发生概率不能忽略的系统中。

由于物料需要两道工序加工而成，且工序分先后。因此物料作废有两种情况，即在第一道工序加工时损坏或第一道工序顺利完成、第二道工序加工损坏。

还需算出在工作多长时间后出故障，以及处理故障的时间。若下次需要此故障 CNC 工作，需要比较故障解决后的时间与下次 RGV 接收到 CNC 信号的时间，取较大的当作当时的 RGV 开始向此 CNC 运动的时间。

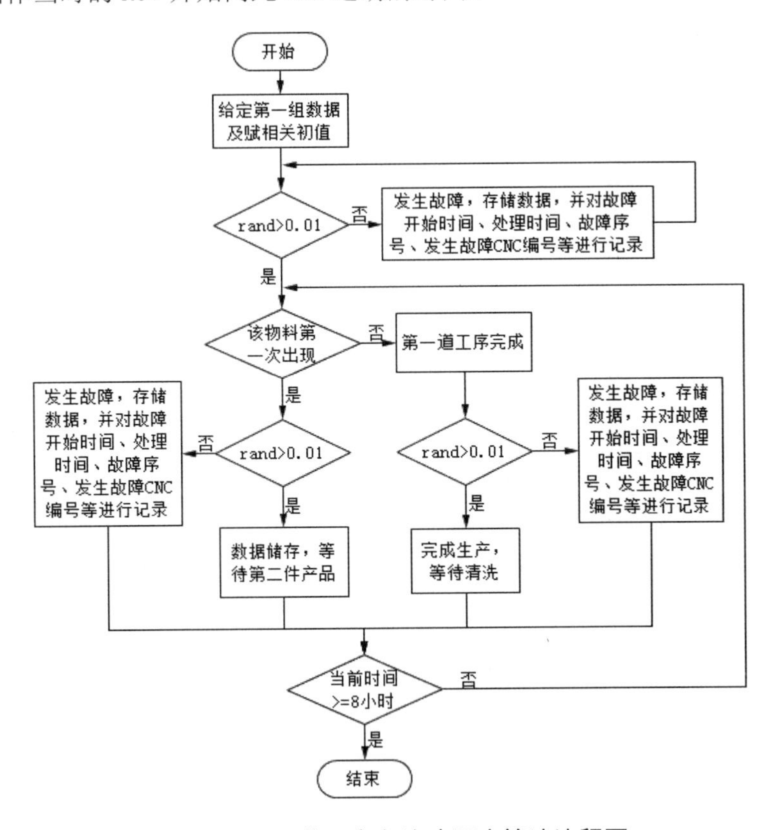

Figure 14 两道工序有故障调度算法流程图

2. 模型的建立

在模型三的条件下，由于增加了随机的故障，因此应增加约束条件。在 RGV 完成给任意一台 CNC 上下料的作业后，CNC 开始工作。从这个时刻起，CNC 都有可能产生故障。因此先判断这台 CNC 是否发生了故障。如果没发生故障，则正常进入下一个状态。如果发生了故障，则必须对 CNC 发生的时间以及 CNC 被修理的时间进行随机化分析，并记录。

由于是两个步骤，并且有先后次序之分。因此，当故障发生在奇数侧 CNC 时，路径就会断，物料不再向下传递。如果偶数侧 CNC 发生故障，那么下次 RGV 到达这台 CNC 并上料时，没有熟料取出，也不用清洗。

这一题的目标函数同问题二的相同，求完成清洗作业被送出系统的物料总数的最大值，可表示成一班次 8 小时内的 RGV 运送总物料件数SUM_t减去因为 CNC 故障而报废的物料件数，即

$$f = -I = -(SUM_t - SUM_b) \tag{15}$$

3. 模型的求解

完成目标函数的编写之后，利用 MATLAB 遗传算法工具箱，代入变量范围、整数指数、遗传种群大小、初始种群、停止准则等参数进行求解，并生成最佳适应度和最佳个体的图形。

其中，为了提高遗传算法求解的效率，我们人为设定一个与第三问模型相同的初始种群。一般主观认为 RGV 从近到远依次为 CNC 进行上下料作业，并且认为由奇数侧的 CNC 直接将半成品给与其正好相对的偶数侧 CNC，编号即为：1、6、11、16。验证此解是一种较为省时的方法，所以可以认为这是一个较优解，并以此为初始种群产生新的子代种群进行优化。

针对题目中给出的加工系统作业的三组数据，代入建立的调度模型进行求解，分别给出一个班次 8 小时中近似最优的加工 CNC 编号、上料开始时间、加工完成时间、下料开始时间和清洗完成时间。

五、 系统作业效率分析

1. 一道工序的效率分析

当程序为一道工序时，对题目中所给的三组不同数据进行分析发现，8 小时内，一道工序无故障所生产的成品数为 356、336、366，一道工序含故障所生产的成品数为 346、326、353。我们用相同时间内所生产的成品数作为评价一道工序生产效率的指标，显然，生产效率由高到低的排序：第 3 组、第 1 组、第 2 组。根据题目中所给条件发现，第二组的 RGV 的移动时间、上下料时间、清洗作业所需时间以及 CNC 加工完成所需时间最多，第 3 组所用时间最小。因此，用每组作业的工作总时间与全部时间的比值作为系统作业效率P_{work}。

即：

$$P_{work1} = \frac{(T_{over} - T_{sum})}{T_{over}} \tag{16}$$

（其中P_{work1}、T_{over}、T_{sum}分别为一道工序的生产效率、工作总时间、每列参数全部遍历一遍的时间）

2. 两道工序的效率分析

当工作情况为两道工序时，8 小时内无障碍时生产的成料数量分别为 66、55、57，有障碍时产生的成料数为 62，53，55。分析得出结果，第一组的工作效率最高。两道工序与仅一道工序结果相比发现，不能简单地用一道工序的系统作业效率来表示两道工序的系统作业效率。由于我们假设 4 台 CNC 做第一道工序，另外 4 台 CNC 做第二道工序，因此，主要影响系统作业效率的因素应该为第一道工序与第二道工序的 CNC 的数量比。因此我们定义，评价多组两道工序生产效率的标准为：在生产物料的一、二两道工序的工作时间之比更接近生产第一道与第二道工序的 CNC 数量比时，生产效率越高。

$$P_{work2} = \frac{\frac{CNC_1}{CNC_2} - \frac{T_1}{T_2}}{\frac{CNC_1}{CNC_2}} \tag{17}$$

（其中P_{work2}、CNC_1、CNC_2、T_1、T_2分别代表两道工序的生产效率、第一道工序 CNC 数量、第二道工序 CNC 数量、CNC 进行第一道工序的时间、CNC 进行第二道工序的时间）

得出结论：在生产中，如果只有一个步骤，我们尽量以缩短每一项步骤的时间来提高生产效率，如果需要两道工序才能完成，可以根据实际问题对工序 1 和工序 2 的 CNC 数量做出改变，来提高生产效率。

§6 评价推广

一、 误差分析

1. 在问题一中求解模型时，我们根据一般经验假定了它的初始序列为“1，2, 3, 4, 5, 6, 7, 8, 1, 2, 3……”，虽然因此缩短了求近似最优解的时间，但带有主观色彩，对模型的遗传方向有一定的影响。
2. 在模型中，为了对遗传算法形成约束，使模型不要过早陷入局部最优解，我们采用了式$\sum_{r=1}^{8} x_{ri} = 1$，$T_{r(i-1)} \le T_{ri}$，以及$f = I, E_i \le 28800$对其进行约束。

二、 模型的优点

1. 本文对问题的求解有合理的假象、假设、计算；
2. 本文利用遗传算法，对结果进行了多次优化尝试，最终选出了我们认为的最优解；
3. 本文结合大量图表辅助说明，使文章更清晰明了；
4. 结合概率生成随机数经过多次拟合得到最优调度模型，具有随机性；
5. 本文利用合适且准确的约束条件与目标函数，结合 MATLAB 代码，利用 MATLAB 遗传算法工具箱来求解最优模型，简化了计算过程；
6. 相对于精确搜索算法，本算法具有并行性，能够在短时间内进行大量迭代，速度较快，占用内存较小；
7. 本文结合题目，采用随机化算法与遗传算法结合的混合遗传算法，保留了精英策略遗传，使算法的收敛速度更快、稳定性更高；

8. 本文的模型构建好以后，又进行了大量重复检验与随之不断的改进，使模型更加精确。

三、　模型的缺点

1. 本文采取遗传算法，所以得出的是模型在 n 代以内相对最优，只是近似最优解，而非实际最优解。若要获得更加精确的解就要增加遗传代数、种群数等等，又会大量增加求解时间；
2. 在问题一中，将目标转化后，选择一个合适的目标完成数量将会比较困难，并且将会造成误差；
3. 模型求解过程中只能根据经验判断可行解是否合理，而没有精确标准，可能存在纳伪错误。

四、　推广应用

1. 横向推广

目前 RGV 多用于仓库管理以及工业自动化生产中，本文所得出的 RGV 调度算法，可用于大型立体仓库中 RGV 的运货策略，仓库规模越大，模型对整体货运效率的改进越高[3]。也完全可以用于有多道工序，多台 RGV 的自动化生产车间，汽车装配车间等等。

2. 纵向推广

本文的利用遗传算法所建立的动态调度模型可用于其他各个行业，特别是用于动态路径规划[4]，而自适应性还可被利用于改进移动 WSN 覆盖方法[5]、实时个性化推荐[6]等等。本文所建的基于遗传算法的调度模型，还可用于工业建筑的健康监测，如识别桥梁损伤程度[7]，以及军工领域航空武器气动参数的辨识[8]等等各种问题。本文所构建的遗传算法模型还易于与其他诸如模拟退火算法、聚类分析等有机结合，满足各种问题的不同需要。

§7 参考文献

[1] 陈华,孙启元.基于 TS 算法的直线往复 2_RGV 系统调度研究[J].工业工程与管理,2015,20 (5):80-88.

[2] 聂峰,程珩.多功能穿梭车优化调度研究[J].物流技术,2008(10):265-267.

[3] 刘万强,周亚勤,杨建国.新型棋盘格密集仓库的出入库货位分配优化[J].东华大学学报(自然科学版),2017,43(4): 496-502.

[4] 邹金茂,徐向荣.基于改进混合遗传算法的无人车路径规划[J].佳木斯大学学报（自然科学版）,2018,36(4): 573-577.

[5] 朱利民,赵丽.基于改进自适应遗传算法的移动 WSN 覆盖方法[J].计算机应用研究,2018,36(5).

[6] 荆玲.基于遗传算法的实时个性化推荐研究[D].重庆:重庆大学,2003.

[7] 毛云霄,王英杰,肖军华. 基于过桥车辆响应的遗传算法桥梁损伤识别[J].振动、测试与诊断，2018,38(4): 696-703.

[8] 杜昌平,周德云,宋笔锋. 基于遗传算法的弹道参数辨识方法研究[J]. 西北工业大学学报，2008,26(3): 373-375.

附录内容扫码查看

附录十三 MATLAB 代码

一、问题一代码

```
function f = objfun(x)
%   一道工序物料加工
%   此处显示详细说明

    W = [32;27];        %RGV 为编号分别为偶奇数的 CNC 一次上下料所需时间
    S = 0;              %RGV 移动所需时间
    C = 25;             %RGV 完成一个物料的清洗作业所需时间
    G = 545;            %CNC 加工完成一个一道工序的物料所需时间
    t = 0;
    flag = 0;

    X = zeros(500,5);
    X(:,1) = x;

    start = 1;
    over = fix((X(1,1) + 1) / 2);
    flag = abs(over - start);

    switch (flag)
        case 0
            S = 0;
        case 1
            S = 18;
        case 2
            S = 32;
        case 3
            S = 46;
    end


    X(1,2) = S;
    X(1,3) = X(1,2) + G + W(int32(mod(X(1,1),2)) + 1);


    for i = 2:500
        i = double(int32(i));

        X(i,2) = X(i-1,2) + W(mod(X(i-1,1),2) + 1);

        start = fix((X(i-1,1) + 1) / 2);
```

```
        over = fix((X(i,1) + 1) / 2);
        flag = abs(over - start);

        switch (flag)
            case 0
                S = 0;
            case 1
                S = 18;
            case 2
                S = 32;
            case 3
                S = 46;
        end

        for j = i-1:-1:1
            if X(j,1) == X(i,1)
                t = j;
            break
            else
               t = 0;
            end
        end

        if (t == 0)         %这个数据是第一次出现

            X(i,2) = X(i-1,2) + W(int32(mod(X(i-1,1),2) + 1)) + S;
            X(i,3) = X(i,2) + G + W(int32(mod(X(i,1),2) + 1));

        else

            if (X(i-1,2) + W(int32(mod(X(i,1),2) + 1)) + C - X(t,3) <
0)      %判断是否工作完成,1 为没有
                X(i,2) = X(t,3) + S;
            else
                X(i,2) = X(i-1,2) + W(int32(mod(X(i-1,1),2) + 1)) + C +
S;
            end

            X(i,3) = X(i,2) + G + W(int32((mod(X(i,1),2)) + 1));

        end

    end
```

```
    for i = 1:500

        for j = i+1:500
            if X(j,1) == X(i,1)
                t = j;
            break
            else
                t = 0;
            end
        end

        if t > 0
            X(i,4) = X(t,2);
        else
            X(i,4) = 10000000000;
        end

        X(i,5) = X(i,4) + W(int32((mod(X(i,1),2)) + 1)) + C;

    end

X;
f = -length(find(X(:,5) <= 28800));
```

二、问题二代码

```
function f = brkfun(x)
%    一道工序物料加工,包括故障
%    此处显示详细说明

    W = [31;28];         %RGV 为编号分别为偶奇数的 CNC 一次上下料所需时间
    S = 0;               %RGV 移动所需时间
    C = 25;              %RGV 完成一个物料的清洗作业所需时间
    G = 560;             %CNC 加工完成一个一道工序的物料所需时间
    a = 0;
    flag = 0;
    breakdown = 0;
    repair = 0;
    sum_brk = 0;
    k = 0;

    Y = zeros(500,4);
```

```
Y(:,1) = [1:500];

X = zeros(500,5);
X(:,1) = x;

start = 1;
over = fix((X(1,1) + 1) / 2);
flag = abs(over - start);

switch (flag)
    case 0
        S = 0;
    case 1
        S = 20;
    case 2
        S = 33;
    case 3
        S = 46;
end

if (rand <= 0.01)
    X(1,2) = S;
    breakdown = G * rand;
    repair = rand * 600 + 600;
    X(1,3) = X(1,2) + breakdown + repair + W(int32(mod(X(1,1),2)) +
1);
    Y(1,2) = 1;
    Y(1,3) = X(1,2) + breakdown + W(int32(mod(X(1,1),2)) + 1);
    Y(1,4) = repair + Y(1,3);
else
    X(1,2) = S;
    X(1,3) = X(1,2) + G + W(int32(mod(X(1,1),2)) + 1);
end

for i = 2:500
    i = double(int32(i));
    X(i,2) = X(i-1,2) + W(mod(X(i-1,1),2) + 1);

    start = fix((X(i-1,1) + 1) / 2);
    over = fix((X(i,1) + 1) / 2);
    flag = abs(over - start);

    switch (flag)
        case 0
```

```
            S = 0;
        case 1
            S = 20;
        case 2
            S = 33;
        case 3
            S = 46;
    end

    for j = i-1:-1:1
        if X(j,1) == X(i,1)
            a = j;
        break
        else
           a = 0;
        end
    end

    if (a == 0)       %这个数据是第一次出现

        X(i,2) = X(i-1,2) + W(int32(mod(X(i-1,1),2) + 1)) + S;

        if (rand <= 0.01)       %机器故障
            breakdown = G * rand;
            repair = rand * 600 + 600;
            X(i,3) = X(i,2) + breakdown + repair +
W(int32(mod(X(i,1),2)) + 1);
            Y(i,2) = 1;
            Y(i,3) = X(i,2) + breakdown + W(int32(mod(X(i,1),2)) +
1);
            Y(i,4) = repair + Y(i,3);
        else
            X(i,3) = X(i,2) + G + W(int32(mod(X(i,1),2) + 1));
        end

    else

        if (X(i,2) + W(int32(mod(X(i,1),2) + 1)) + C  - X(a,3) <
0)      %判断是否工作完成,1 为没有
            X(i,2) = X(a,3) + S;
        else
            X(i,2) = X(i-1,2) + W(int32(mod(X(i-1,1),2) + 1)) + C +
S;
        end
```

```
            if (rand <= 0.01)       %机器故障
                breakdown = G * rand;
                repair = rand * 600 + 600;
                X(i,3) = X(i,2) + breakdown + repair +
W(int32(mod(X(i,1),2)) + 1);
                Y(i,2) = 1;
                Y(i,3) = X(i,2) + breakdown + W(int32(mod(X(i,1),2)) +
1);

                Y(i,4) = repair + Y(i,3);
            else
                X(i,3) = X(i,2) + G + W(int32((mod(X(i,1),2)) + 1));
            end

        end

    end

    for i = 1:500
        for j = i+1:500
            if X(j,1) == X(i,1)
                a = j;
            break
            else
                a = 0;
            end
        end
        if a > 0
            X(i,4) = X(a,2);
        else
            X(i,4) = 10000000000;
        end
        X(i,5) = X(i,4) + W(int32((mod(X(i,1),2)) + 1)) + C;
    end

    for i = 1:500
        if X(i,5) <= 28800
            sum_brk = sum_brk + Y(i,2);
            k = k + 1;
        end
    end

X
Y
```

```
f = -(k - sum_brk);
```

三、问题三代码

```
function f = bpobjfun(x)
%   两道工序物料加工
%   单数序号表示第一道工序，偶数序号表示第二道工序。

    W = [32;27];          %RGV 为编号分别为偶奇数的 CNC 一次上下料所需时间
    C = 25;               %RGV 完成一个物料的清洗作业所需时间
    T_prd1 = 455;         %CNC 加工完成一个两道工序物料的第一道工序所需时间
    T_prd2 = 182;         %CNC 加工完成一个两道工序物料的第二道工序所需时间
    flag = 0;
    a = 0;
    product = 0;          %序号出现的集合
    times = 0;            %序号出现次数的集合
    product1 = 0;         %CNC 第一次加工出熟料的集合
    times1 = 0;           %CNC 加工出的熟料集合
    product2 = 0;         %CNC 第二次加工出成料的集合
    times2 = 0;           %CNC 加工出的成料集合
    TPmax1 = 0;           %第一道工序 CNC 的最大时间
    TPmax2 = 0;           %第二道工序 CNC 的最大时间

    i = 0;
    i = double(int32(i));
    a = ones(400);
    result = ones(400,8);

    X = zeros(400,8);
    X(:,1) = x;        %x 为序号 1-16

    %编号和 CNC 之间的转换关系:
    for i = 1:400
    if (X(i,1)==1)||(X(i,1)==2)||(X(i,1)==3)||(X(i,1)==4)
        X(i,2) = 1;
    else if (X(i,1)==5)||(X(i,1)==6)||(X(i,1)==7)||(X(i,1)==8)
            X(i,2) = 3;
        else if (X(i,1)==9)||(X(i,1)==10)||(X(i,1)==11)||(X(i,1)==12)
                X(i,2) = 5;
            else if
(X(i,1)==13)||(X(i,1)==14)||(X(i,1)==15)||(X(i,1)==16)
                    X(i,2) = 7;
```

```
                end
            end
        end
    end
    if (X(i,1)==1)||(X(i,1)==5)||(X(i,1)==9)||(X(i,1)==13)
        X(i,5) = 2;
    else if (X(i,1)==2)||(X(i,1)==6)||(X(i,1)==10)||(X(i,1)==14)
            X(i,5) = 4;
        else if (X(i,1)==3)||(X(i,1)==7)||(X(i,1)==11)||(X(i,1)==15)
                X(i,5) = 6;
            else if
(X(i,1)==4)||(X(i,1)==8)||(X(i,1)==12)||(X(i,1)==16)
                    X(i,5) = 8;
                end
            end
        end
    end
    end


    %计算距离
    flag = abs(fix((X(1,2) + 1) / 2) - 1);

    switch (flag)
        case 0
            S = 0;
        case 1
            S = 18;
        case 2
            S = 32;
        case 3
            S = 46;
    end


    X(1,3) = S;
    X(1,4) = X(1,3) + T_prd1;
    X(1,6) = X(1,4) + W(int32(mod(X(1,2),2)) + 1);
    X(1,7) = X(1,6);
    X(1,8) = X(1,2);

    for i = 2:400
        i = double(int32(i));
```

```
%计算每个步骤的距离，上面几台机器的状况
product = find(X(1:i-1,1)==X(i,1));
times = length(product);

product1 = find(X(1:i-1,2)==X(i,2));
times1 = length(product1);

product2 = find(X(1:i-1,5)==X(i,5));
times2 = length(product2);

%先确定每一步的位置
if times1 == 0       %第一次一次也没出现过（不进行二，在一停止）
    flag = abs(fix((X(i,2) + 1) / 2) - fix((X(i-1,8) + 1) / 2));

    switch (flag)
        case 0
            S = 0;
        case 1
            S = 18;
        case 2
            S = 32;
        case 3
            S = 46;
    end

    X(i,3) = S + X(i-1,6);
    X(i,4) = X(i,3) + T_prd1;
    X(i,8) = X(i,2);       %当前位置为第一次出现的位置
    X(i,6) = X(i,3) + W(int32(mod(X(i,2),2)) + 1);
    X(i,7) = X(i,6);
    X(i,8) = X(i,2);

else       %经历超过两次，正常运行,所以每次都要清洗。

    flag = abs(fix((X(i,2) + 1) / 2) - fix((X(i-1,8) + 1) / 2));

    switch (flag)
        case 0
            S = 0;
        case 1
            S = 18;
        case 2
            S = 32;
        case 3
```

```
        S = 46;
end

for j = 1:i
    if  X(j,4) >= TPmax1;
        TPmax1 = X(j,4);
    end
end
for j = 1:i
    if X(j,7) >= TPmax2;
        TPmax2 = X(j,7);
    end
end

flag = abs(fix((X(i,2) + 1) / 2) - fix((X(i-1,8) + 1) / 2));

switch (flag)
    case 0
        S = 0;
    case 1
        S = 18;
    case 2
        S = 32;
    case 3
        S = 46;
end

if (times2 <= 1)
    C = 0;
else
    C = 25;
end

if (TPmax1 > X(i-1,3))
    X(i,3) = TPmax1 + S;
else
    X(i,3) = X(i-1,6) + S + C;
end
    X(i,4) = X(i,3) + T_prd1;

flag = abs(fix((X(i,2) + 1) / 2) - fix((X(i,5) + 1) / 2));

switch (flag)
    case 0
```

```
                S = 0;
            case 1
                S = 18;
            case 2
                S = 32;
            case 3
                S = 46;
        end

        if (times2 <= 1)
            C = 0;
        else
            C = 25;
        end
        if (TPmax2 > X(i,3) + W(int32(mod(X(i,2),2)) + 1))
            X(i,6) = TPmax2 + S;
        else
            X(i,6) = X(i,3) + W(int32(mod(X(i,2),2)) + 1) + S;
        end
            X(i,7) = X(i,6) + T_prd2;
            X(i,8) = X(i,5);
    end
end

for i = 1:400
    if (X(i,7) - X(i,6) < T_prd2)
        X(i,6) = 0;
        X(i,7) = 0;
    end
end

result(:,1) = [1:400];

for i = 1:400
    result(i,2) = X(i,2);
    result(i,3) = X(i,3);
    for j = i+1:400
        if ((X(j,2) == X(i,2)) && X(j,6))
            result(i,5) = X(j,5);
            result(i,6) = X(j,6);
            break
        end
    end
end
```

```
    for i = 1:400
        for j = i+1:400
            if (result(j,2) == result(i,2))
                result(i,4) = result(j,3);
                break
            end
        end
    end

    for i = 1:400
        flag = 100000000;
        for j = i:400
            if ((result(j,5) == result(i,5)) && (result(j,6) >
result(i,6)) && (flag > result(j,6)))
                flag = result(j,6);
            end
        end
        result(i,7) = flag;
        result(i,8) = result(i,7) + C + W(int32(mod(result(i,5),2)) +
1);
    end

    X
    result

f = -length(find(result(:,8) <= 28800));
```

四、问题四代码

```
function f = bpbrkfun(x)
%   两道工序物料加工,包括故障
%   单数序号表示第一道工序，偶数序号表示第二道工序。

    W = [32;27];        %奇偶数 CNC 工作时间
    C = 25;            %清洗作业的时间
    T_prd1 = 455;       %第一道工序工作时间
    T_prd2 = 182;       %第一道工序工作时间
    flag = 0;
    a = 0;
    product = 0;        %序号出现的集合
    times = 0;          %序号出现次数的集合
```

```
product1 = 0;        %CNC 第一次半成品的集合
times1 = 0;          %CNC 生产的半成品次数
product2 = 0;        %CNC 第二次生产成品的集合
times2 = 0;          %CNC 第二次成品的集合
TPmax1 = 0;          %第一道工序 CNC 的最大时间
TPmax2 = 0;          %第二道工序 CNC 的最大时间
sum_brk = 0;
k = 0;

i = 0;
i = double(int32(i));
a = ones(400);

result = zeros(400,5);
for i = 1:400
    result(i,1) = i;
end

output = ones(400,8);

X = zeros(400,9);
X(:,1) = x;        %x 为序号 1-16

%编号和 CNC 之间的转换关系：
for i = 1:400
    if (X(i,1)==1)||(X(i,1)==2)||(X(i,1)==3)||(X(i,1)==4)
        X(i,2) = 1;
    else if (X(i,1)==5)||(X(i,1)==6)||(X(i,1)==7)||(X(i,1)==8)
            X(i,2) = 3;
         else if
(X(i,1)==9)||(X(i,1)==10)||(X(i,1)==11)||(X(i,1)==12)
                 X(i,2) = 5;
              else if
(X(i,1)==13)||(X(i,1)==14)||(X(i,1)==15)||(X(i,1)==16)
                      X(i,2) = 7;
                   end
              end
         end
    end
if (X(i,1)==1)||(X(i,1)==5)||(X(i,1)==9)||(X(i,1)==13)
    X(i,5) = 2;
else if (X(i,1)==2)||(X(i,1)==6)||(X(i,1)==10)||(X(i,1)==14)
         X(i,5) = 4;
     else if (X(i,1)==3)||(X(i,1)==7)||(X(i,1)==11)||(X(i,1)==15)
```

```
                X(i,5) = 6;
            else if
(X(i,1)==4)||(X(i,1)==8)||(X(i,1)==12)||(X(i,1)==16)
                    X(i,5) = 8;
                end
            end
        end
    end
    end


    %计算距离

    flag = abs(fix((X(1,2) + 1) / 2) - 1);

    switch (flag)
        case 0
            S = 0;
        case 1
            S = 18;
        case 2
            S = 32;
        case 3
            S = 46;
    end

    if (rand() > 0.01)
        X(1,3) = S;
        X(1,4) = X(1,3) + T_prd1;
        X(1,6) = X(1,3) + W(int32(mod(X(1,2),2)) + 1);
        X(1,7) = X(1,6);
        X(1,8) = X(1,2);
    else
        X(1,3) = S;
        X(1,4) = X(1,3) + T_prd1 * rand + (rand * 600 + 600);
        X(1,6) = X(1,4);
        X(1,7) = x(1,6);
        X(1,8) = X(1,2);
        result(1,2) = 1;
        result(1,3) = X(1,2);
        result(1,4) = X(1,3);
        result(1,5) = X(1,4);
    end
```

```
for i = 2:400
    i = double(int32(i));

    %计算每个步骤的距离，上面几台机器的状况
    product = find(X(1:i-1,1) == X(i,1));
    times = length(product);

    product1 = find(X(1:i-1,2) == X(i,2));
    times1 = length(product1);

    product2 = find(X(1:i-1,5) == X(i,5));
    times2 = length(product2);


    %先确定每一步的位置

    if times1 == 0       %第一次一次也没出现过（不进行二，在一停止）
        flag = abs(fix((X(i,2) + 1) / 2) - fix((X(i-1,8) + 1) / 2));

        switch (flag)
            case 0
                S = 0;
            case 1
                S = 18;
            case 2
                S = 32;
            case 3
                S = 46;
        end

        if (rand > 0.01)
            X(i,3) = S + X(i-1,6);
            X(i,4) = X(i,3) + T_prd1;
            X(i,8) = X(i,2);       %当前位置为第一次出现的位置
            X(i,6) = X(i,3) + W(int32(mod(X(i,2),2)) + 1);
            X(i,7) = X(i,6);
        else
            X(i,3) = S + X(i-1,6);
            X(i,4) = X(i,3) + T_prd1 * rand + (rand * 600 + 600);
            X(i,8) = X(i,2);
            X(i,6) = X(i,3) + W(int32(mod(X(i,2),2)) + 1);
            X(i,7) = X(i,6);
            result(i,2) = 1;
```

```
        result(i,3) = X(i,2);
        result(i,4) = X(i,3);
        result(i,5) = X(i,4);
    end

else        %经历超过两次，正常运行,所以每次都要清洗。

    flag = abs(fix((X(i,2) + 1) / 2) - fix((X(i-1,8) + 1) / 2));

    switch (flag)
        case 0
            S = 0;
        case 1
            S = 18;
        case 2
            S = 32;
        case 3
            S = 46;
    end

    for j = 1:i
        if  X(j,4) >= TPmax1;
            TPmax1 = X(j,4);
        end
    end
    for j = 1:i
        if X(j,7) >= TPmax2;
            TPmax2 = X(j,7);
        end
    end

    flag = abs(fix((X(i,2) + 1) / 2) - fix((X(i-1,8) + 1) / 2));

    switch (flag)
         case 0
             S = 0;
         case 1
             S = 18;
         case 2
             S = 32;
         case 3
             S = 46;
    end
```

```
        if (times2 <= 1)
            C = 0;
        else
            C = 25;
        end

        if (TPmax1 > X(i-1,3))
            X(i,3) = TPmax1 + S;
        else
            X(i,3) = X(i-1,6) + S + C;
        end

        if(rand <= 0.01)
            %完成第一道工序的过程中故障
            X(i,4) = X(i,3) + T_prd1 * rand + (rand * 600 + 600);
            result(i,2) = 1;
            result(i,3) = X(i,2);
            result(i,4) = X(i,3);
            result(i,5) = X(i,4);
            X(i,6) = X(i,4);
            X(i,7) = X(i,4);
            X(i,8) = X(i,2);
        else
            X(i,4) = X(i,3) + T_prd1;
            flag = abs(fix((X(i,2) + 1) / 2) - fix((X(i,5) + 1) /
2));

            switch (flag)
                case 0
                    S = 0;
                case 1
                    S = 18;
                case 2
                    S = 32;
                case 3
                    S = 46;
            end

            flag = 1;
            if (X(i,8) == X(i,2))
                flag = 0;
            end

            if  ((TPmax2 > X(i,3) + W(int32(mod(X(i,2),2)) + 1)) &&
```

```
flag)
                X(i,6) = TPmax2 + S;
            else
                X(i,6) = X(i,3) + W(int32(mod(X(i,2),2)) + 1) + S;
            end

            if(rand > 0.01)
                X(i,7) = X(i,6) + T_prd2;
                X(i,8) = X(i,5);
            else
                X(i,7) = X(i,6) + rand * T_prd2 + (rand * 600 + 600);
                X(i,8) = X(i,5);
                result(i,2) = 1;
                result(i,3) = X(i,5);
                result(i,4) = X(i,6);
                result(i,5) = X(i,7);
            end
        end
    end
end

for i = 1:400
    if (X(i,7) - X(i,6) < T_prd2)
        X(i,6) = 0;
        X(i,7) = 0;
    end
end


output(:,1) = [1:400];

for i = 1:400
    output(i,2) = X(i,2);
    output(i,3) = X(i,3);
    for j = i+1:400
        if ((X(j,2) == X(i,2)) && X(j,6))
            output(i,5) = X(j,5);
            output(i,6) = X(j,6);
            break
        end
    end
end

for i = 1:400
```

```
        for j = i+1:400
            if (output(j,2) == output(i,2))
                output(i,4) = output(j,3);
                break
            end
        end
    end

    for i = 1:400
        flag = 100000000;
        for j = i:400
            if ((output(j,5) == output(i,5)) && (output(j,6) >
output(i,6)) && (flag > output(j,6)))
                flag = output(j,6);
            end
        end
        output(i,7) = flag;
        output(i,8) = output(i,7) + C + W(int32(mod(output(i,5),2)) +
1);
    end

    for i = 1:400
        if output(i,8) <= 28800
            sum_brk = sum_brk + result(i,2);
            k = k + 1;
        end
    end

    X;
    result
    output


f = -(k - sum_brk);
```

6.作业车间动态调度问题研究

摘要

本文针对作业车间调度的问题，对三种一般情况进行了多方分析，分别给出了合理的调度方案，使系统加工效率达到最大化。我们运用了动态调度的相关理论，借鉴了排队论、模拟仿真的思想、根据遗传算法等启发式算法的原理，综合运用了 JAVA、Excel 等软件编程求解，得出了单工序加工物料作业以 1、2、3、4、5、6、7、8 的顺序开始加工工作为最优解、双工序加工中加工时间短的机床位置集中分布对系统工作效率有正向影响、合理预测故障问题来评估加工风险等结论，最后结合实际给出合理化建议。

针对任务一拆分为如下三种问题：一道工序柔性作业问题、二道工序部分柔性作业问题、考虑故障发生的加工作业问题，对于三种类型的问题我们进行了具体分析。针对单工序柔性作业，基于排队论的观点，我们从完成任务数最多、智能车移动路径最短和空闲时间最短三个目标建立多目标规划模型，通过模拟仿真的思想，运用 JAVA 软件编程，代入任务二的数据求解模型；针对双工序部分柔性作业问题，我们围绕优化的中心思想化繁为简，以机床加工时间最长为目标，采用遗传算法、模拟退火等启发式算法的原理，使用 JAVA 软件编程，寻找全局最优解；针对故障发生的加工作业问题，我们发现故障发生对于单双工序的影响是近似相同的，基于此，我们引入故障因子，遵循概率随机的思想，采用随机遍历抽样的方式对前两个模型进行修正，得出考虑故障情况的加工生产结果，并精准定位故障发生位置，降低系统风险。

最后，我们对所建立的三种模型进行客观全面的评价，并对模型进行一些展望，建议读者能够从多 RGV 动态调度、轨道形状、智能车预测机制入手来分析提高系统加工效率的方法，对问题进行更深层次的研究。

【关键词】 动态调度；模拟仿真；多目标规划；启发式算法

作　　者：李辉，王一名，樊帆

指导教师：罗虎明

奖　　项：2018年全国大学生数学建模竞赛国家二等奖

1 问题重述

1.1 智能加工系统概述

智能加工系统由8台计算机数控机床(Computer Number Controller,机床)、1辆轨道式自动引导车(Rail Guide Vehicle,RGV)、1条RGV直线轨道、1条上料传送带、1条下料传送带等附属设备组成。RGV是一种无人驾驶、能在固定轨道上自由运行的智能车。它根据指令能自动控制移动方向和距离,并自带一个机械手臂、两只机械手爪和物料清洗槽,能够完成上下料及清洗物料等作业任务。

具体工作流程为:智能车首先在数控机床操作台放置未加工物料,等到该数控机床加工时间完毕后,向智能车发出需求信号,智能车接受信号,经过判断后,按照预订时间,移动至该数控机床操作台完成上下料工作,机床开始执行新加工任务,如果智能车取出的物料为熟料,则智能车立即进行熟料清洗工作。完成清洗工作后的智能车将等待下一个数控机床需求信号的处理。

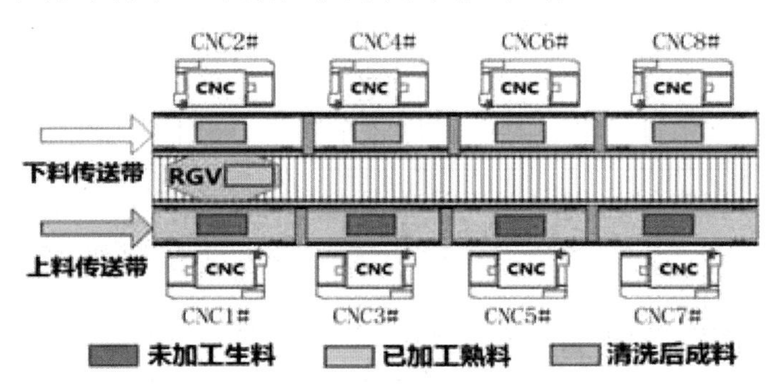

1.2 需解决的问题

针对下面的三种具体情况:

(1)一道工序的物料加工作业情况,每台机床安装同样的刀具,物料可以在任一台机床上加工完成;

(2)两道工序的物料加工作业情况,每个物料的第一和第二道工序分别由两台不同的机床依次加工完成;

(3)机床在加工过程中可能发生故障(据统计:故障的发生概率约为1%)的情况,每次故障排除(人工处理,未完成的物料报废)时间介于10~20分钟之间,故障排除后即刻加入作业序列。要求分别考虑一道工序和两道工序的物料加工作业情况。

请你们团队完成下列两项任务:

任务1:对一般问题进行研究,给出RGV动态调度模型和相应的求解算法;

任务2:利用系统作业参数的3组数据分别检验模型的实用性和算法的有效性,给出RGV的调度策略和系统的作业效率。

2 符号说明

表 1 部分符号说明

符号	含义
M_1	智能车移动 1 单位距离花费的时间
M_2	智能车移动 2 单位距离花费的时间
M_3	智能车移动 3 单位距离花费的时间
C	智能车清洗熟料花费的时间
T_w	智能车处于待工状态的时间
W	机床完成一道工序的时间
M	智能车完成一次上下料的时间
i	第 i 个任务点的编号
n	机床的编号（n=1、2、…、8）
z	智能车的位置编号
α	优先服务因子
w_1	机床加工完成一个两道工序物料的第一道工序所需时间
w_2	机床加工完成一个两道工序物料的第二道工序所需时间

3 基本假设

（1）在一个班次结束时，如果存在正在加工的物料，则该物料的加工视为无效，不计入完工产品数量中，将机床和智能车对该在产品之前的加工过程等同于空闲状态；

（2）物料在传送带上移动的时间忽略不计；

（3）智能车在总的工作时间内不会出现故障，系统能够持续正常运转；

（4）假定机床故障发生后的修复时间服从期望为 15，标准差为 1 的正态分布。

4 问题分析

对于任务一的三种情况，我们可以把其划分为四种类型进行处理：（1）不考虑机床加工过程出现故障的一道工序的物料加工作业；（2）不考虑机床加工过程出现故障的两道工序的物料加工作业；（3）考虑机床加工过程出现故障的物料加工作业。针对不同的类型我们分别进行如下分析：

4.1 类型一分析：

一道工序的物料加工作业可分为两个阶段：加工阶段和换料阶段。每完成一次加工任务，机床将立即发出需求信号，智能车接受需求信号后，将进入判断环节：如果智能车正在进行换料或者清洗工作，那么该需求信号将进入排队状态，等到智能车完成当前工作后，再进行处理；如果智能车处于空闲状态，那么智能车将立即移动至该机床，进行换料和清洗工作；如果智能车处于刚完工状态，那么智能车将对当前排队的需求信号按照排队规则进行排序，优先处理排在第一位置的需求信号。

通过对比题目给出的相关数据，我们发现加工过程的时间相同，而对于不同编号的机床，智能车为偶数编号机床一次上下料所需时间要大于为奇数编号机床一次上下料所需时间。在给定时间下（8 个小时），我们要达到三个目标：完工物料数量最多、所有设备处于空闲状态的时间最少，智能车移动的路径单位最短，基于此，对于排队的需求信号，我们以换料时间最短、距离最近作为优先处理的原则进行建模，通过减少智能车移动的时间和所有设备空闲状态的时间来提高系统的工作效率。

4.2 类型二分析

两道工序物料加工过程可分为两次一道工序物料加工的组合，即一个任务的完成需要经历“一次加工——换料——二次加工——换料”。但是，和一道工序物料加工相比，差别在于两次工序所使用的刀具是不同的，每个机床只能安装一种刀具，这就导致两道工序物料加工需由两台不同机床按顺序完成，根据三组参数来看，两种机床的加工时间长短也有差异。加工时间短的机床，使用频率比加工时间长的机床更高，根据“短板效应”理论，应当持有更多的加工时间长的机床来提升系统加工效率。由于加工时间长的机床使用频率低，应当分布于车间两端，而加工时间短的机床，使用频率高，适合分布于车床中央，从而可以节省智能车的移动时间，提高智能车的工作效率。

每种机床数量的确定以及位置的分布将直接影响任务完成的数量，因此进行两道工序动态调度模型建立时，需要将不同类型的机床排列组合作为重点考虑因素。

4.3 类型三分析

考虑机床在加工过程中出现故障的情况，显然使动态调度问题更加复杂化，一方面需要考虑发生故障的可能性对系统工作效率的影响，另一方面由于故障发生时，修复的时间是不确定的，更加使发生故障对系统造成的延误性难以估计。

既然是在机床加工阶段产生故障，那么故障的产生所造成的停工滞留时间必然会造成该次加工时间的加长，而每一次加工过程中均会存在 1%可能性的故障，因此，过增加故障概率和故障影响程度来对原有两种类型进行修正，便可以获得在考虑故障发生时的结果。

5 模型建立与求解

5.1 一道工序多目标规划模型

5.1.1 模型建立

基于排队论的观点，我们可以把一道工序的实现加工阶段和换料阶段拆分为三个过程：输入过程、排队规则和服务过程。智能车相当于单服务台，8 个机床相当于有限源顾客，顾客相互独立，可以是一个一个到达也可以是成批到达。系统采用等待制，一次处理顾客需求的最大容量为 8，当成批到达时，系统采用优先服务的原则接受最优先的顾客需求，而非先到先得。

智能车开始位于 1#机床与 2#机床中间的轨道位置处，所有的车床处于空闲状态，均向智能车发出需求信号，遵循换料时间最短和距离最短原则，智能车在初始阶段将按照 1#、2#、3#、4#、5#、6#、7#、8#的顺序依次处理换料工作。智能车上下料时，根据机床的编号 n 不同，上下料时间为：

$$m_n=\begin{cases}t_o,(n\text{为奇数})\\t_e,(n\text{为偶数})\end{cases}$$

智能车在结束所有上料工作后，将处于空闲状态，直到有机床完成加工过程，需要智能车进行换料工作。机床发出需求信号，智能车接收信号后开始移动到该机床处，进行换料工作。完成换料后，机床开始新的加工任务，同时，智能车开始进行清洗熟料的工作。

在此过程中，智能车的位置 z 可由智能车停止或工作时，对应的机床的编号 n 来确定：

$$z=\begin{cases}\dfrac{n}{2},(n\text{为偶数})\\[2ex]\dfrac{n+1}{2},(n\text{为奇数})\end{cases}\quad n\in[1,8],n\text{为整数}$$

智能车在接受到编号为n' 的机床发出的需求信号后，预计移动的距离为 M_l， 其中：

$$l=\begin{cases}\left|\dfrac{n'+1}{2}-z\right|,(n'\text{为奇数})\\[2ex]\left|\dfrac{n'}{2}-z\right|,(n'\text{为偶数})\end{cases}$$

记机床从开始加工到加工任务完成，智能车换料后，才算做一次任务完成，因此，编号为 n 的机床完成第 i 任务所花费的时间为：

$$t_{in} = W + M_{li} + m_{ni} + C$$

当开始执行第 b 个任务时，整个系统运行的时间为T'，此时，智能车已经运行的时间为：

$$t^{rgv}{}_{b} = \sum_{i=1}^{b-1} t_{in} - (b-1)W$$

智能车处于空闲状态的时间为：

$$T_w = T' - t^{rgv}{}_{b}$$

当j($j \in (1,8]$)个机床发出需求信号时，智能车排序的依据为优先服务因子：

$$\alpha_j = \frac{M_{lj} + m_{nj}}{\sum_j (M_{lj} + m_{nj})}$$

优先服务因子α 为智能车到达每个排队的机床预期的时间和上下料的时间占所有排队的机床预期的时间和上下料的时间之和的比率。显然，α_j越小，说明对应的机床在所有排队等待的机床中就其换料和与智能车的距离综合来看，更能节省时间。

为使智能车效率最大化，我们需要达到在单位时间内完成的任务数量最大、智能车移动的路径最短和智能车处于空闲状态的时间最短，为实现三个目标，我们设定目标函数为：

$$\begin{aligned} &Max\ i \\ &Min \sum_i M_{li} \\ &Min\ T_w \end{aligned}$$

在这三个目标中，智能车的空闲状态时间越短，就意味着它在给定时间范围内的工作时间越长，完成的任务数也就越多。所以，单位时间内完成任务数量最大和智能车处于空闲状态时间最短的相关性很强，并不独立，只选取一个即可；而智能车移动路径最短从某种程度上并不能明显体现出智能车效率最大化，可作为次要目标保留。

综上所述，我们对一道工序加工作业动态调度建立多目标规划模型：

$$\begin{aligned} &\text{main } Max\ i \\ &Min \sum_i M_{li} \\ &s.t. \begin{cases} \sum_{i=1}^{b-1} t_{in} - (b-1)W < T_{总} \\ \min[\alpha_j = \frac{M_{lj} + m_{nj}}{\sum_j (M_{lj} + m_{nj})}] \\ n \in [1,8], n \in N^* \\ i \in N^* \\ z \in [1,4], z \in N^* \\ l \in [1,3], l \in N^* \end{cases} \end{aligned}$$

由于上述模型的约束条件复杂，无法使用一般的线性规划方法进行求解，而且由于涉及到与时间相关，用现代优化解法获得的结果误差太大，因此，我们根据模拟仿真的思想做到完全柔性式加工生产，并通过以下流程图编程求解的算法，求解使单位时间内完成的任务数量达到最大。流程图如图 1 所示：

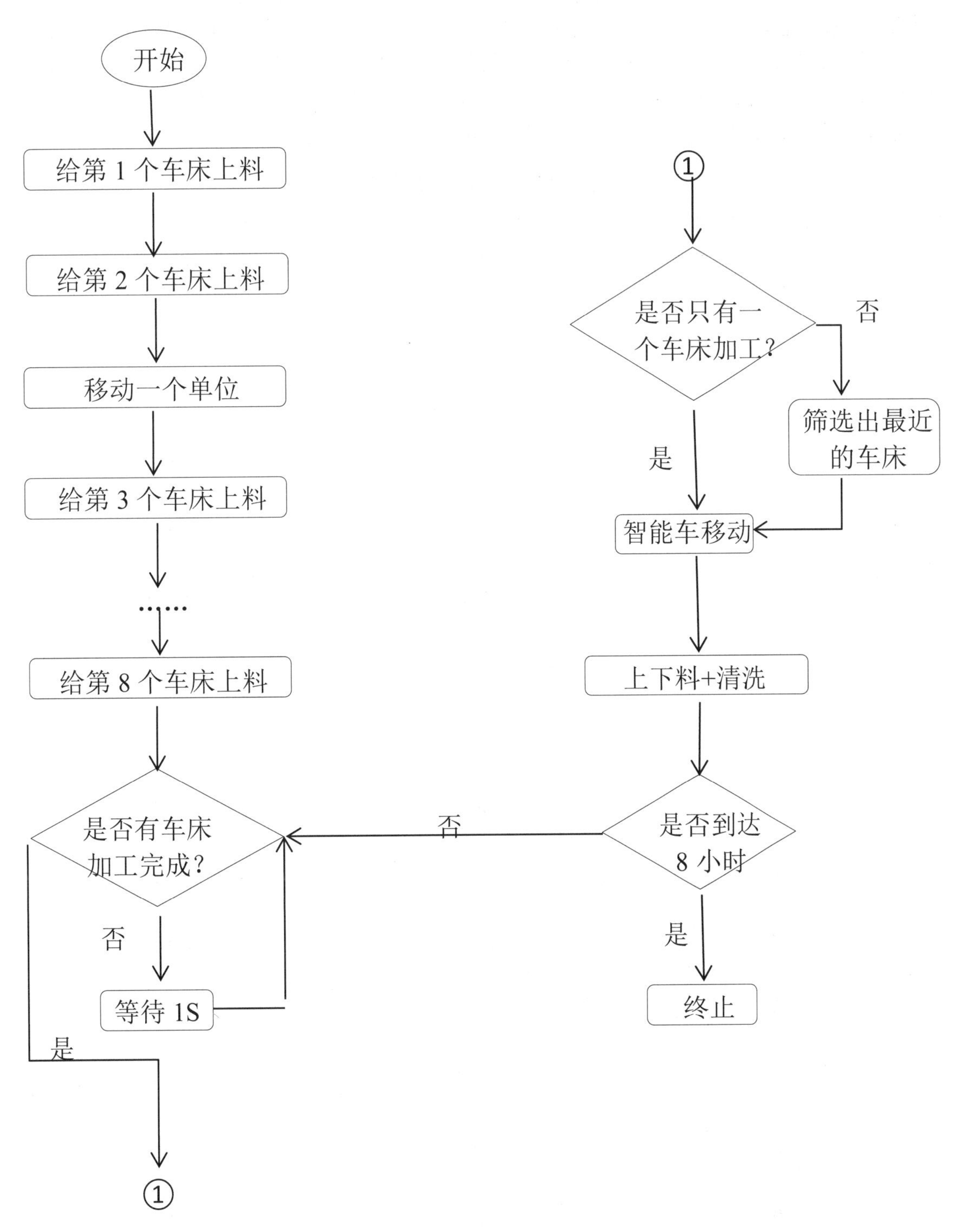

图 1 一道工序加工过程流程图

5.1.2 模型求解

通过编程，我们发现，最优的顺序是以 1#、2#、3#、4#、5#、6#、7#、8#的顺序进行的，第一组最终完成任务 290 个、第二组最终完成任务 312 个、第三组最终完成任务 304 个。从加工时间上看，三组参数从大到小排列为：第 2 组>第 1 组>第 3 组，符合加工时间越长，最终完成任务越低的理论，结果具有一定的可信性，表明我们所构建的模型具有一定的使用性，算法具备有效性。求解的部分结果如表 2 所示（以第 3 组情况为例，具体详见附件）：

表 2 第 3 组加工情况部分结果

加工 CNC 编号	上料开始时间(s)	下料开始时间(s)
1	0	618
2	28	788
3	78	867
4	106	985
5	156	1064
6	184	1182
7	234	1261
8	262	1379
1	727	1425
2	829	1504
3	913	1622
4	1015	1701
5	1099	1819
6	1201	1898
7	1487	2016
8	1571	2095

5.2 两道工序动态调度模型

5.2.1 模型建立

两道工序物料加工，由于加工工序所用的加工时间不同，将会影响到任务完成的数量。一般来说，工序加工时间长的机床使用频率要低于加工时间短的机床，如果两种机床持有量相等，则会造成半成品数量拥堵的现象，不仅会降低产量还可能会增加仓储成本。因此我们将根据工序加工时间的长短来确定两种机床的数量。每种机床的数量为：

$$N_k = [\frac{8w_k}{w_1 + w_2}] \quad （k=1、2，\ N_k \in N^*\ ）$$

如果工序加工时间具有明显差异，那么工序加工时间长的机床使用低频性将会导致智能车移动至该类机床的次数将明显低于工序加工时间短的机床，为降低智能车移动耗费的时间，我们建议将工序加工时间长的机床安放在车间两端，增强工序加工时间短的机床的集中度，从而减少智能车移动的路径。不同编号的机床上下料所用的时间虽有差异，而且差异并不明显，但是我们无法排除这种差异对结果的影响。

由于两道工序加工比一道工序加工需要增加考虑加工时间不同、机床的分布情况、不同类型机床的具体数量等因素，因此，限于比赛时间的紧迫，我们无法在短时间内通过模拟仿真的方法解决两道工序加工的作业类型，我们只能进行部分环节柔性处理，将采用遗传算法的思想进行求解，即在适应度函数和约束条件下寻找最优的加工机床调度编排序列，实现部分柔性加工生产。

各个加工机床因为编号的不同导致上下料的取用时间不同，因为种类的不同，导致加工时间的不同，但是上下料取用时间和加工时间是已知的，而且两道工序有先后之分，这就对机床的加工次序形成了约束。只有确定与加工约束条件相符的机床，并在这些机床中进行筛选出最佳方案，才能使加工性能各指标达到最优，也即使单位时间内加工产量尽可能多。单位时间内加工的产量足够多，那么同样的，需要付出的加工时间也将会更多，所以，系统工作的效率可以通过 8 个机床工作的时间来体现，因此，我们定义系统的加工（作业）效率为：

$$\phi = \frac{\sum W_i}{8T_{总}}$$

其中，W_i 表示序列中所有编号机床的加工时间，用 $n(i)$ 表示序号 i 所对应的机床编号，则有：

$$w_i=\begin{cases}w_1,n(i)\in\{第一道工序车床编号\}\\ w_2,n(i)\in\{第二道工序车床编号\}\end{cases}$$

我们的目标是使所有机床的加工时间达到最大，因此适应度函数为：

$$\max\ \sum w_i$$

常见的车间调度问题一般需要考虑三个方面的约束：(1)工艺方面约束；(2)机器方面约束；(3)时间方面约束；本题中两道工序的加工作业问题的约束条件，归纳起来一共包含：（1）一个机床在同一时间只能加工一个工件或者等待加工工件；（2）每个机床仅能实现一道加工工序；（3）加工工序必须按照顺序进行，即未经过第一道工序加工的生料，不能直接进行第二道加工工序；（4）智能车同一时间只能进行移动或清洗或上下取料或清洗熟料工作中的一个，而且仅有一个智能车。 转化为逻辑表达式为：

$$if\ \ n(i+1)=n(i),\ \ then\begin{cases}\sum(w_{i+1}-w_i)>w_1,n(i)\in\{第一道工序机床\}\\ \sum(w_{i+1}-w_i)>w_2,n(i)\in\{第二道工序机床\}\end{cases}$$

$$b_i=\begin{cases}-1,n(i)\in\{第一道工序机床\}\\ 1,n(i)\in\{第二道工序机床\}\end{cases}$$

$$\left|\sum_{j=0}^{7}b_{i+j}\right|\le|N_2-N_1|$$

$$\sum w_i<T_{总}$$

最理想的加工生产状况是每个加工车间都在进行加工生产，而且智能车没有移动的时间损耗，在这种理想状态下，（考虑上下取料的时间）完成一个加工周期的时间为 $\max\{w_1,w_2\}+\min\{N_1,N_2\}(t_o+t_e)$ 秒，生产 $\min\{N_1,N_2\}$ 个工件。规定时间内，能够生产的最大产量为:

$$P_{\max}=[\frac{T_{总}}{\max\{w_1,w_2\}+\min\{N_1,N_2\}(t_o+t_e)}\bullet\min\{N_1,N_2\}]$$

最差的加工生产状况是每个加工车间虽然仍进行生产，但是除了上下取料的时间外，智能车具有空闲状态和移动所花费的时间。在这种状况下，完成一个加工周期的时间为 $time'$ ，生产 $\min\{N_1,N_2\}$ 个工件。在规定时间内，能够生产的最大产量为 $P_{\min}$ 。

$$time' = \max\{w_1, w_2\} + \min\{N_1, N_2\}(t_o + t_e) + \frac{7}{3}\sum_{l=1}^{3} m_l + \mu T_{总}$$

$$P_{\min} = [\frac{T_{总}}{time'} \bullet \min\{N_1, N_2\}]$$

其中，μ 为智能车发生空闲状态的时间概率，取值应根据具体的车间分配状况决定，$\mu T_{总}$ 为智能车处于空闲状态的期望值。因此，初始化种群中一个染色体序列的长度应该所处的范围为：

$$\max(i) \in [P_{\min}, P_{\max}]$$

以下我们将根据遗传算法等启发式算法原理进行编程求解，遗传算法的步骤如下：

Step1：种群初始化，对遗传个体进行编码，采用随机数量随机生产若干机床编码序列；

Step2：选择，通过约束条件和目标函数计算每个个体的目标函数值，即每个可行序列所对应的生产加工总时间；

Step3：筛选，我们评估每个个体的适应度，遵循适应度越高，选择概率越大的原则，采用轮盘赌算法，从种群中选择两个个体作为父方和母方；

Step4：抽取双方个体的子代，进行交叉，产生子代；

Step5：对子代染色体进行变异；

Step6：重复 2、3、4、5，直到新种群的产生。

5.2.2 模型求解

我们根据三组数据中的第一道工序和第二道工序加工时间的比重，确定了 3 组情况各自的两种机床的数量。同时，我们根据两种车床数量的差异，按照差异越大，智能车空闲状态的概率越大的原则，设置了三组值；并代入具体的参数，计算出最大产量和最低产量的值，获得下表：

表 3 二道工序三种加工情况相关参数

	第 1 组	第 2 组	第 3 组
N_1:N_2	1:1	3:5	3:1
空闲状态概率 μ	0.1	0.15	0.2

可行解序列的长度范围（P_{min}~P_{max}）	30~181	29~165	30~157

根据加工时间长的工序对应的机床分布于车间两端，加工时间短的工序对应的机床分布于车间中间的原则，我们对三组参数的车间分布情况如下图所示：

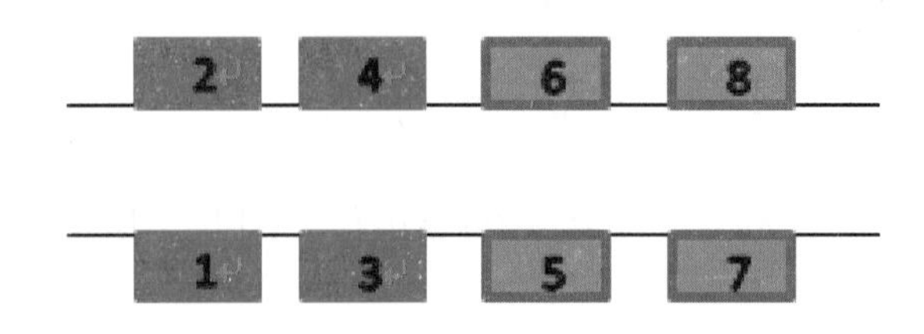

图 2 第 1 组情况车间机床编排图

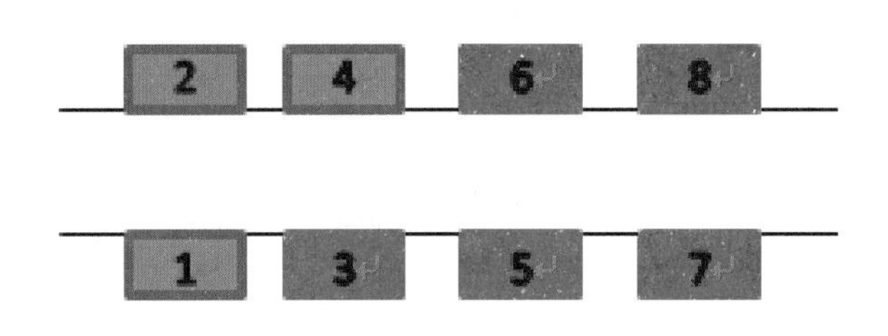

图 3 第 2 组情况车间机床编排图

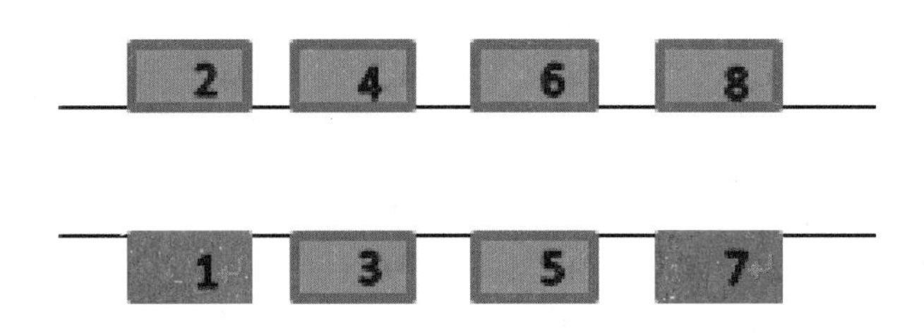

图 4 第 3 组情况车间机床编排图

*其中，蓝色表示第一道工序的机床；

红色表示第二道工序的机床

将相关参数代入代码中，通过 10000 次迭代后，筛选出最优的序列编号，使加工时间达到最大。经计算可得，三组情况的加工效率分别为：

表 4 双工序加工效率

	第一组	第二组	第三组
加工效率	99.40%	98.06%	99.85%

从表 4 可知，机床的数目和排布的方式不同将会影响加工的效率，但是影响差异不大，从三组情况的加工效率就能明显看出。目前来看，根据加工时间的长短来决定不同加工工序的机床带来的加工效率均能达到 99%左右，结果可观，表明我们建立的模型和算法具有良好的效果。（篇幅有限，具体结果详见附件中

的支持材料 Case-2-result.xls，在此不再赘述。）

5.3 故障修正的动态调度模型

5.3.1 故障因子设置

机床在加工过程中将会出现故障的情况，每台机床故障发生的概率为 1%，出现故障后的机床将中断加工过程，无法继续加工工作，需要立即进行维修，但是维修完成的时间也不是固定的，介于 10~20 分钟之间。为了简化模型，我们将维修时间假定为服从 μ=15，σ=1 的正态分布。因此，为衡量故障对每台车间上下取料的影响，我们对每台机床的加工时间引入故障概率因子，则考虑故障情况下，每台机床的工作时间为：

$$w' = w + 60a \cdot \exp\{\frac{-(x-15)^2}{2}\}$$

其中，a 表示故障是否发生，在算法中可以用随机遍历抽样算法进行模拟发生概率为 1%的故障的发生。将以上两个模型中的工序加工时间均通过以上方式添加故障因子，重新代入两个模型中，输出结果，便可以得出考虑机床在加工阶段出现故障情况下的加工结果。

5.3.2 一道工序故障模型求解

引入故障因子后，对模型一编制的代码进行稍作修正后，重新运行，获得的工作效率情况如下：

表 5 引入故障因子的单工序加工效率

	第一组	第二组	第三组
加工效率	93.25%	91.60%	94.45%

从上表来看，引入故障因子后，对单工序加工效率的影响较为严重，但也更符合客观实际。因此，企业在进行加工生产时，需要预防机床在生产过程中出现故障的情况，不仅降低生产效率，而且还会增加维修成本，对企业经济效益带来很大危害。企业可以通过定期维修检测设备，及时更换老化设备等手段降低风险的影响程度。（具体结果详见附件中的支持材料 Case-3-result1.xls，在此不再赘述。）

5.3.3 两道工序故障模型求解

同理，在引入故障因子后，对模型二进行相同处理，运行结果如下：

表 6 引入故障因子的双工序加工效率

	第一组	第二组	第三组
加工效率	97.20%	96.77%	95.32%
降低率	2.2%	1.29%	4.53%

根据表 6 的可知，考虑故障因素后，双工序加工效率有所降低，但是没有单工序加工降低的剧烈。原因在于，单工序加工由于加工频率高，故障发生的频数也相应增加，对系统造成的影响更大，符合高收益高风险的定律。而对于双工序加工作业，故障风险对第二组的影响最小，而对第三组的影响最大，这是因为不同工序的机床数量上差异越大，也会使故障风险的效应向两极划分，由于双工序作业本身的工序分级，第一道工序对第二道工序形成限制，如果加工第一道工序的机床多余加工第二道工序的机床，那么第一道工序发生故障时的的替补能力更强，而生产所具有的短板效应增强了第二道工序的重要性，加工第二道工序的机床发生故障将会使生产停滞，这也就是故障对第三组影响更大的原因。

因此，对于双工序作业生产，企业应该抱有风险意识，不能完全凭借每道工序的加工时间来确定不同种类机床的数量，应当结合工序的重要性，以及故障风险对每种机床造成的不同影响来衡量不同种类机床的数量和分布，确保使风险降低至最小化。（具体结果详见附件中的支持材料 Case-3-result2.xls，在此不再赘述。）

6 模型的评价

6.1 优点

1. 在对一道工序加工作业处理时，引入排队论和模拟仿真的思想，设计合理的流程图，并依据流程图编写相应算法，最大程度实现全面柔性车间加工过程。
2. 在对两道工序加工作业的处理中，充分考虑影响加工效率的三大约束来源，并尽可能多的增加约束条件来提高结果的精确度，采用遗传算法的思想，避免陷入局部最优解的风险。
3. 考虑故障对两种情况的影响，引入故障因子修正模型，考虑了故障的发生概率对加工过程的影响。

6.2 缺点

1. 在对两道工序加工作业的求解时，没有通过模拟仿真，而是通过求解最优序列来获得结果，对机床的分布也没有进行深入思考而且通过遗传算法每次获得的结果均不同，存在的误差比较大。
2. 在对故障考虑时，仅考虑了加工过程中机床发生故障的状况，没有考虑在机床其他阶段是否出现故障的情况。

附录

模型一代码（JAVA 编程）：

1.机床

```
public class cheChuang
{
    Boolean flag_shangLiao = false;
    Boolean        flag_xiaLiao = false;
    Boolean        flag_call = false;
    Boolean flag_work = false;

    int work_time_has = 0;
    int work_time_one = 560;
    int work_time_two_one = 400;
    int work_time_two_two = 378;

    //设定车床的 CNC 代码
    int num = 0;
    int flag = 0;
    void init(int number)
    {
        this.num = number;
        //刚开始建立时候就请求上料
        this.raise_flag_for_RGV(0);
    }

    void work_done()
    {
        if(this.work_time_has >= this.work_time_one)
            raise_flag_for_RGV(1);

    }

    void raise_flag_for_RGV( int number)
    {
        if(number == 0)
        {
            this.flag_call = true;
            //需要上料
            this.flag_shangLiao = true;
```

```
            this.set_flag_for_RGV(0);
        }

        if (num == 1)
        {
            this.flag_call = true;
            //需要下料
            this.flag_xiaLiao = true;
            this.set_flag_for_RGV(1);
        }

    }

    void set_flag_for_RGV(int num)
    {
        this.flag = num;
    }

    int return_flag_for_RGV()
    {
        return this.flag;
    }


}
```

2.工作

```
public class main
{
    public static void main(String [] args)
    {
        RGV rgvTest = new RGV();
        cheChuang cc1 = new cheChuang();
        cc1.init(1);
        cheChuang cc2 =new cheChuang();
        cc2.init(2);
        cheChuang cc3 =new cheChuang();
        cc3.init(3);
        cheChuang cc4 =new cheChuang();
        cc4.init(4);
        cheChuang cc5 =new cheChuang();
        cc5.init(5);
```

```
        cheChuang cc6 = new cheChuang();
        cc6.init(6);
        cheChuang cc7 = new cheChuang();
        cc7.init(7);
        cheChuang cc8 =new cheChuang();
        cc8.init(8);
        cheChuang cheChuangList[] = {cc1, cc2, cc3, cc4, cc5, cc6, cc7, cc8};


        main_work mainJob = new main_work();
        mainJob.init(rgvTest, cc1, cc2, cc3, cc4, cc5, cc6, cc7, cc8);
        rgvTest.set_list(cheChuangList);
        mainJob.first_work();
        mainJob.cycle();
    }
}
```

3.核心加工工作

```
import java.util.Vector;

public class main_work
{
    cheChuang chechuangList[] = new cheChuang[8];
    cheChuang cc1 = null;
    cheChuang cc2 = null;
    cheChuang cc3 = null;
    cheChuang cc4 = null;
    cheChuang cc5 = null;
    cheChuang cc6 = null;
    cheChuang cc7 = null;
    cheChuang cc8 = null;
    RGV rgv = null;

    void  init(RGV rgvTest,cheChuang cc1, cheChuang cc2,cheChuang
cc3,cheChuang cc4,cheChuang cc5,cheChuang cc6,cheChuang cc7,cheChuang cc8)
    {
        System.out.println("开始 main 函数创建");
        this.rgv = rgvTest;
        this.cc1 = cc1;
        this.cc2 = cc2;
        this.cc3 = cc3;
        this.cc4 = cc4;
```

```
        this.cc5 = cc5;
        this.cc6 = cc6;
        this.cc7 = cc7;
        this.cc8 = cc8;
        this.chechuangList[0] = (this.cc1);
        this.chechuangList[1] = (this.cc2);
        this.chechuangList[2] = (this.cc3);
        this.chechuangList[3] = (this.cc4);
        this.chechuangList[4] = (this.cc5);
        this.chechuangList[5] = (this.cc6);
        this.chechuangList[6] = (this.cc7);
        this.chechuangList[7] = (this.cc8);

        System.out.println("结束 main 函数创建");
    }

    void first_work()
    {
        System.out.println("开始初始化");
        rgv.receive(this.cc1);
        rgv.receive(this.cc2);
        rgv.move_to(3);
        System.out.println("rgv 的位置: " + rgv.location);
        rgv.receive(this.cc3);
        rgv.receive(this.cc4);
        rgv.move_to(5);
        System.out.println("rgv 的位置: " + rgv.location);

        rgv.receive(this.cc5);
        rgv.receive(this.cc6);
        rgv.move_to(7);
        System.out.println("rgv 的位置: " + rgv.location);

        rgv.receive(this.cc7);
        rgv.receive(this.cc8);
        System.out.println("结束初始化");
    }


    void cycle()
    {
        while(this.rgv.time_has_work < this.rgv.time_end_work)
```

```
            {
                System.out.println("开始循环");
                System.out.println("运行时间" + this.rgv.time_has_work);
                for(int i = 0; i < 8; i++)
                {
                    this.chechuangList[i].work_done();
                    if(this.rgv.flag_busy)
                        this.rgv.receive(this.chechuangList[i]);
                    else
                    {
                        for(int j = 0; j < 8; j++)
                            this.rgv.add_Time_relax(this.chechuangList[j]);
                    }
                }
            }
        }
}
```

4.智能车

```
import java.util.Vector;
import java.math.*;
public class RGV
{
    int time_move_one_distance = 20;
    int time_move_two_distance = 33;
    int time_move_three_distance = 46;
    int time_shangXiaLiao_Qishu = 28;
    int time_shangXiaLiao_Oushu = 31;
    int time_qingXi = 25;
    int time_start_work = 0;
    int time_has_work = 0;
    int time_end_work = 28800;

    Boolean flag_busy = false;

    //start Location
    int location = 1;
    Vector<Integer> waitList_num = new Vector<Integer>();
    Vector<Integer> waitList_action = new Vector<Integer>();

    //计算完成的工件数目
    int count_Finish = 0;
```

```
cheChuang cheChuang_List[] = new cheChuang[8];

void  init()
{

}

void set_list( cheChuang []li)
{
    for(int i = 0; i < 8; i++)
    {
        cheChuang_List[i] = li[i];
    }
    for(int i = 0; i < 8; i++)
        System.out.println(this.cheChuang_List[i].num);
}


void receive(cheChuang chuang)
{
    System.out.println("接收信号");
    int num = chuang.num;
    int action = chuang.return_flag_for_RGV();
    //接收信号时候判断是否忙碌
    if(waitList_num.size() == 0) {
        flag_busy = false;
    }
    else {
        flag_busy = true;
    }


    if(!this.flag_busy)
    {
        this.flag_busy = true;
        System.out.println("存在 1111111111 个信号");
        this.move_to(num);
        this.action(num, action);
        this.flag_busy = false;
    }

    if(this.flag_busy)
```

```
				{
					System.out.println("存在多个信号");
					this.waitList_num.add(num);
					this.waitList_action.add(action);
					//计算最短的路径并执行
					int                              lowestDistance                              =
this.count_distance(this.waitList_num.get(0));
					for(int i = 1; i < this.waitList_num.size(); i++)
					{
						if(this.count_distance(this.waitList_num.get(0))                    <
lowestDistance)
						{
							lowestDistance                                                  =
this.count_distance(this.waitList_num.get(i));
							num = this.waitList_num.get(i);
							action = this.waitList_action.get(i);
						}
					}
					//计算最短的路径并执行
					this.flag_busy = true;
					this.move_to(num);
					this.action(num, action);
					this.flag_busy = false;
				}

			}


			void action(int num,int flag_requese)
			{
				System.out.println("执行信号");
				if(flag_requese == 0)
				{
					if(num % 2 == 0)
					{
						this.time_has_work += this.time_shangXiaLiao_Oushu;
						for(int i = 0; i < this.cheChuang_List.length; i++)
						{
							this.add_Time_Work(this.cheChuang_List[i],
 this.time_shangXiaLiao_Oushu);
						}
					}
```

```
            if(num % 2 == 1)
            {
                this.time_has_work += this.time_shangXiaLiao_Qishu;
                for(int i = 0; i < this.cheChuang_List.length; i++)
                {
                    this.add_Time_Work(this.cheChuang_List[i],
this.time_shangXiaLiao_Oushu);
                }
            }

        }

        if(flag_requese == 1)
        {
            if (num % 2 == 0)
            {
                this.time_has_work += this.time_shangXiaLiao_Oushu;
                this.count_Finish += 1;
                this.time_has_work += this.time_qingXi;
                this.time_has_work += this.time_shangXiaLiao_Oushu;
                for(int i = 0; i < this.cheChuang_List.length; i++)
                {
                    this.add_Time_Work(this.cheChuang_List[i],
this.time_shangXiaLiao_Oushu);
                    this.add_Time_Work(this.cheChuang_List[i],
this.time_shangXiaLiao_Oushu);
                    this.add_Time_Work(this.cheChuang_List[i],
this.time_qingXi);
                }
            }

            if (num % 2 == 1)
            {
                this.time_has_work += this.time_shangXiaLiao_Qishu;
                this.count_Finish += 1;
                this.time_has_work += this.time_qingXi;
                this.time_has_work += this.time_shangXiaLiao_Qishu;
                for(int i = 0; i < this.cheChuang_List.length; i++)
                {
                    this.add_Time_Work(this.cheChuang_List[i],
this.time_shangXiaLiao_Qishu);
```

```
                        this.add_Time_Work(this.cheChuang_List[i],
this.time_shangXiaLiao_Qishu);
                        this.add_Time_Work(this.cheChuang_List[i],
this.time_qingXi);
                    }
                }
            }
        }


        //函数返回值存在正负，越大为正，越小为负
        int count_distance(int num)
        {
            int distance;
            //处理偶数
            if(num % 2 == 0)
            {
                distance = Math.abs((this.location) - (num / 2));
                System.out.println("增加的距离" + distance);
                this.location += distance;
                System.out.println("现在的位置" + this.location);
                return distance;
            }
            //处理奇数
            if(num % 2 == 1)
            {
                distance = Math.abs((this.location) - ((num + 1) / 2));
                System.out.println("增加的距离" + distance);
                this.location += distance;
                System.out.println("现在的位置" + this.location);
                return distance;
            }
            return 0;
        }


        void move_to(int num)
        {
            int time_move = 0;
            int distance = this.count_distance(num);
            if(distance == 1)
                time_move = this.time_move_one_distance;
```

```
		else if(distance == 2)
			time_move = this.time_move_two_distance;
		else if(distance == 3)
			time_move = this.time_move_three_distance;
		this.time_has_work += time_move;
		this.cheChuang_List[num - 1].flag_work = true;
		for(int i = 0; i < this.cheChuang_List.length; i++)
			this.add_Time_Work(this.cheChuang_List[i], time_move);
	}



	//小车运动的同时增加正在工作的机床的时间
	void add_Time_Work(cheChuang cheChuang, int time)
	{
		if(cheChuang.flag_work)
			cheChuang.work_time_has += time;
	}
	void add_Time_relax(cheChuang cheChuang)
	{
		cheChuang.work_time_has += 1;
	}
}
```

模型二&故障修正：

```
import java.util.Vector;

public class Main
{
	public static void main(String []args)
	{
		int Max = 0;
		int Max_temp = 0;
		//一次产生十个随机序列
		int time = 100000;
		int w1 = 400;
		int w2 = 378;
		for(int num = 0; num < time; num++)
		{
			//产生随机长度
			double length = Math.random() * 151 + 30;
```

```
int length2 = new Double(length).intValue();
/*System.out.println("长度为: ");
System.out.println(length2);*/
//产生随机序列数组
int data[] = new int[length2];
int data2[] = new int[length2];
Boolean flag[] = new Boolean[time];
for (int i = 0; i < time; i++)
	flag[i] = true;
//定义数组
int first[] = {1, 2, 3, 4};
int second[] = {5, 6, 7, 8};

int begin_for_create_data = 0;
for (int num_two = 0; num_two < length2; )
{
	int number = new Double(Math.random() * 8 +
1).intValue();

	if(number != begin_for_create_data)
	{
		{
			data[num_two] = number;
			num_two++;
			begin_for_create_data = number;
		}
	}
}

/*for (int te: data
	) {
	System.out.print(te + " ");
}
System.out.println("\n\n");*/
//剔除不合理的序列
int sum;
for (int temp = 0; temp < data.length - 7; temp++)
{
	sum = 0;
	//第一道检验
	int b[] = new int[8];
	flag[temp] = true;
```

```
            if (data[temp] == data[temp + 1])
            {
                System.out.println("第一检验错误");
                flag[num] = false;
                break;
            }

            //第二道检验
            for (int j = 0; j < 8; j++) {
                if (contains(first, data[j + temp]))
                    b[j] = -1;
                if (contains(second, data[j + temp]))
                    b[j] = 1;
            }
            for (int index : b
                    ) {
                sum += index;
            }
            if ((sum) > 0)
            {
                //System.out.println("第二检验错误");
                flag[num] = false;
                break;
            }
            //第二道检验

        }

        //求和
        if (flag[num])
        {
            Max_temp = 0;
            for (int temp = 0; temp < data.length; temp++)
            {
                    //System.out.println("序号: " + temp + " " +
data[temp]);
                    if (contains(first, data[temp]))
                        Max_temp += w1;
                    if (contains(second, data[temp]))
                        Max_temp += w2;
            }
            if (Max < Max_temp && Max_temp < 28800)
```

```
                    {
                        Max = Max_temp;
                        System.out.print("\n\n\n\n\n\n\n\n");
                        for(int temp2 = 0; temp2 < data.length; temp2++)
                        {
                            data2[temp2] = data[temp2];
                        }
                        System.out.println("最大的数组: " + data2.length);
                        System.out.println(" 故 障 存 在 的 位 置 "  +  new
Double(Math.random() * data2.length).intValue());
                        int trouble_time = new Double(Math.random() * 10 +
10).intValue();
                        System.out.println("处理故障的时间" + trouble_time);
                        System.out.print("最大的数组的数值: ");
                        System.out.println(Max + trouble_time);
                        System.out.println("数组数据: ");
                        for (int temp3: data2
                            ) {
                            System.out.print(temp3 + " ");
                        }
                    }
                }

                /*if(num == (time - 1))
                {
                    System.out.println("最大的数组: ");
                    for(int temp2 = 0; temp2 < data2.length; temp2++)
                    {
                        System.out.print(data2[temp2] + " ");
                    }
                }*/
            }
            System.out.println();
            System.out.println("迭代次数: " + time);
            System.out.println("最大的: " + Max);
        }

        static Boolean   contains(int[] data, int num)
        {
            for(int temp = 0; temp < data.length; temp++)
            {
                if (num == data[temp])
```

```
                    return true;
                }
                return false;
            }
        }
```

参考文献

[1] 王进峰, 阴国富, 雷前召,等. 基于改进遗传算法的柔性作业车间调度[J]. 现代制造工程, 2013(5):50-53.

[2] 吴焱明, 刘永强, 张栋,等. 基于遗传算法的 RGV 动态调度研究[J]. 起重运输机械, 2012(6):20-23.

[3] 陶泽, 谢里阳, 郝长中,等. 基于混合遗传算法的车间调度问题的研究[J]. 计算机工程与应用, 2005, 41(18):19-22.

[4] 陈振同. 基于改进遗传算法的车间调度问题研究与应用[D]. 大连理工大学, 2007.

[5] 韩明. 遗传算法在作业车间调度问题中的应用[D]. 吉林大学, 2015.

[6] 钱晓龙, 唐立新, 刘文新. 动态调度的研究方法综述[J]. 控制与决策, 2001, 16(2):141-145.

7.机场出租车的司机决策行为与上车点、优先权方案安排研究

摘要：本文研究了机场出租车司机的决策行为、出租车“上车点”的优化方案制定和出租车“优先权”方案安排。基于经济人假设，给出了以收益最大化为原则的出租车司机决策模型和以收益均衡为原则的短途载客“优先权”发放方案；结合排队理论，提出了以安全性原则为先、总乘车效率最高的“上车点”设置，并确定单次进入乘车区的出租车合理数量和乘客的分流设计。

针对任务一，首先，定义了出租车载客循环、乘客平均目的地和出租车空车返回方向并给出了相应的证明和解释。随后，基于经济人假设，将探究影响司机决策行为因素转化为探讨出租车司机收益率的影响因素，并从利益最大化原则出发，分别建立了A、B两种方案对应的出租车司机的收益率模型。之后引入“出租车运力需求”这一概念，确定出租车运力需求阈值。当蓄车池内已有车辆数超过阈值时，司机应选择放空返回城区拉客。反之，司机需要对比两种方案中自身的收益率，当A方案的收益率大于B方案的收益率时，司机应选择方案A进入蓄车池排队，反之选择方案B。

针对任务二，选取常州市奔牛国际机场进行分析，将相关数据代入问题一中的出租车司机决策模型，得出机场地区出租车运力需求阈值和应当采取方案A 决策时的蓄车池已有车辆阈值。此外，使用MATLAB对决策结果进行数学和图像描述，可直观判断出租车运力需求的阈值，以及随着时间点、季度的变化，A、B两种方案出租车司机收益率的变化。之后，通过比较仅依靠个人进行随机选择时出租车司机群体的平均收益率和使用出租车司机决策模型进行选择时的收益率，验证模型的合理性。结果显示，本文建立的决策模型较为合理，符合经济人假设。最后，利用经济学中的比较静态分析方法，定量探究司机决策对拥堵延时指数等13个因素的依赖程度。

针对任务三，首先对现有常见双车道案例进行分析，发现其在安全性方面仍存在问题。据此进行改进设计，包括增加隔离带、设计蛇形队列和设置双侧上车点。其次，根据“短板效应”和排队理论，改进上车点位置、双侧单次进入乘车区的出租车总量和乘客分流模式，使得总乘车效率最高。最后，以常州机场为例，进行了实例分析，结果显示，双侧单次进入乘车区的出租车总量为10；道路两侧乘客数量之比与其离开出站口到上车的整个过程花费的时间成反比。

针对任务四，首先对短途载客行为进行界定，当出租车司机两次返回机场的时间差小于某一定值时，将其认定为短途载客出租车。然后，基于收益率均衡原则，分别对短途载客和非短途载客司机的收益率进行度量。最后，建立收益均衡模型，使用MATLAB对模型进行求解，得出“优先权”发放的时间阈值，当出租车司机两次返回机场的时间差小于该时间阈值时，应当给予司机“优先权”。

关键词：运力需求阈值出租车司机决策模型排队论收益率均衡模型

作　　者：谷满仓，徐梦玥，郭颖

指导教师：胡小宁

奖　　项：2019年全国大学生数学建模竞赛国家一等奖

一、问题重述

1.1 问题背景

目前，我国民航产业迅速发展，乘客数量不断攀升。机场已成为出租车载客的主要去处。对于送客至机场的单个司机而言，由于机场离市区较远，此时的决策行为对其收益影响很大。

1.2 出租车接机规则

有打车意愿的乘客在下飞机后需到“乘车区”排队，由管理人员指挥分批乘车；此时排队等候载客的出租车会分批定量进入“乘车区”载客。

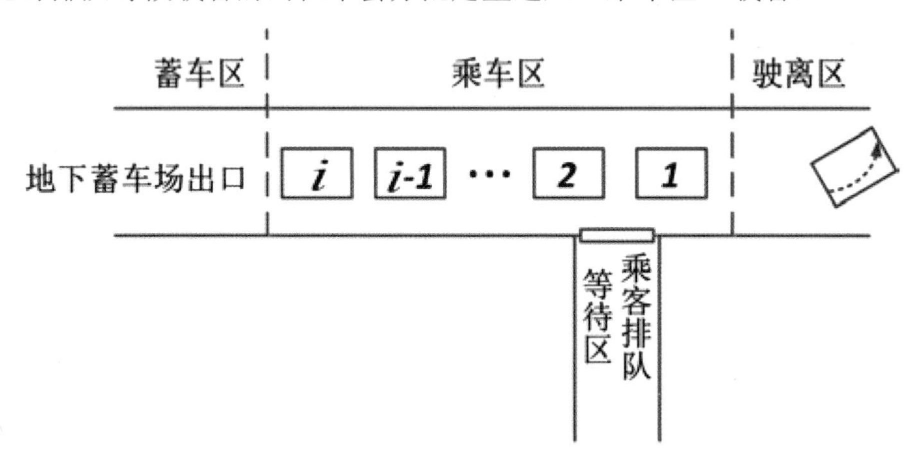

图 1 出租车接机示意图

在进入“蓄车池”之前，出租车司机可观测到在一定时间内抵达的航班数和“蓄车池”内的排队车辆数。同时，司机还可通过个人经验对抵达航班数和乘客数量的多寡进行判断。

1.3 两种选择——送客至机场后司机的决策

（1）驶入到达区，在指定的“蓄车池”进行排队，依次载客返回市区，此时需付出时间成本；

（2）以放空状态返回市区进行拉客，此时需付出空载费用，损失载客的潜在收益。

1.4 四项任务——题目的要求

任务一：结合机场的乘客数量变化及出租车司机的成本效益，探究影响司机决策的显著因素，给出决策模型及选择策略。

任务二：针对国内某机场及所在城市的数据，为司机提供针对性决策方案，并分析模型是否合理和整个模型对相关因素的依赖性。

任务三：对于具有两条并行车道的“乘车区”，给出合理的出租车和乘车安排方案，以达到最高的乘车效率；

任务四：分析如何安排对短途载客并返回的出租车的“优先”安排，以尽可能实现不同载客情况出租车的收益均衡。

二、符号说明

符号	意义	单位
t^0	一个循环中出租车空载状态下行驶的时间	小时
t^1	一个循环中出租车载客到达目的地的行驶时间	小时
d^0	一个循环中出租车空载状态下行驶的里程数	公里
d^1	一个循环中出租车载客到达目的地的行驶里程数	公里
B	出租车司机单位载客周期收入	元
t^0	出租车平均排队时间	小时
t_i^1	$i \in \{A,B\}$,出租车司机选择决策i时的期望载客时间	小时
d_i^1	$i \in \{A,B\}$，出租车司机选择决策i时的期望里程数	公里
RR_j	$j \in \{A,B,1,2\}$；当 $j \in \{A,B\}$时，表示选择决策的出租车司机的预期收益率；当 $j \in \{1,2\}$时，表示取得“优先权”的短途司机与非短途司机的预期收益率	—

此处仅列出论文中的通用符号，用于特定模型的符号会在首次使用时加以说明。

三、模型假设

1、出租车司机仅凭个人经验接客，不考虑网络订单。

2、出租车司机符合经济学中的经济人假设，即每一个从事经济活动的人采取的经济行为都是力图以自己最小的经济代价去获得资金最大的经济利益。

3、出租车以燃烧车用天然气为动力。

4、出租车往返机场的路程和时间不变。

5、乘客组从出站口到“乘车区”近似服从 Posion 分布。

6、出租车到“乘车区”接客近似服从 Posion 分布。

7、假设“乘车区”的每个停车位的乘客组平均上车时间相同。

8、并行车道为直道。

9、绿色通道内不会产生拥堵。

10、不考虑由于实行“优先权”政策导致部分司机使用非法手段取得“优先权”及其产生的一系列影响。

四、问题分析

2.1 任务一分析

任务一要求分析出租车司机在多种因素影响下的决策行为，并探究司机在不同实际情况下的选择策略。由于出租车司机被假定为经济人，因此其决策行为必将是以在单位时间内获取最大收益为目标的。由此，本文将探究影响司机决策行为的因素这一问题转化为探讨影响出租车司机收益率的因素。

基于司机个人的利益最大化原则，本文确定了方案 A 和方案 B 的时空变量及成本、效益参数，从而建立起出租车司机的收益率模型。之后引入“出租车运力需求阈值”这一概念，分别探讨出租车司机可能面临的三种不同情况时的理性选择，由此建立起出租车司机决策模型。

2.2 任务二分析

任务二要求以某机场作为数据样本，为该机场的出租车司机提供针对性的决策方案，并讨论模型的合理性和对各个影响因素的依赖性。

通过查阅资料，本文以常州市奔牛国际机场为例，将机场和城市的相关数据代入任务一中的出租车司机决策模型，并利用 MATLAB 进行决策结果分析，直观判断出出租车运力需求的阈值及出租车司机收益率最大时所处的季度和时段，并给出出租车司机对 A、B 两种方案进行决策时的临界状态。

通过对比出租车司机群体仅依赖个人经验进行决策和根据本文构建模型进行决策的平均收益率，验证模型的合理性。对影响临界状态的各个因素进行比较静态分析，确定模型对各个影响因素的依赖性。

2.3 任务三分析

任务三要求在遵循安全性原则的前提下，分析两条并行车道的“上车点”优化设计，从而获得最高的乘车效率。首先对现行常见案例进行分析，找出其存在的问题和缺陷，在此基础上提出有效的改进方案。由于本题要考虑出租车及乘客的安排问题，本文考虑采用排队论的方法，对单次可进入乘车区的出租车数量和乘客分流设计，并给出“上车点”设置的示意图。

2.4 任务四分析

任务四要求为短途载客的出租车司机制定优先权安排方案，以保证各个司机收益率的均衡。可以认为，由于通常情况下“短途”出租车司机两次返回机场的时间差较小，若这一时间差小于某一定值，则应当给予司机“优先权”。基于收益率均衡原则，分别对短途载客和非短途载客司机的收益率进行度量，之后建立收益均衡模型，求出“优先权”发放的时间阈值，从而给出合理的优先权安排方案。

五、模型的建立与求解

5.1 任务一模型的建立

5.1.1 定义与指导原则

在模型建立之前，本文对出租车的载客情况进行了分析，基于此进行模型构建。对于任务一，提出以下定义及指导原则：

定义 1（出租车载客循环）：将出租车由空驶状态，到载客状态并将乘客送至目的地称为一个循环（见图 2）。

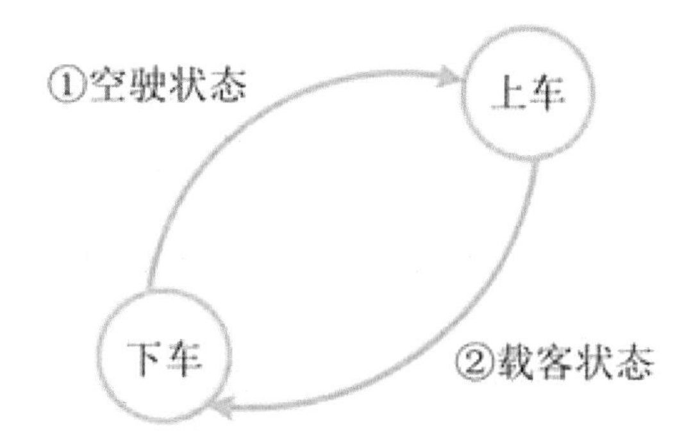

图 2 出租车载客循环

可证：出租车空驶状态将影响紧随其后的载客状态，例如，空驶状态出租车的行驶速度、行驶方向等因素将影响出租车在何时何地搜索到乘客。因此，为方便起见，本文将出租车从空驶状态开始到载客状态结束合称为一个“出租车载客循环”。

定义 2（乘客平均目的地）：司机预期在机场搭乘出租车的乘客平均目的地为市中心。

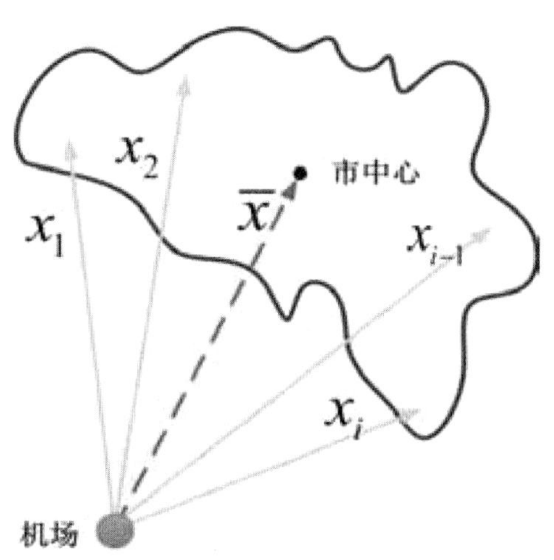

图 3 乘客平均目的地

机场的出租车司机在进行决策时，无法判断乘客最终目的地，只能依据以往经验得出从机场搭载出租车的乘客将乘坐的平均里程数。通常情况下，乘客乘坐的里程数如上图所示为 x_i，其平均里程数即为从机场到市中心的距离 $\overline{x}$。因此，本文定义司机认为在机场搭乘出租车的乘客平均目的地为市中心。

原则 1（利益最大化原则）：出租车司机按照个人利益最大化的角度出发进行决策。基于经济学的理性人假设，出租车司机无论是选择在机场等候载客还是直接返回市区，都将是从自身利益出发，即力图付出最小的经济代价以获取最大的收益。

定义 3（出租车空车返回方向）：出租车空车返回市区时，司机朝着市中心行驶。

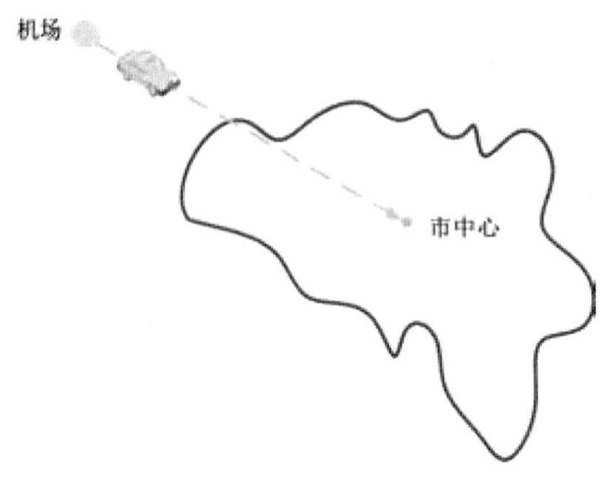

图 4 司机空车返回方向

通常情况下，城市中心附近区域人流量大，乘车人数较多。司机作为经济人，将在个人利益最大化的驱使下理性选择前往市中心区域，寻找乘客。

5.1.2 单位载客周期时空变量及成本、效益参数的刻画

（1）时空变量的刻画

在一个周期内，出租车只有空驶和载客两种情形，因此构造行驶时间和行驶距离的 0-1 变量：

$$\alpha=\begin{cases}0, & \text{空驶状态}\\1, & \text{载客状态}\end{cases} \tag{1}$$

- 时间变量的刻画

$$T=\sum_{\alpha=0}^{1}t^{\alpha} \tag{2}$$

其中，T 为单位循环的总时间，t^0 为一个循环中出租车空载状态行驶的时间，t^1 为该次循环中出租车载客到达目的地的行驶时间。

- 空间变量的刻画

$$D=\sum_{\alpha=0}^{1}d^{\alpha} \tag{3}$$

其中，D 为单位循环的总里程数，d^0 为一个循环中出租车空载状态行驶的里程数，d^1 为该次循环中出租车载客到达目的地的行驶里程数。

（2）成本、效益参数的刻画

为更好地度量出租车司机在一次循环中的收益情况，本文选取单位载客周期收入、单位时间租金和单位里程的燃气费三个参数。

- 单位载客周期收入参数的刻画

根据相关政策，通常情况下，出租车在夜间加收 λ 的费用。为确定出租车司机该时段的准确收入，引入“收入系数”来表示出租车司机真实收入和里程收入之比。根据出租车运营时间的不同，可将夜间和白天的收入系数定义为：

$$\varphi \triangleq \begin{cases} 1, & \text{白天} \\ 1+\lambda, & \text{夜间} \end{cases} \tag{4}$$

其中，φ为收入系数，λ为政府规定的出租车司机夜晚加收费用比例。

因此，出租车的收入为：

$$B = \begin{cases} \varphi \times B_{sf}, & d^1 \le d^s \\ \varphi \times \left[B_{sf} + P \times (d^1 - d^s) \right], & d^1 > d^s \end{cases} \tag{5}$$

其中，B_{sf}为出租车的起步价，d^s为起步里程数，P为超出起步里程数后的单位里程价。

- 租金参数的刻画

为提高出租车的使用效率，一辆出租车通常由两个司机交接班。查阅文献可得，出租车司机通常按月缴纳租金，且每位司机一个月（按 30 天算）内工作 15 天，每天平均运营时间为 18 小时。

$$t = 15 \times 18h = 16200 \min \tag{6}$$

$$c^r = \frac{x}{t} \tag{7}$$

其中，c^r为单位时间内的租金，x为出租车司机需缴纳的月租金，t为出租车每月平均运营的时间。

- 单位里程的车用天然气成本参数的刻画

目前许多城市的出租车为降低成本，已逐步以车用天然气作为主要动力来源，因此本文以下叙述均只对车用天然气进行分析。

根据相关报道[1]出租车每百公里平均消耗 9.83 立方米车用天然气。因此，每公里的车用天然气成本为c^f，即：

$$c^f = \frac{n \times \rho}{y} \tag{8}$$

其中，y为出租车在一次循环中行驶的距离，n为出租车行驶y公里消耗的车用天然气量，ρ为车用天然气的单位价格。

5.1.3 出租车司机收益率模型

送客至机场后，出租车司机需要对等候载客或空车返回进行决策，期间须考虑机场航班数量、乘客数量多寡、等待或返回的时间成本以及空载费用成本等一系列影响因素，由此本文建立出租车司机决策模型。

为满足收益率最大化原则，本文从成本—收益角度，定义出租车司机的净收益模型如下：

$$Profit \triangleq B - \left(c^r \times T + c^f \times D \right) \tag{9}$$

其中，$Profit$为出租车司机的净收益。

通常情况下，出租车司机希望减少空载时间并延长利润距离，以获得更高的收入。由此，本文将单个出租车载客循环中出租车司机所获取的运营收益定义为收益率，以此来衡量司机感知到的利润。

$$RR \triangleq \frac{Profit}{T} \tag{10}$$

其中，RR表示在一次循环中出租车司机的收益率。

对于A、B两种不同的选择方案，其具体时空参数的刻画如下：

（1）A方案时空参数的刻画

1）时间参数的刻画

司机在排队等候进入乘车区时，需要耗费t^0的时间成本，主要由出租车数量、行李搬运时间等因素导致。引入“服务速度”对t^0进行测算。计算如下：

$$\delta = \frac{a}{\mu *} \tag{11}$$

$$t_A^0 = \delta \times n \tag{12}$$

其中，$\mu *$为服务强度，即单位时间内用于为乘客提供服务的平均时间；a为平均每个出租车搭载的乘客数；δ为出租车的服务速度，即单个出租车搭载乘客离开乘车区消耗的时间。

根据定义1，出租车司机在进入蓄车池时的期望载客时间t^1为机场到市中心所需时间，即：

$$t_A^1 = \varepsilon t_{center} \tag{13}$$

其中，ε为城市道路拥堵延时指数，t_{center}为从机场到达市中心的自由流（畅通）旅行时间

2）空间参数的刻画

司机需要在指定的“蓄车池”进行排队，等候进入乘车区载客。在排队时，由于相对出租车载客里程数而言，出租车移动距离过短，因此本文近似认为出租车的行驶距离为0，即：

$$d_A^0 = 0 \tag{14}$$

根据定义1，出租车司机在进入蓄车池时的预期期望里程数d^1为机场到市中心的距离，即：

$$d_A^1 = d_{center} \tag{15}$$

其中，d_{center}为机场到市中心的距离。

（2）B方案时空参数的刻画

当出租车司机选择放空返回市区时，根据定义3，出租车司机的返回方向为市中心方向。

1）时间参数的刻画

司机选择放空返回市区时，其载客状态和空驶巡航状态时间之比由空驶率表示，因此放空返程时预期空驶时间为：

$$t_B^0 = \gamma t_{center} \tag{16}$$

其中，γ为空驶率。

对于出租车司机而言，离开机场后接到的第一单的期望消耗时间与其个人的经验判断有关，因此：

$$t_B^1 = \varepsilon t_{ave} \tag{17}$$

其中，ε为城市道路拥堵延时指数，t_{ave}为在“自由流旅行”情况下，出租车司机每单消耗的平均时间

2）空间参数的刻画

由于单位载客周期中，出租车司机在城市中行驶的平均速度基本一致，因此其空载巡航里程为：

$$d_B^0 = \gamma d_{center} \tag{18}$$

同理，d_B^1为该城市出租车司机平均运每单的里程数。

5.1.4 出租车司机决策模型

对于出租车司机而言，选择 A、B 两种方案都将付出一定的成本，出租车司机作为经济人，将遵循“利益最大化原则”，以最小的经济代价去获取最大的经济利益作为最终目标进行经济决策。因此，出租车司机必将选择收益率较大的方案。其决策模型如下：

Step 1： 将蓄车池已有出租车数量和该时段出租车运力需求进行对比。

在本文中，引入“出租车运力需求”的概念，表示该时段内最大出租车运力需求。

$$N = [\frac{k\varpi}{\beta}] \tag{19}$$

其中，N为该时段的出租车运力需求阈值，k为该时间段内乘客总人数，ϖ为选择乘坐出租车的乘客比例，β为出租车的载客率，$[\bullet]$表示取整。

出租车司机将通过比较当前排队车辆数m与出租车运力需求直接进行决策。

- $m \geq N$

当$m \geq N$时，“蓄车池”中的车辆数量已经达到了出租车运力需求阈值，机场出租车运力需求已得到基本满足。此时不必再进行 Step i（$i>1$）即可确定司机的最优决策为放空返回市区拉客。

- $m \leq N$

当$m \leq N$时，即“蓄车池”中已有出租车数量尚未达到阈值，机场出租车运力需求尚未完全满足，进入机场蓄车池等待可能存在盈利空间。此时转 step2 继续进行计算以确定最优决策。

Step 2： 带入数据，计算该季节、该时段对应的方案 A 预期收益率。

Step 3： 带入数据，计算该季节、该时段对应的方案 B 预期收益率。

Step 4： 比较方案 A 和方案 B 的预期收益率并进行决策。

- $RR_A \geq RR_B$，出租车司机选择方案 A，即进入蓄车池排队等待。
- $RR_A \leq RR_B$，出租车司机选择方案 B，即直接放空返回市区拉客。

5.2 任务二模型的建立

5.2.1 数据分析

本文采用常州市奔牛国际机场作为数据样本，从高德地图、期刊等得到以下相关数据。

（1）常州机场相关数据

本文使用用国内最大的航空公司——南方航空公司的 6 架常用机型[2]（波音 B777A、B777B、B737；空客 A319、A380；ERJ190）的最大客容量取平均值作为到达机场的飞机的最大客容量，为 256 人；本文取 5 家航空公司（东航、国航、南航、春秋、吉祥）的平均客座率做为到达机场的飞机的客座率，为 85.27%；选取入港航班（到达城市是常州的航班），用到达机场的飞机的平均最大客容量与客座率的乘积作为一次入港航班到达机场的旅客数，具体数据见附录 2。

（2）常州市区相关数据[3]

常州市机场到市中心的距离为 24.4 公里，出租车平均行驶 48 分钟。根据“高德地图”发布的《中国主要城市季度交通报告—常州市》可知，各季度常州交通拥堵延时指数、常州高峰期交通拥堵延时指数如下：

表 1 常州市拥堵延时指数

季度数	高峰拥堵延时指数	全天拥堵延时指数
第一季度	1.454	1.2554
第二季度	1.457	1.32
第三季度	1.442	1.31
第四季度	1.4492	1.277

注：由于“高德地图”近三年未发布第四季度的拥堵延时指数，因此综合考虑节假日天气等因素，认为第四季度的高峰拥堵延时指数为 $\varepsilon_4 = 0.6\varepsilon_1 + 0.4\varepsilon_3$

（3）常州市出租车相关数据

根据岳喜展[4]对某国际机场的调查统计数据分析，离港乘客选择乘坐出租车的比例为 39.7%。

目前，常州市出租车的起步里程数为 3 公里，起步价为 10 元/3 公里，当超出起步里程后，超出部分每单位里程收取 2.1 元[5]。

胡浩[6]实证考察得出，单个出租车平均一次可搭载 1.5 位乘客，其服务强度平均为 0.625 人/min。

景雪琴[7]对常州 2010 年 11 月份的出租车运营数据进行了分析，得到常州出租车各时段空载率（单位：%）如下：

表 2 常州市出租车空载率

时间段	01~02	03~04	05~06	07~08	09~10	11~12	13~14	15~16	17~18	19~20	21~22	23~24
空载率	46.57	42.99	45.93	45.86	42.34	39.87	39.90	39.67	39.80	40.04	40.95	39.91

从出租车的成本来看，近日常州市宣布取消出租车的月租金，有利于降低其成本。在车用天然气方面，其单位价格为 4.19 元/立方米，经计算，每单位车用

天然气的成本为 0.4119 元/公里[8]。

5.2.2 决策结果分析

（1）出租车运力需求阈值

出租车司机将当前排队车辆数 *m* 与出租车运力需求相对比直接进行决策。

表 3 不同时间段内出租车运力需求阈值数

时段	0:00	1:00	2:00	3:00	4:00	5:00	6:00	7:00
出租车阈值数	58	29	0	0	0	0	0	0
时段	8:00	9:00	10:00	11:00	12:00	13:00	14:00	15:00
出租车阈值数	58	87	145	58	145	29	116	0
时段	16:00	17:00	18:00	19:00	20:00	21:00	22:00	23:00
出租车阈值数	116	87	145	58	174	145	29	58

由表 3 可知，出租车运力需求阈值集中于夜间，20:00 达到峰值，为 174 辆，之后呈递减趋势。

（2）方案 A

通过分析常州市 2018 年各个季度机场出租车的相关数据，利用 MATLAB 绘制出各个季度、不同时间段内该机场中“蓄车池”已有出租车数量与司机收益率的关系图如下：

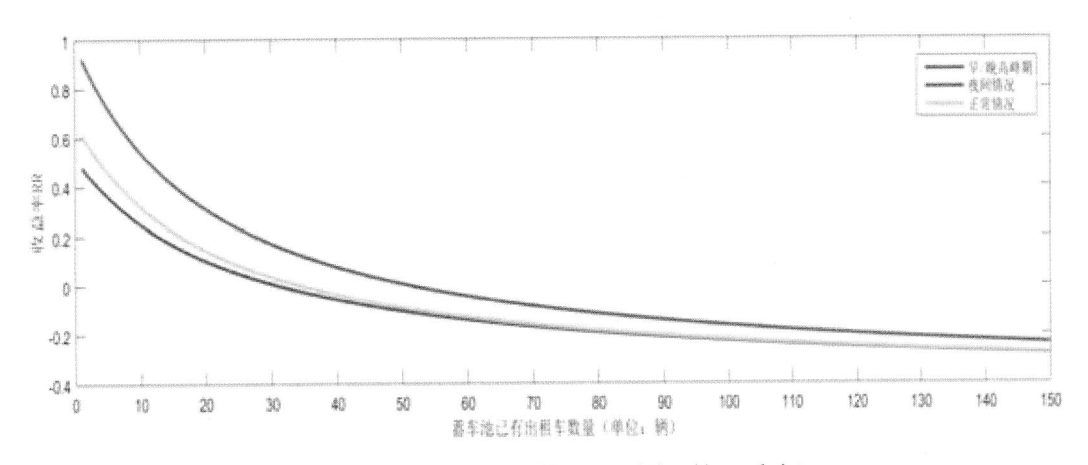

图 5 方案 A 不同时段的 n-RR 图（第一季度）

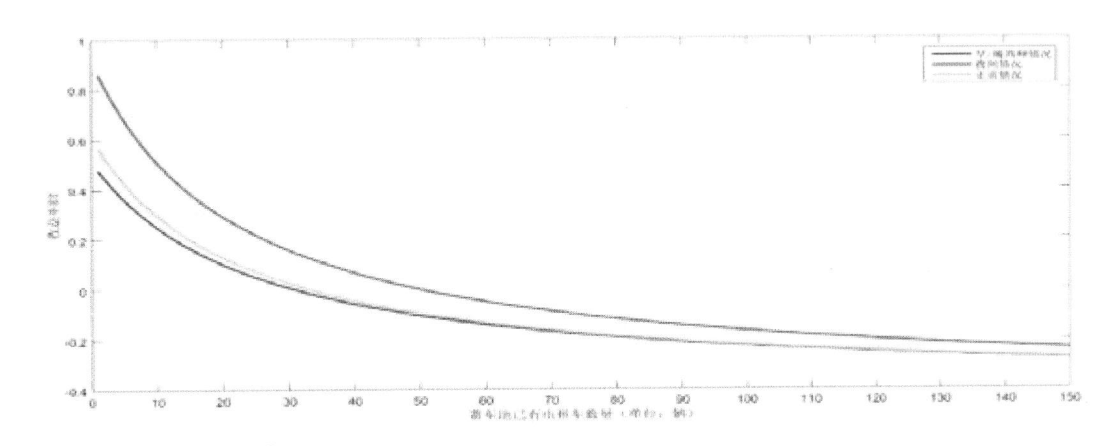

图 6 方案 A 不同时段的 n-RR 图（第二季度）

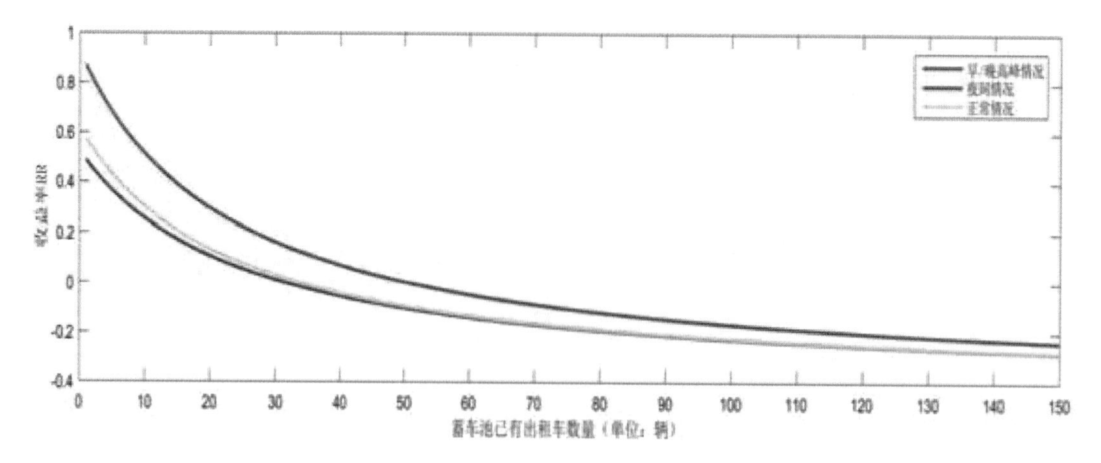

图 7 方案 A 不同时段的 n-RR 图（第三季度）

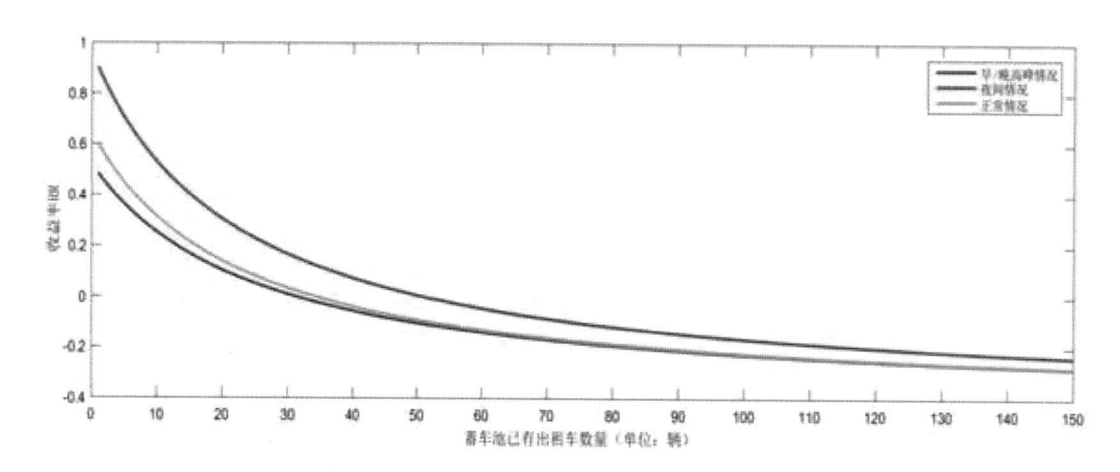

图 8 方案 A 不同时段的 n-RR 图（第四季度）

由常州机场四个季度的 n-RR 图可知：

1）横向来看，随着蓄车池内出租车数量的增长，收益率呈递减趋势，无限趋近于一个值，且在早/晚高峰期司机的收益率最先为 0。

2）纵向来看，在四个季度中，夜间载客的收益率最高，早/晚高峰期间出租车司机收益率相对偏低。

（3）方案 B

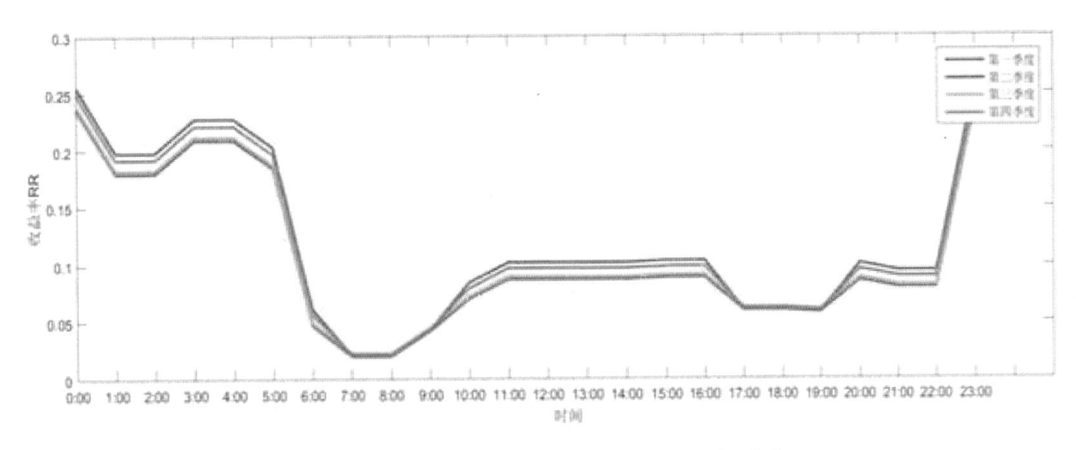

图 9 方案 B 不同时段的 t-RR 图（四个季度）

由图 9 可知：

1）横向来看，不同时段内出租车司机的收益率波动较大，夜间的收益率最大，峰值在 0.25 以上；

2）纵向来看，在第一季度时，出租车司机的收益率在任意时段均位居首位，第二季度司机的收益率紧随其后。

（4）司机决策模型

当出租车司机送客到机场后，若蓄车池已有出租车数量大于蓄车池已有车辆阈值，司机选择 B 方案。反之，司机应选择 A 方案进入蓄车池。以常州机场为例，在分别计算出不同季度、不同时段内两种方案的预期收益率后，进行比较，得出蓄车池已有车辆阈值如下表所示。

表 4 不同季度、不同时段内蓄车池已有车辆阈值

时间	0:00	1:00	2:00	3:00	4:00	5:00	6:00	7:00
第一季度	23	27	27	25	25	27	27	28
第二季度	23	28	28	25	25	27	27	28
第三季度	23	28	28	25	25	27	27	28
第四季度	23	27	27	25	25	27	27	28

时间	8:00	9:00	10:00	11:00	12:00	13:00	14:00	15:00
第一季度	28	25	25	23	23	23	23	23
第二季度	28	25	25	24	24	23	23	23
第三季度	28	25	25	23	23	23	23	23
第四季度	28	24	25	23	23	23	23	24

时间	16:00	17:00	18:00	19:00	20:00	21:00	22:00	23:00
第一季度	23	24	24	24	23	24	24	23
第二季度	23	23	23	24	24	24	24	23
第三季度	23	24	24	24	23	24	24	23
第四季度	24	24	24	24	23	24	24	23

5.2.3 模型的合理性分析

根据司杨和关宏志[9]在北皋机场、首都国际机场停车场等地进行实地调研获得的数据统计发现，出租车司机愿意进入蓄车池的比例为 68%。因此，在出租车

司机仅依靠个人经验进行随机选择时，出租车司机群体的平均收益率为：

$$\overline{RR} = 68\% RR_A + (1-68\%) RR_B \tag{20}$$

由于 $\overline{RR}$ 恒小于 $\max\{RR_A, RR_B\}$，所以出租车司机通过本文所述的决策模型进行选择，可以使个人利益实现最大化，满足“理性人”假设。

5.2.2 对相关因素的依赖性分析

根据以上分析，可知出租车司机的决策行为受到许多因素的影响，当考虑模型对相关因素的依赖性时，即在其他因素保持不变的条件下，一种因素变动对决策变量所产生的影响。同时，由于出租车司机载客的最终目的是实现个人的收益最大化，即达到均衡状态。基于以上分析，本文引入经济学中的比较静态分析方法，定量探究各个因素对司机决策行为的影响效果。

Step 1：求蓄车池已有车辆阈值。

令：

$$RR_A = RR_B \tag{21}$$

运用 MATLAB 编程求解，解得蓄车池已有车辆阈值 n。

Step 2：对 n 进行比较静态分析，探究各个因素对司机决策行为的影响效果。

$$\xi_{\mathrm{X}} = \frac{\partial n}{\partial \mathrm{X}} \tag{22}$$

其中，X 为单个因素，ξ_{X} 为因素 X 对司机决策行为的影响效果。

例如：

拥堵延时指数 ε 对司机决策行为的影响效果为：

$$\xi_\varepsilon = \frac{\partial n}{\partial \varepsilon} = -a\mu^*[t_{center} + \frac{t_{ave}(B_{sf}\varphi - c^f d_{center} - d^s P\varphi + d_{center}P\varphi)}{(c^f d_{ave} - B_{sf}\varphi + c^f d_{center}\gamma + d^s P\varphi - d_{ave}P\varphi)}] \tag{23}$$

其他因素对司机的决策行为影响效果 ξ_{X} 的见附录 3。

5.3 任务三模型的建立与求解

5.3.1 定义与指导原则

在大多数情况下，司机和乘客均需要排队等候，在保证安全的前提下如何合理地安排“上车点”以提高乘车效率，已成为机场管理人员亟待解决的问题。本部分将针对两条并行车道的乘车区进行上车点的设计，以达到上述要求。对于任务三，提出以下定义及指导原则：

定义 4（乘客组）：共乘一辆出租车的乘客即为乘客组。

原则 2（安全性原则）：上车点的设计必须遵循安全性原则，以确保车辆和乘客的安全，避免人员伤亡、设备及财产损失等意外事件的发生，这在考虑上车点的数量及所在位置时尤为重要。

5.4.2 并行车道案例

目前，对于两条并行车道的出租车接机服务，许多国际化机场（如里斯本国际机场）已建立了较为完善合理的“上车点”，如下图所示：

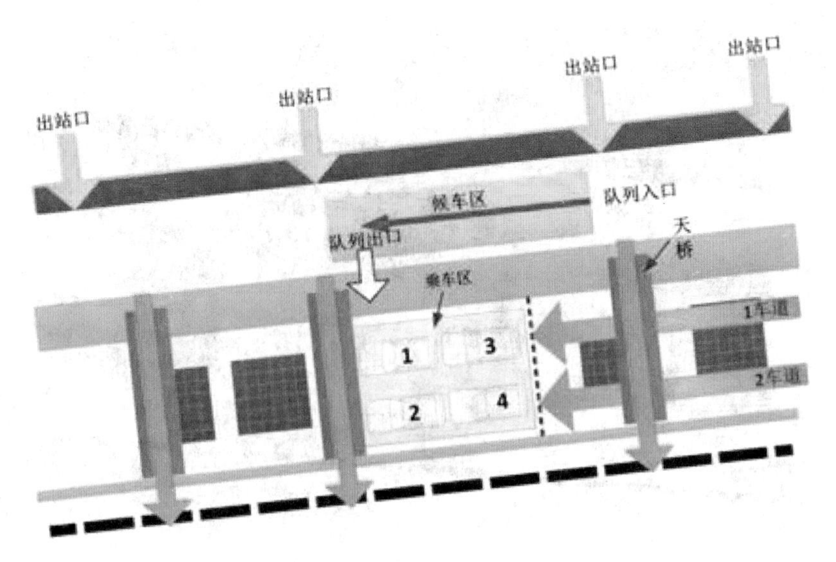

图 10 国外机场“上车点”设计示意图

由图 10 可知，“上车点”设计在乘车区靠近出站口的一侧，乘客既可以在候车区排队等候乘车，也可通过天桥到道路的另一侧乘车。出租车从 1、2 车道并行驶入“蓄车池”，而后分批进入乘车区载客。

值得注意的是，这种设计方案固然提高了乘车效率，但却忽视了更为重要的安全问题。在经济社会中，乘客和司机均为经济人，都希望付出最小的时间成本完成乘车或载客。因此会有部分乘客不会从天桥到达道路另一侧，而是选择直接横穿乘车区乘坐 3、4 号出租车；当斜前方出现空位时，部分司机会选择插队以减少等待时间。在这些情况下，极易发生交通事故，危害乘客和车辆的安全。（如图 3）

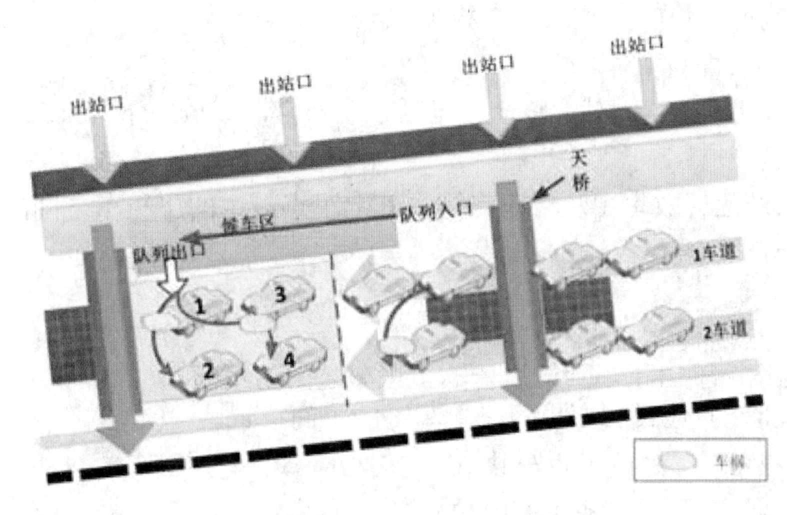

图 11 上述设计存在的安全问题

5.4.3 并行车道设计

本部分在已知现有案例的问题和缺陷的基础上，对并行车道上车点进行了设计（图 4），以保证安全性，并通过合理的安排机场离港乘客和出租车的运行，提高了出租车的接机效率。

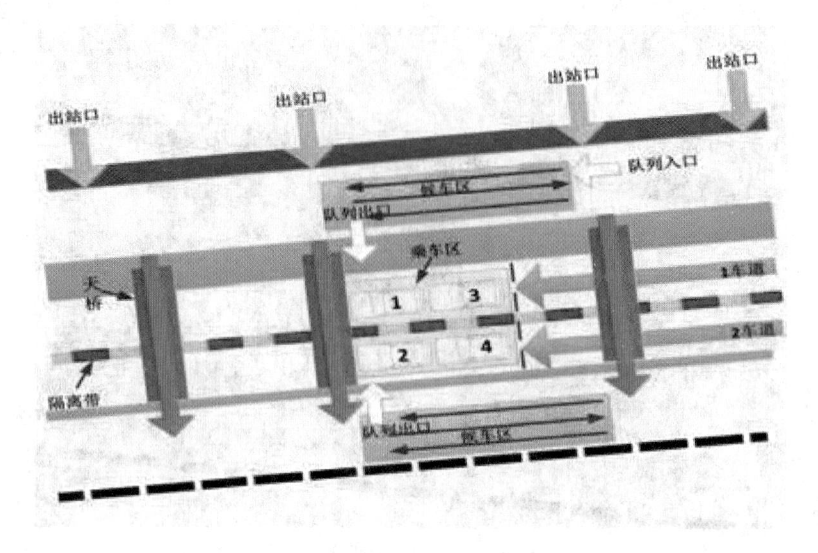

图 12 乘车区改进设计

1）隔离带的加入

在道路中央加入隔离带，使乘客只有原地排队等候或通过天桥乘车两种选择，从而大幅度减少了横穿乘车区的行为，以外在约束保障了乘客的安全。

2）蛇形队列的设计

与直线型队列相比，蛇形队列最大限度地缩短了一个客户退出队伍到下一个客户出现之间的时间，从而提高出租车服务的整体性能，进一步提高乘客的乘车效率。

3）双侧上车点的设置

本文在道路另一侧设置了候车区，由出站口的管理人员按一定比例将乘客分散到道路两侧，从而提高乘车效率。

5.3.4 基于排队理论的上车点设置和出租车、乘客安排

（1）上车点设置

借鉴“短板效应”，对于整个上车区系统而言，其放入下一批次出租车的时间由最慢载客驶离上车区的出租车决定。由于乘客上车的平均速度基本一定，因此上车区系统运转效率由从上车点到距其上车点最远的出租车之间的距离 $x_{\max}$ 决定。由于：

$$x_{\max} \geq \frac{1}{2}L \tag{24}$$

其中，L 为第一辆出租车到最后一辆出租车之间的距离。

因此，上车点应设置在出租车上车区的中间位置。

（2）单次进入乘车区的出租车数量

对于机场而言，除考虑上车点位置外，还需考虑乘车区单次允许进入的车辆数、两侧候车区乘客数量之比等因素。根据假设，出租车到达“乘车区”和乘客到达“乘车区”都服从近似的泊松分布。

同时，由于从“蓄车池”到“上车区”出口之间都加装了隔离带，因此对于隔离带一侧而言，单次可进入乘车区的出租车数量和到达出租车车流构成了“单路排队多通道服务系统”（M/M/N）。

对于出租车而言，该系统指到达的出租车均按先后顺序进入“乘车区”，且按照由远至近的停靠原则，以保证停车位尽可能得到充分利用，即只要有停车位空闲，出租车就可以进入“服务台”接受服务；对于乘客而言，该系统指到达的旅客均以乘客组形式按先后顺序进入“乘车区”，且按照由远至近的上车原则，以保证出租车得到尽可能充分利用，即只要有出租车空闲，乘客就可以进入“服务台”接受服务，乘车离开。

基于此，本文根据排队理论构建模型如下：

Step 1：认为出租车排队进入服务台（乘车区）接受服务，求解服务台（乘车区）个数。

从出租车角度出发，引入标准的 M/M/N 模型。认为出租车排队进入乘车区接受“服务”，单次可进入乘车区的出租车数量即为“服务台”数量，各个出租车位上的乘客平均上车时间相同，平均服务率相同，即 $\mu_1=\mu_2=\cdots=\mu_N=\mu$。

设出租车的平均到达率为 λ，则服务强度为：

$$\rho=\frac{\lambda}{N\mu} \tag{25}$$

根据排队理论[10-11]可知：

系统中没有出租车的概率为：

$$P_0=\left[\sum_{k=0}^{N-1}\frac{1}{k!}\left(\frac{\lambda}{\mu}\right)^k+\frac{1}{N!}\cdot\frac{1}{1-\rho}\cdot\left(\frac{\lambda}{\mu}\right)^N\right]^{-1} \tag{26}$$

系统中有 k 辆车的概率为：

$$P_k=\begin{cases}\frac{1}{k!}\left(\frac{\lambda}{\mu}\right)^k P_0 & (k\le N)\\ \frac{1}{N!N^{k-N}}\left(\frac{\lambda}{\mu}\right)^k P_0 & (k>N)\end{cases} \tag{27}$$

系统中平均出租车数量为：

$$\bar{n}=\frac{(N\rho)^N\rho}{N!(1-\rho)^2}P_0+\frac{\lambda}{\mu} \tag{28}$$

平均排队长度为：

$$\bar{q}=\bar{n}-\rho \tag{29}$$

若 λ,μ,N 为已知数据，则根据上述公式可求得排队的平均车辆数 $\bar{n}$，但由于需要考虑多出租车进入停留时间不确定的影响，本文不能直接选取 $\bar{n}$ 作为单次可进入乘车区的出租车数量。

对于多出租车进入情况，参考公交车站点设计处理方式[]引入有效单次可进入乘车区的出租车数量作为约束条件，来处理乘客位置和出租车驶入速度的不均匀性。

由有效单次入区车辆数系数可知，当平均车辆数满足如下条件时，可认为N满足成为推荐设计的单次进入乘车区的出租车数量数的条件。

$$\begin{aligned}&\bar{n}\le N_b\\&P(n\le N)>M\end{aligned} \tag{30}$$

其中，N_b为有效出租车进入数量，M为系统中车辆数不超过单次出租车可进入数量的置信度。

由上式可知，单次出租车可进入数量可通过试算确定，在满足式（30）时停止。算法如下：

步骤 1：确定初始值N^0。

$$N_0=\frac{\lambda}{\mu} \tag{31}$$

以常州机场为例，$N^0=2$。

步骤 2：计算N^0情况下的$P(0)$、$\bar{n}$，判断是否满足条件。若满足条件则停止，若不满足条件，进行步骤 3。

以常州机场为例，此时$P(0)=0.0870$，$\bar{n}=$[illegible]，不满足条件，进行步骤 3。

步骤 3：取$N^0=N^0+1$，计算该情况下的$P(0)$、$\bar{n}$，判断是否满足条件。若满足条件则停止，若不满足条件，则重复步骤 3。

以常州机场为例，$N^0=3$则$P(0)=0.1699$，$\bar{n}=2.0683$，满足条件，停止。

Step 2：认为乘客排队进入乘车区接受服务，求解乘客到达后必须等待的概率。若概率$P>5\%$，认为会影响到乘车效率。则试增加单次进入乘车区的出租车（服务台）数量，重复 Step 2。最终确定单次进入乘车区的出租车数量N^{0*}。

从乘客角度出发，引入标准的 M/M/c 模型。本文认为，此时，乘客作为“顾客”进入乘车区接受“服务”，此时出租车为该模型中的“服务台”。

以常州机场为例。根据上述计算公式可求得，乘客到达后必须等待（即乘车区已有 3 名乘客，即各出租车均已载客）的概率为：

$$P(n\ge 3)=0.7133 \tag{32}$$

在这种情况下，乘车需等待的概率较大，影响乘车区的乘车效率，无法使得乘车效率达到最大。增加一个进入乘车区的出租车数，计算得到乘客到达后必须等待（即乘车区已有 4 名乘客，即各出租车均已载客）的概率为：

$$P(n\ge 4)=0.0932 \tag{33}$$

此时，乘车需等待的概率依然较大，影响乘车区的乘车效率。再增加一个进入乘车区的出租车数，计算得到乘客到达后必须等待（即乘车区已有 5 名乘客，即各出租车均已载客）的概率为：

$$P(n \geq 5) = 0.0180 \quad (34)$$

此时，乘车需等待的概率可以被接受。

Step 3： 将 N^{0*} 带入 Step 1 的算法中，检验 $P(0)$ 和 $\bar{n}$，判断是否满足条件。

以常州机场为例，算得 $P(0)=0.1858$，$\bar{n}=1.6958$，可以被接受，故而单次进入乘车区的出租车数量为 5，双侧单次进入乘车区的出租车数量为 10。

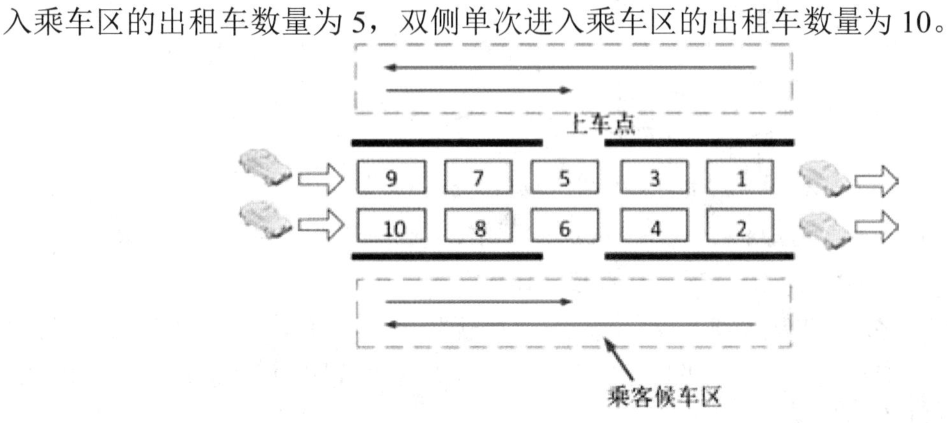

图 13 上车区示意图

（2）双侧候车区乘客比例的刻画

管理人员在指挥乘客分散到道路两侧时，需按照一定的比例进行。由于在道路对侧等候乘车的乘客要多花费通过天桥的时间为：

$$t = \frac{L}{v} \quad (35)$$

因此本文认为，应尽量调整两侧候车区的乘客比例为：

$$n_1 : n_2 = (\bar{t} + t) : \bar{t} \quad (36)$$

其中，n_1 和 n_2 分别为靠近和远离出站口的候车区，$\bar{t}$ 为乘客的平均上车时间。

5.4 任务四模型的建立与求解

通常情况下，“短途”出租车司机两次返回机场的时间差较小，因此本文认为，若这一时间差小于某一定值，则应当给予司机“优先权”。

5.4.1 定义与指导原则

对于任务四，本文提出以下定义及指导原则：

定义 5（出租车司机机场载客周期）： 将出租车司机进入蓄车池排队到第二次将离港乘客送达目的地的运营周期称为“出租车司机机场载客周期”，其具体情况如图 14 所示。过程①为从蓄车池到上车区，过程②为从上车区到第一个乘客下车点，过程③为从第一个乘客下车点到返回蓄车池，过程④为再次从蓄车池到上车区，过程⑤为从上车区到第二位乘客的下车点。

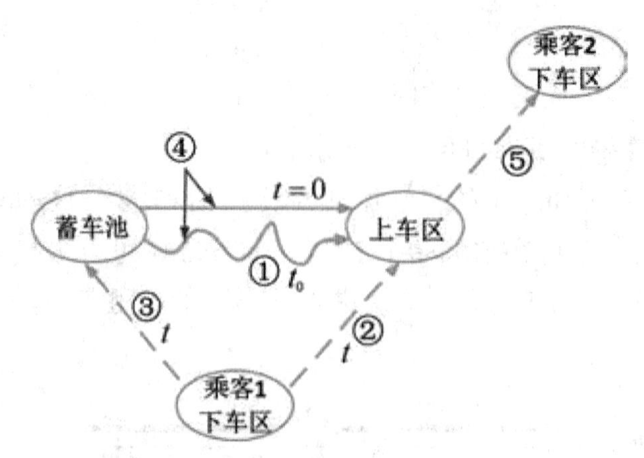

图 14 出租车司机机场载客周期示意图

定义 6（短途）：驶离上车区到回到蓄车池的时间为$2t$，当$2t$小于某临界时间t_p，则将该次出租车司机的运营称为“短途”。

原则 3（收益率均衡原则）：利用短途返程优先配客方案，弥补短途载客司机在该次循环中遭受的损失，力图使各个司机的收益率达到均衡状态，以实现使各个出租车司机收益均衡的目的。

5.4.2 时空参数的刻画

（1）时间参数的刻画

对于获得“优先权”的出租车司机而言，在单个“出租车司机机场载客周期”中，其工作时间由过程①消耗的时间t_0，过程②消耗的时间t过程③消耗的时间t、过程④消耗的时间$t_{0-second}$和过程⑤消耗的时间t_1共同决定。

在设计“优先权”方案的时候，为使各个出租车司机的收益实现均衡，因此需使用从蓄车池到上车区平均时间来表示过程①的时间，即：

$$t_0 = \overline{t_0} \tag{37}$$

同理，过程⑤的时间为：

$$t_1 = \overline{t_1} \tag{38}$$

根据定理 2 可知，$\overline{t_1}$为机场到市中心所需的时间。

此外，本文假设出租车不需在绿色通道排队，因此，对于获得“优先权”的出租车司机而言，过程④的时间为 0，即：

$$t_{0-second} = 0 \tag{39}$$

（2）空间参数的刻画

同理，对于获得“优先权”的出租车司机而言，在单个“出租车司机机场载客周期”中，其里程数由过程①到过程⑤的里程数之和共同决定。

由于相对出租车载客里程数而言，出租车移动距离过短，因此本文近似认为过程①中出租车的行驶距离为 0，即：

$$d_0 = 0 \tag{40}$$

同理，过程④行驶的里程数$d_{0-second} = 0$。

对于过程②和过程③，由于通常情况下可测算出城市内车辆行驶的平均速度v，因此过程②和过程③车辆行驶的距离为：

$$d = vt \tag{41}$$

对于过程⑤，根据定理 2 可知，d_1 为从机场到市中心的距离。

5.4.3 收益均衡模型的构建与求解

（1）获“优先权”的短途司机收益率计算

根据公式（5）和公式（9），过程①和过程②的收益为：

$$\begin{cases} Profit_1 = B_1 - \left[c^r \times (t^0 + t) + c^f \times d\right] \\ B_1 = \varphi \times [B_{sf} + P \times (d - d^s)] \end{cases} \tag{42}$$

过程③的收益为：

$$Profit_2 = -c^r \times t - c^f \times d \tag{43}$$

过程④和过程⑤的收益为：

$$\begin{cases} Profit_3 = B_3 - \left[c^r \times t_1 + c^f \times d_1\right] \\ B_3 = \varphi \times [B_{sf} + P \times (d_1 - d^s)] \end{cases} \tag{44}$$

在整个“出租车司机机场载客周期”中，出租车司机的收益率为：

$$RR_1 = \frac{Profit_1 + Profit_2 + Profit_3}{t_0 + 2t + t_1} \tag{45}$$

（2）非短途载客出租车司机平均收益率计算

同理，非短途载客出租车司机的收益为：

$$\begin{cases} Profit_4 = B_4 - \left[c^r \times (t_0 + t_1) + c^f \times d_1\right] \\ B_4 = \varphi \times [B_{sf} + P \times (d_1 - d^s)] \end{cases} \tag{46}$$

在单个出租车载客循环中，出租车司机的平均收益率为：

$$RR_2 = \frac{Profit_4}{t_0 + t_1} \tag{47}$$

（3）收益均衡模型

当短途载客和非短途载客司机的收益率达到均衡时，“优先权”措施的实施目的得以实现。因此，建立收益均衡模型如下：

$$RR_1 = RR_2 \tag{48}$$

（4）“优先权”发放的时间阈值求解

使用 MATLAB 对收益均衡模型进行求解，得到过程②和过程③的用时为：

$$t = \frac{\varphi[B^{sf}(t_1 + t_0) - Pd^s(t_1 + t_0)]}{v(t_1 + t_0)(2c^f - \varphi P) + 2(\varphi B^{sf} - d_1 c^f + \varphi d_1 P - \varphi P d^s)} \tag{49}$$

六、模型的优缺点

6.1 模型的优点

1、模型的建立基于对现实生活中机场出租车的实际情况的理性分析和合理推导，对某些重要概念进行定义，对求解原则进行深刻全面的分析讨论，使模型具有足够的理论支持

2、算法简单易懂，编程实现较为简单，操作性强，具有普适性，易于推广

3、在分析影响出租车司机决策时，利用经济学专业知识，较为全面的考虑了相关影响因素，保证了模型的实用性和完整性。

6.2 模型的缺点

1、由于难以获得微观数据，本文使用部分相关网站数据来源，非国家统计局等专业机构数据，数据来源有限，一定程度上影响了实证分析结果的准确性。

2、选取的影响司机决策的因素之间难免互相影响，未考虑因素之间的相互影响，影响模型的有效性。

七、模型的改进和推广

1、存在可深入研究的问题。如结合实际情况对斜列式出租车道、多列并行车道的情况进行进一步探讨。

2、可考虑获得大量连续微观数据后，运用智能算法对模型进行改进，以提高模型精度。

3“优先权”可能会导致恶性竞争。例如，上海浦东机场通过为载短途乘客出租车司机发放“短途票”的方式进行“优先权”设计。然而，部分司机篡改短途票上的时间，以达到“免排队”的目的，由此导致短途虚高问题；部分司机为取得“短途票”超速驾驶，进而带来了安全隐患。为防止“优先权”被恶意抢夺，可以考虑从制度等方面入手，如对出租车司机进行监督，采取严格的顾客评价制度，接到有关投诉就永久取消“优先权”资格等方式对上述现象进行控制。

4、本文是对机场出租车安排、调度等方面进行了分析与评价，其中引入的出租车载客循环、运力需求阈值等概念、思想可以被进一步转化推广到多种情况（例如火车站或汽车站对出租车和乘客的安排）中，具有很强的现实意义。

附录 1：参考文献

[1] http://jsnews2.jschina.com.cn/system/2014/06/05/021107731.shtml

[2]http://www.csair.com/cn/index.shtml?WT.mc_id=sem-baidu-pbzctitle0828&utm_source=19SEM&utm_campaign=sem-baidu-pbzc-title0828

[3] https://trp.autonavi.com/detail.do?city=320400

[4]岳喜展. 机场陆侧交通需求预测及集散道路方案设计[D].吉林大学,2011.

[5] http://js.people.com.cn/n2/2018/1126/c360302-32329041.html

[6]胡浩.大型铁路客运枢纽与周边交通系统衔接方式研究[D].南京林业大学,2016.

[7] 景雪琴.出租车营运数据仓库的设计与实现[J]. 交通信息与安全,2012,30(05):139-142.
[8]www.jgtong.com.cn/ggsy/ggsyindex.aspx?lbid=25b2585b-ef0d-4e89-9033-7a10a45f87ab&lbbm=003
[9]司杨,关宏志.计划行为理论下出租车驾驶员寻客行为研究[J]. 交通运输系统工程与信息,2016,16(06):147-152+175.
[10] 胡运权,郭辉煌.运筹学教程（第四版）. 北京：清华大学出版社，2012.
[11]吕林.城市公交站点优化设计方法研究[D].东南大学,2006.

附录 2：常州机场每时刻到港旅客数

表 1 每时段到港旅客数

旅客到达时间段		到达旅客数量
起始	终止	
0:00	1:00	437
1:00	2:00	219
2:00	3:00	0
3:00	4:00	0
4:00	5:00	0
5:00	6:00	0
6:00	7:00	0
7:00	8:00	0
8:00	9:00	437
9:00	10:00	655
10:00	11:00	1091
11:00	12:00	437
12:00	13:00	1092
13:00	14:00	219
14:00	15:00	874
15:00	16:00	0
16:00	17:00	874
17:00	18:00	655
18:00	19:00	1092
19:00	20:00	437
20:00	21:00	1310
21:00	22:00	1092
22:00	23:00	219
23:00	0:00	437

附录3：各个因素对ξ_X的影响效果

$$\xi_{t_{center}}=\frac{\partial n}{\partial t_{center}}=-a\frac{\mu^*\left[\varepsilon+\gamma(B_{sf}c^f d_{center}-d^s P\varphi+d_{center}P\varphi)\right]}{c^f d_{ave}-B_{sf}\varphi+c^f d_{center}\gamma+d^s P\varphi-d_{ave}P\varphi}$$

$$\xi_a=\frac{\partial n}{\partial a}=\frac{a^2\mu^*\left[\varepsilon t_{center}+(\gamma t_{center}+\varepsilon t_{ave})(B_{sf}\varphi-c^f d_{center}-d^s P\varphi+d_{center}P\varphi)\right]}{c^f d_{ave}-B_{sf}\varphi+c^f d_{center}\gamma+d^s P\varphi-d_{ave}P\varphi}$$

$$\xi_{\mu^*}=\frac{\partial n}{\partial \mu^*}=\frac{-a\left\{\varepsilon t_{center}+\left[\left(\gamma t_{ave}+\varepsilon t_{ave}\right)\left(B_{sf}\varphi-c^f d_{center\_}-d^s P\varphi+d_{center}P\varphi\right)\right]\right\}}{c^f d_{ave}-B_{sf}\varphi+c^f d_{center}\gamma-d_{ave}P\varphi}$$

$$\begin{cases} g=\mu^*(\gamma t_{center}+\varepsilon t_{ave}) \\ G=(c^{f2}d_{ave}-d^s B_{sf}^2\varphi^2+d_{ave}B_{sf}^2\varphi^2-B_{sf}c^f\varphi+B_{sf}P\varphi^2 \\ \qquad +d^s c^f P\varphi-2c^f d_{ave}P\varphi+B_{sf}c^f\gamma\varphi-d^s c^f\gamma\varphi)) \\ H=\left[a\left(c^f d_{ave}-B_{sf}\varphi\right)+c^f d_{center}\gamma+d^s P\varphi-d_{ave}P\varphi\right]^2 \\ \xi_{d_{center}}=\dfrac{\partial n}{\partial d_{center}}=\dfrac{gG}{H} \end{cases}$$

$$\begin{cases} l=\dfrac{1}{a(c^f d_{ave}-B_{sf}\varphi+c^f d_{center}\gamma+d^s P\varphi-d_{ave}P\varphi)^2} \\ L=-c^f\mu^*(\gamma t_{center}+\varepsilon t_{ave})(B_{sf}d_{ave}-B_{sf}t_{center}+B_{sf}d_{center}\gamma \\ \qquad -d^s d_{ave}P+d^s d_{center}P+d_{center}{}^2P\gamma-d^s d_{center}P\gamma) \\ \xi_\varphi=\dfrac{\partial n}{\partial\varphi}=lL \end{cases}$$

$$\begin{cases} R=(B_{sf}d_{ave}\varphi-B_{sf}t_{center}\varphi-d^s c^f d_{ave}+d^s c^f d_{center}+c^f d^2{}_{center}\gamma-d^s c^f d_{center}\gamma) \\ \xi_{B_{sf}}=\dfrac{\partial n}{\partial B_{sf}}=\dfrac{-\mu^*\varphi(\gamma t_{center}+\varepsilon t_{ave})R}{a(c^f d_{ave}-B_{sf}\varphi+c^f d_{center}\gamma+d^s P\varphi-d_{ave}P\varphi)^2} \end{cases}$$

$$\xi_{d^s}=\frac{\partial n}{\partial d^s}=\frac{P\mu^*\varphi(\gamma t_{center}+\varepsilon t_{ave})(c^f d_{ave}-c^f d_{center}+c^f d_{center}\gamma-d_{ave}P\varphi+d_{center}P\varphi)}{a(c^f d_{ave}-B_{sf}\varphi+c^f d_{center}\gamma+d^s P\varphi-d_{ave}P\varphi)^2}$$

$$\xi_{c^r}=\frac{\partial n}{\partial c^r}=0$$

$$\begin{cases} Q=(c^f d_{ave}-c^f d_{center}+c^f d_{center}\gamma-d^s d_{ave}P+d^s d_{center}P+d^2{}_{center}P\gamma-d^s d_{center}P\gamma) \\ \xi_{c^f}=\dfrac{\partial n}{\partial c^f}=\dfrac{\mu^*\varphi(\gamma t_{center}+\varepsilon t_{ave})Q}{a(c^f d_{ave}-B_{sf}\varphi+c^f d_{center}\gamma+d^s P\varphi-d_{ave}P\varphi)^2} \end{cases}$$

$$\xi_{t_{ave}}=\frac{\partial n}{\partial t_{ave}}=-\frac{\varepsilon\mu^*\left(B_{sf}\varphi-c^f d_{center}-d^s P\varphi+d_{center}P\varphi\right)}{a\left(c^f d_{ave}-B_{sf}\varphi+c^f d_{center}\gamma-d_{ave}P\varphi\right)}$$

$$\begin{cases} F=\mu^*(B_{sf}\varphi-c^f d_{center}-d^s P\varphi+d_{center}\mathrm{P}\varphi)(B_{sf}t_{center}\varphi \\ \quad -c^f d_{ave}d_{center}+c^f d_{center}\varepsilon t_{ave}-d^s Pt_{center}\varphi+d_{ave}P\varphi) \\ f=\dfrac{1}{a(c^f d_{ave}-B_{sf}\varphi+c^f d_{center}\gamma+d^s P\varphi-d_{ave}P\varphi)^2} \\ \xi_\gamma=\dfrac{\partial n}{\partial\gamma}=Ff \end{cases}$$

$$\xi_{d_{ave}}=\frac{\partial n}{\partial d_{ave}}=\frac{\mu^*(c^f-B_{sf}\varphi)(\gamma t_{center}+\varepsilon t_{ave})(B_{sf}\varphi-c^f d_{center}-d^s P\varphi+d_{center}P\varphi)}{a(c^f d_{ave}-B_{sf}\varphi+c^f d_{center}\gamma+d^s P\varphi-d_{ave}P)}$$

$$\begin{cases} W=\left(B_{st}d_{ave}\varphi-B_{st}d_{center}\varphi-d^s c^f d_{ave}+d^s c^f d_{center}+c^f d^2{}_{center}\gamma-d^s c^f d_{center}\gamma\right) \\ \xi_P=\dfrac{\partial n}{\partial P}=\dfrac{-\left[\mu^*\varphi\left(\gamma t_{center}+\varepsilon t_{ave}\right)W\right]}{a\left(c^f d_{ave}-B_{st}\varphi+c^f d_{center}\gamma+d^s P\varphi-d_{ave}P\varphi\right)^2} \end{cases}$$

附录 3：

附录程序运行环境：MATLAB 版本: 8.6.0.267246(R2015b) MATLAB

许可证编号: 123456

操作系统: Microsoft Windows 10 家庭中文版 Version 10.0 (Build 17134)

```
%各时段阈值数
%p 各时段飞机场离港人数
%以一小时为单位计算
clc
clear all
p=437
m = zeros(24,1);
a=0.397;
k=1.5;
for j=1:24
    l=0;
    l=a*p(j);
    m(j)= ceil(l/(k*2));
end

% A 方案 正常情况
% a 前方出租车数量
% d0 空车行驶距离
% d1 到市中心距离
% n 已有等待车辆数
% v 服务速度
% u 服务强度
% k 载客率
% p 拥堵指数
% t 正常行驶速度
% t1 期望载客所花时间
% B 期望乘客支付车费
% b0 常州市出租车起步价
% b1 常州市出租车单位价格
% d 出租车起步里程
% i 第 i 季度
% cf 出租车燃气成本
a=150
```

```
RR=zeros(4,a);
for i=1:4
    for n=1:a;
        d1= 24.6;
        d0=0;
        t = 41;
        u=5/8;
        k=1.5;
        v = 1.5/(5/8);
        t0 = v*n;
        if i == 1
            p=1.2554;
        elseif i == 2
            p=1.32;
        elseif i == 3
            p=1.31;
        else
            p=1.277;
        end
        t1= p*t;
        b0 = 10;
        d = 3;
        b1 = 2.1;
        B = b0+b1*(d1-d);
        cf = (9.83*4.19)/100;
        c1=cf*(t1+t0);
        cr = 0;
        c2 = cr*(d1+d0);
        profit=B-c1-c2;
        RR(i,n)=profit/(t1+t0);
    end
end

%A 方案 早晚高峰
% d0 空车行驶距离
% d1 到市中心距离
% n 已有等待车辆数
% v 服务速度
```

```
% u 服务强度
% k 载客率
% p 拥堵指数
% t 正常行驶速度
% t1 期望载客所花时间
% B 期望乘客支付车费
% b0 常州市出租车起步价
% b1 常州市出租车单位价格
% d 出租车起步里程
% i 第 i 季度
% cf 出租车燃气成本
a=150
RR=zeros(4,a);
for i=1:4
    for n=1:a;
        d1= 24.6;
        d0=0;
        t = 41;
        u=5/8;
        k=1.5;
        v = 1.5/(5/8);
        t0 = v*n;
        if i == 1
       p=1.454;
          elseif i == 2
   p=1.457;
          elseif i == 3
   p=1.442;
          else
   p=1.4492;
          end
        t1= p*t;
        b0 = 10;
        d = 3;
        b1 = 2.1;
        B = b0+b1*(d1-d);
        cf = (9.83*4.19)/100;
        c1=cf*(t1+t0);
```

```
            cr = 0;
            c2 = cr*(d1+d0);
            profit=B-c1-c2;
            RR(i,n)=profit/(t1+t0);
        end
end

% A 方案 夜间情况
% d0 空车行驶距离
% d1 到市中心距离
% n 已有等待车辆数
% v 服务速度
% u 服务强度
% k 载客率
% p 拥堵指数
% t 正常行驶速度
% t1 期望载客所花时间
% B 期望乘客支付车费
% b0 常州市出租车起步价
% b1 常州市出租车单位价格
% d 出租车起步里程
% i 第 i 季度
% cf 出租车燃气成本
a=150
RR=zeros(4,a);
for i=1:4
        for n=1:a;
            d1= 24.6;
            d0=0;
            t = 41;
            u=5/8;
            k=1.5;
            v = 1.5/(5/8);
            t0 = v*n;
            if i == 1
                p=1.2554;
            elseif i == 2
                p=1.32;
```

```
        elseif i == 3
            p=1.31;
        else
            p=1.277;
        end
        t1= p*t;
        b0 = 10;
        d = 3;
        b1 = 2.1;
        B = (b0+b1*(d1-d))*1.3;
        cf = (9.83*4.19)/100;
        c1=cf*(t1+t0);
        cr = 0;
        c2 = cr*(d1+d0);
        profit=B-c1-c2;
        RR(i,n)=profit/(t1+t0);
    end
end

% B 方案 正常情况
% k 空载率
% p 拥堵率
RR=zeros(4,24);
t00= 41;
d00= 24.6;
for j=1:24
    k=0;
    k=K(j)/100;
    for i=1:4
        if i == 1
            p=1.2554;
        elseif i == 2
            p=1.32;
        elseif i == 3
            p=1.31;
        else
            p=1.277;
        end
```

```
            t0=(k*t00)*p;
            d0=k*d00;
            d1=(6.89+7.29)/2;
            t1=(15.2+16.1)/2;
            b0 = 10;
            d = 3;
            b1 = 2.1;
            B = b0+b1*(d1-d);
            cf = (9.83*4.19)/100;
            c1=cf*(t1+t0);
            cr = 0;
            c2 = cr*(d1+d0);
            profit=B-c1-c2;
            RR(i,j)=profit/(t1+t0);
    end
end

% B 方案    早晚高峰
% k 空载率
% p 拥堵率
RR=zeros(4,24);
t00= 41;
d00= 24.6;
for j=1:24
    k=0;
    k=K(j)/100;
    for i=1:4
        if i == 1
                p=1.454;
          elseif i == 2
                p=1.457;
            elseif i == 3
                p=1.442;
        else
                p=1.4492;
            end
          t0=(k*t00)*p;
          d0=k*d00;
```

```
            d1=(6.89+7.29)/2;
            t1=(15.2+16.1)/2;
            b0 = 10;
            d = 3;
            b1 = 2.1;
            B = b0+b1*(d1-d);
            cf = (9.83*4.19)/100;
            c1=cf*(t1+t0);
            cr = 0;
            c2 = cr*(d1+d0);
            profit=B-c1-c2;
            RR(i,j)=profit/(t1+t0);
        end
    end

% B 方案    夜间情况
% k 空载率
% p 拥堵率
RR=zeros(4,24);
t00= 41;
d00= 24.6;
for j=1:24
    k=0;
    k=K(j)/100;
    for i=1:4
        if i == 1
            p=1.2554;
        elseif i == 2
            p=1.32;
        elseif i == 3
            p=1.31;
        else
            p=1.277;
        end
          t0=(k*t00)*p;
          d0=k*d00;
          d1=(6.89+7.29)/2;
          t1=(15.2+16.1)/2;
```

```
            b0 = 10;
            d = 3;
            b1 = 2.1;
            B = (b0+b1*(d1-d))*1.3;
            cf = (9.83*4.19)/100;
            c1=cf*(t1+t0);
            cr = 0;
            c2 = cr*(d1+d0);
            profit=B-c1-c2;
            RR(i,j)=profit/(t1+t0);
        end
end

% 求解 n 的偏导数
%B 收益
%z 收入系数
%K 起步价
%k 单位价格
%c 起步距离
%dcenter 到市中心距离
%tcenter 到市中心自由流时间的时间
%m 拥堵延时指数
%P 净收益
%x 服务强度
%y 单位时间服务人数
%l 空驶率
%d0 空驶里程
%d1 运营里程
%t0 空驶时间
%t1 运营时间
%tave 平均运营时间
%dave 平均运营里程
% 分别对：m tcenter y x   dcenter z K k   c cr cf   tave l dave 求偏导
syms t1a m tcenter t0a y x n d1a dcenter Ba z K k d1a c Pa cr cf RRa t0b t1b tave d0b
l d1b dave Bb Pb RRb S s
t1a=m*tcenter;
t0a=y/x*n;
d1a=dcenter
```

```
Ba=z*(K+k*(d1a-c));
Pa=Ba-cr*(t1a+t0a)-cf*d1a;
RRa=Pa/(t1a+t0a);
t0b=l*tcenter;
t1b=m*tave;
d0b=dcenter*l;
d1b=dave;
Bb=z*(K+k*(d1b-c));
Pb=Bb-cr*(t1b+t0b)-cf*(d0b+d1b);
RRb=Pb/(t1b+t0b);
s=solve(RRa/RRb==1,n)
S=simplify(s);
dfa=diff(S,m);
SS=simplify(dfa)

% 第三题
% c 服务台数量
% a 平均到达率
% b 平均服务率
syms k
c=4;
a=0.7;
b=(5/8)/1.5;
d=a/(c*b);
f1=(1/factorial(k))*(a/b)^k
y1=symsum(f1,k,0,c-1);
y2=(1/factorial(c))*(1/(1-d))*(a/b)^c;
p0=1/(y1+y2)
n=a/b+((((c*d)^c)*d)/(factorial(c)*(1-d)^2))*p0

syms k
c=4;
a=7;
b=3;
d=a/(c*b);
f1=(1/factorial(k))*(a/b)^k
y1=symsum(f1,k,0,c-1);
y2=(1/factorial(c))*(1/(1-d))*(a/b)^c;
```

```
p0=1/(y1+y2)
P=(((c*d)^c*p0*d)/(factorial(c)*(1-d)^2))/c

%第四问
% a 到市中心时间
% A 到市中心距离
% b 短途时间
% B 短途距离
% d 到市中心时间
% F 收益
% K 起步价
% k 单位距离价格
% c 起步距离
% P 净收益
% cr 单位租金成本
% cf 单位燃气成本
% e 站内平均等待时间
% m 收入系数
syms a A b B v K k c cr cf e m Fdt Fszx Pdtcz Pdthc Pszxbd Pszxyd RRzc RRyxq
B = b*v;
Fdt = m*(K + k*(B-c));
Fszx = m*(K + k*(A-c));
Pdtcz = Fdt - cr*(e+b) -cf*B;
Pdthc = -cr*b - cf*B;
Pszxbd = Fszx - cr*a - cf*A;
Pszxyd = Fszx - cr*(e+a) - cf*A;
RRzc = Pszxyd/(e+a);
RRyxq = (Pdtcz+Pdthc+Pszxbd)/(e+b+b+a);
solve(RRzc==RRyxq,b)
Y=collect(y)
```

8. “新时代”机场与出租车互利双赢计划

摘要：本文研究的是机场出租车资源配置和乘客运输的问题，随着国内航空运输业的不断发展，国内的各机场枢纽不断改造扩建，机场客流吞吐量不断增加，其交通运输能力与客运需求之间的矛盾日益显著，因此此类问题的解决具有重大意义。在本文的研究过程中，为了使出租车和机场综合效益最大化，主要采用了基于非线性规划模型的经济收益方程思想和基于排队论的枢纽上车点数目优化等策略，对问题进行分析和解决。

针对问题一，结合出租车的决策的首要决定因素——相对收益率，通过研究收益与抽象的等待时间成本等影响因素间关系，得到一个收益的变化规律，从而将其作为最终决策的判据。通过从机场官方网站和查阅文献得到大量航班信息，得到时段与客运量对应关系折线图。为了研究波动规律，采用季节指数法，得到近年来客流量与各季度的相关关系。通过研究大量航班数据，使用区间分段法，得到24 小时内8 个时段的人流密度变化规律。进而通过综合季度与日时段，得到乘客数量的变化规律。基于已有的乘客数目变化规律，结合其他重要因素，得到了各因素与等待时间的关系模型，进而以此为传递量，得出一个综合考虑各主要因素的净利润决策模型，为到达机场的出租车提供了一个合理的决策方案。

针对问题二，基于问题一得到的模型，采集并重点分析整理了上海虹桥机场与一日上海市出租车时刻位置经纬度的大量数据。通过历年季度客流量，预测得到未来各个季度客流量，通过分析各时段航班数量，得到虹桥机场航班时刻分异结构表。最终得到一张基于上海虹桥机场数据F 函数计算值与季度和航段关系表，实现了清晰地判断各季度各航段Wa 与Wb 的效益大小关系，达到了帮助上海市出租车司机做出更合理的决策的目的。此外，通过分析模型的合理性和相关因素的依赖性，我们得出采用该模型可以使出租车司机平均收益率提高11.6%，期望里程和出租车总数量之和对决策函数F 的贡献率超过60%，与另外因素形成明显差异，从而验证了模型的准确性和科学性。

针对问题三，为使总乘车效率最高，基于基本的排队理论，分别研究了旅客排队上车模型和出租车排队模型，得出M/M/c 模型更适合机场旅客的运输的结论。从而基于旅客M/M/c 排队模型，对双车道并行路段上车点的数目设置进行究，得到最佳乘车点数目与乘客等待时间成本的函数关系，从而使总的乘车效率最高，顾客的等待时间也更加均匀，以上海机场为例，其“短时等待”顾客数降低10.3%左右，“长时等待”顾客提高5.5%左右，平均队列长降低21%左右。

针对问题四，为了使出租车收益均衡，保证乘客乘车安全，结合我国科技“新时代”大数据技术，建议机场部门采用利用智能GPS 轨迹识别定位系统，对出租车真实目的地进行定位。通过关于长途单与短途单经济效益相对平衡的约束方程，得到一个以机场为中心，“短程上限距离”为半径的圆域，同时得到与之对应的相对合理的往返时间，使拥有短程票多次往返的出租车司机收益与接一次长途单收入差别在10%以内，满足相对车程5%的缓冲时间，总体使得这些出租车的收益均衡。

关键词：互利双赢 非线性规划模型 经济收益方程 基于排队论的枢纽上车点数目优化。

作　　者：尹晓，田雨，常亚栋

指导教师：胡小宁

奖　　项：2019年全国大学生数学建模竞赛国家二等奖

一、问题重述

1.1 问题背景

随着中国航空业的发展，大型国际机场的旅客吞吐量急剧上升，根据中国民用航空局官方网站给出的统计数据，2018年民航旅客运输量达到6.12亿人次，同比增长10.9%。旅客运输量的激增，将对机场周边道路交通及接续运输组织带来极大的影响。其中出租车作为一种重要的机场接续运输交通方式，具有约为23.7%的交通分担率，并在强化枢纽机场与周边综合交通网的衔接、保障机场交通运输高效有序进行方面发挥着重大的作用。机场通常将送客、接客分为两个不同的通道。为确保公平，提高运输效率，机场管理部门通常设置蓄车池，使等待接客的出租车在蓄车池中排队等候，并按照排队顺序，先来的出租车先上客、先驶离，同时，想要打车的乘客将会到达指定“乘车区”排队，同样按照先后顺序上车。在此种管理模式下，对于送客至机场出租车司机，共有两种选择：前往到达区等待载客回市区即在“蓄车池”排队等候，和直接放空返回市区拉客。其中，在“蓄车池”中排队等待有一定的时间成本，取决于出租车与乘客的数量，而空载返回市区将承担空载费用以及损失潜在的载客收益。

出租车司机的在两种选择中的决策结果取决于许多因素，其中，某时段内抵达的航班数量以及“蓄车池”中已有的车辆数是司机可观测到的确定信息，同时，司机的个人经验，例如某个季节、某个时段航班数量及可能乘客数量的多寡等，对于司机的决策也有一定影响。在实际中，还有很多影响出租车司机决策的确定与不确定因素，并具有不同的关联关系和影响效果。

1.2 问题提出

结合实际情况，建立数学模型研究下列问题：

(1) 分析研究与出租车司机决策相关因素的影响机理，综合考虑机场乘客数量的变化规律和出租车司机的收益，建立出租车司机选择决策模型，并给出司机的选择策略。

(2) 收集国内某一机场及其所在城市出租车的相关数据，给出该机场出租车司机的选择方案，并分析模型的合理性和对相关因素的依赖性。

(3) 在某些时候，经常会出现出租车排队载客和乘客排队乘车的情况。某机场“乘车区”现有两条并行车道，管理部门应如何设置“上车点”，并合理安排出租车和乘客，在保证车辆和乘客安全的条件下，使得总的乘车效率最高。

(4) 机场的出租车载客收益与载客的行驶里程有关，乘客的目的地有远有近，出租车司机不能选择乘客和拒载，但允许出租车多次往返载客。管理部门拟对某些短途载客再次返回的出租车给予一定的“优先权”，使得这些出租车的收益尽量均衡，试给出一个可行的“优先”安排方案。

二、问题分析

2.1 问题一分析

问题一要求我们分析与出租车司机决策相关的影响机理，建立出租车司机选择决策模型，给出司机的选择策略。首先，需要对司机决策的相关因素进行分析，司机将依靠自己的经验，对某季节、某时段的客流量有一个大致的估计，并结合可观测的“蓄车池”中已有的出租车数量，预计排队等候时间，做出两种选择：

A.排队等候,B.到市区接客。考虑到蓄车池车辆密集以及通行的单向性，排队等候的出租车无法中途停止排队进程，并且不考虑驶离机场后中途折返的情况。因此，考虑司机只做一次选择，并且只有A、B两种选择。两种选择相比，选择A可能会付出一定的时间成本，在等待排队阶段将无法得到收入，但机场管理部门将会为其安排乘客上车。选择B相比A节省了排队等待时间，但空载回到市区将承担以油费为主要影响因素的空载费用。因此，对于等待时间的预估将在司机的决策中发挥极其显著的作用，简单地分析，等待时间段越短，排队等候的时间成本就越少，空载回市区的决策将逐渐减少其经济性，出租车司机的决策则更偏向A;而等待时间越长，排队耗费的时间成本越大，司机决策更向B偏移。而预估等待时间通常由乘客数量和排队出租车数量决定，影响乘客量的因素有很多，例如时段、季节、天气等。此外，空载成本也是一个重要的影响因素，其主要的影响条件有机场至市区的里程数、油价等。为综合考虑等待时间成本和空载成本及其他成本，我们将建立决策综合经济收益模型，将时间成本和空载成本量化为经济成本，考虑接客所带来的经济收入，综合考虑两种决策下的经济收益，以收益值作为决策的依据，即两种决策下综合经济收益值较大的决策较优。除却经济影响因素外，另有一些因素例如司机的个人计划、城市道路交通拥堵情况、司机健康状况都将对司机的决策产生一定程度的影响，相比上述因素而言，其影响较小，在此我们对其忽略不计。

2.2 问题二分析

问题二要求收集国内某一机场及其所在城市的相关数据，并根据数据判定司机应该选择的决策。用实际的数据作为问题一所建模型的输入，用模型的输出作为决策的判据。对模型的合理性验证以及对相关因素依赖性分析，可以理解为对问题一所建立的最优决策模型的优越性分析以及敏感性分析。在数据的搜集过程中，数据类型及其采样频率可能与模型输入所需数据不符，因此，对数据进行处理以及对模型进行数据适应性改进在此环节中将发挥极其重要的作用。

2.3 问题三分析

问题三从机场管理部门角度出发，对乘车排队模型进行优化，在保证车辆和乘客安全性的情况下提高总体乘车效率。由于机场建设已经完成，乘车区道路为并行两车道且不能改变，蓄车池的面积大小、出站口距乘车区的距离等因素也已经确定，因此，在问题三综合乘车排队优化模型的建立中，我们着重考虑设置乘车点的个数以及乘客排队等车的队列形式。利用排队论的有关知识，在车速不超速情况下，系统等待时间越短，总体乘车效率越高。

2.4 问题四分析

问题四针对机场出租车不愿意接待短途乘客的现象，要求设计一个合理的、能促进长短途载客收益均衡的管理方案。在机场的出租车载客过程中，司机往往要花费很多的时间成本，一般需要等待两三个小时才能拥有接待乘客的机会，并且无论长途短途一般都有高速路费成本，如果接到短途乘客，司机的盈利将大幅度减少，甚至有时“入不敷出”，严重影响出租车司机拉载短途客的积极性。机场管理部门要保障长短途接客的公平性，不可能改变计价规则，增加短途出租车的利润，只能通过减少短途出租车候客等待时间，来给予短途出租车一定的优先权。在模型的建立与求解部分，本文将着重研究短途载客优先权的确定。

三、模型假设与约定

1、假设排队等候的出租车不能中途离开蓄车池、停止等候进程。
2、假设空载返回市区的出租车不会中途折返。
3、假设出租车在开车过程中车速保持恒定。

四、符号说明及名词定义

符号	说明
W_a	决策 A 条件下的综合经济收益值
W_b	决策 B 条件下的综合经济收益值
C_a	决策 A 条件下的油耗成本(元)
C_b	决策 B 条件下的油耗成本(元)
Po	油价(元/升)
L	单位距离的油耗(升/公里)
a	出租车起步价(元)
b	超过起步价内公里后,每多增一公里出租车收取的费用(元)
T	在“蓄车池”中的预估等待时间(小时)
v	出租车运行的平均速度(千米/小时)
β	市区内出租车的空载率
y	单位时间内平均载客收入(元/小时)
yi	第 i 季度机场航班到港量(架)
Mi	季节指数
$T1$	蓄车池内出租车的基础等待时间
i	司机进行决策时所处的时间单元,$i=1,2,3,\cdots 48$
η	航板到港量对等待时间的增益系数。
p_n	队长 N 的概率分布
λ	乘客到达率
μ	服务率,即单位时间内完成乘车服务的顾客数量
ρ	服务强度
L_i	排队系统的平均队长
W_s	平均滞留时间(小时)
W_q	平均等待时间
c	服务台个数
L_q	平均队长
ρ'	服务机构的繁忙率
α	每个乘客单位时间的等待时间成本

五、模型的建立与求解

5.1.1 问题一分析

问题一的核心问题即为出租车司机的选择决策问题。其中，影响司机决策的主要因素为：等待时间成本、空载成本，等待时间成本取决于乘客数和“蓄车池”内出租车数，其中乘客数受到时段、季节影响较大；空载成本受到油价、机场与市区之间距离的影响。为了衡量决策影响因素对决策的影响，我们建立了决策经济收益分析模型，将各种成本及收入都量化为经济量，并将计算得到的经济收益值作为决策的判据，忽略司机个人因素及道路交通情况对决策的影响。

5.1.2 模型的建立

5.1.2.1 决策综合经济收益模型的建立

为了为两类决策选取合适的判据，我们建立了决策综合经济收益模型，通过计算某段时间内两种决策情况下的经济利润，选取判定较优的决策。针对时间尺度的选取，如果选取的时间段过短，将无法对出租车的整个运行过程和总体收益情况进行全面了解，进而得出的判断可能存在不合理性，通俗来讲，二者在短时间内不具备可比性，如果选取的时间段过长，将增大分析的困难和计算的复杂程度，因此我们选取出租车在飞机场送客完成作为时间起点，将决策B情况下出租车完成将机场乘客送入市区后作为时间终点，并分析分别在决策A、决策B情况下，此时间段内的综合经济收益值W_a、W_b。

5.1.2.2 决策A情况下综合经济收益值W_a的计算

决策A条件下的综合经济收益值W_a的表达式为：

$$W_a = E_a - C_a \tag{1}$$

由于选择排队等候乘客上车，在此条件下，综合经济收益值的计算中的成本主要为返回市区的油耗成本C_o，其计算公式为：

$$C_o = S \cdot P_o \cdot L(1+\lambda) \tag{2}$$

其中，Po为油价，S为机场到市区的距离，L为单位距离的油耗，λ为载人后油耗的增加系数。

同时，决策A条件下的经济收入主要为载客收入，主要指将乘客从机场送至市区所得到的报酬，公式为：

$$E_a = a + (S - x)b \tag{3}$$

其中a为出租车起步价，b为超过起步价内公里后，每多增一公里出租车收取的费用。综上，A决策情况下综合经济收益值W_a的计算公式为：

$$W_a = a + (S - x)b - S \cdot P_o \cdot L(1+\lambda) \tag{4}$$

5.1.2.3 决策B情况下综合经济收益值W_b的计算

与A类似，在决策B情况下综合经济收益值W_b的表达式为：

$$W_b = E_b - C_b \tag{5}$$

决策B情况下，经济成本主要为空载回市区，以及研究时间内在市区中运行油耗成本C_b，假设出租车一直处于运行状态，计算公式为：

$$C_b = T \cdot v \cdot P_o \cdot L \tag{6}$$

其中，T为在“蓄车池”中的预估等待时间，即T为研究时段，v为出租车运行的平均速度。

决策B情况下的收入主要为返回市区之后，出租车接待乘客所得到的报酬E_b，假设两种决策下车速都恒定为v，则如果以决策A情况下出租车回到市区为时间结束标志，决策B条件下出租车在市区运营的总时间即为预估等待时间T，E_b公式如下：

$$E_b = T(1-\beta)y \tag{7}$$

其中β为市区内出租车的空载率，y为单位时间内平均载客收入。

5.1.2.4 预估等待时间T的计算

在出租车司机的决策中，预估等待时间T的计算至关重要，而等待时间是一个难以预测的量，受到多种因素的影响，其主要取决于乘客数量以及“蓄车池”中排队等待的出租车的数量。其中，出租车数量是司机可观测到的量，然而乘客数量将受到许多因素影响而产生波动，例如季节因素（旅游）、时间因素、天气因素（降雨、降雪、大雾天气影响航班飞行情况）、城市周边重大活动等。为简化模型，在此只考虑季节因素与时间因素,并建立非线性模型：

(1) 研究乘客数量与季度的关系，建立指数预测模型，用指数预测模型的计算结果，除以季度的天数（120）得到某个季度的日均客流量。

(2) 根据某一典型日的机场中各时段的到港航班数推算出各时段乘客数量在每日乘客总量所占权重。

Step1:建立乘客数量季度指数预测模型

季节指数预测法是根据以往数据呈现的季节性规律，对未来数据进行合理预测的方法，其形式通常由指数形式表示，根据历史统计数据建立季节指数法预测模型：

$$\begin{cases} y_i = (c+dt)M_i \\ c = \dfrac{\sum y_i}{n} \\ d = \dfrac{\sum y_i t}{\sum t^2} \end{cases} \tag{8}$$

其中，y_i为第i季度机场航班到港量，c、d为待测系数，Mi为季节指数，由机场历年第i季度到港量与全年平均到港量之比定义。

Step2:计算各时段对乘客数量所占权重：

由于不同航班空载率较低且相差无几，本文将以航班到港量估计乘客数量。由上式求得不同季节航班到港量，计算得到该季度内每日平均到港量，选取机场某一典型日，每个半小时统计航班到港量，得到一个长度为48的时间序列，并除以该日内总航班数，可得到每个时间单元客流量所占权重 $g_1, g_2, g_3 \cdots g_{48}$。

Step3:T的计算

结合前两步的计算结果，可得出预估等待时间的计算公式：

$$\begin{cases} T = T_1 + \eta \cdot \dfrac{H_i}{d} \\ H_i = \dfrac{y_t}{120} g_i \end{cases} \tag{9}$$

其中，T_1为蓄车池内出租车的基础等待时间，i为司机进行决策时所处的时间单元，$i = 1,2,3,\cdots 48$；H_i为该时间段内航班到港数量，η为航班到港量对等待时间的增益系数。

5.1.2.5 模型总结

对以上分离建模过程进行总结，可得到最优决策模型函数：

$$F(W_a, W_b) = \frac{W_a}{W_b} \tag{10}$$

其中F为W_a、W_b的比值，反映了两种决策下效益的比较。

因此，出租车司机的选择策略为：计算得到$F(W_a, W_b)$的取值，若大于1，即选择决策A，若小于1，即选择决策B。

5.2.1 问题二分析

问题二要求收集国内某一机场的以及机场所在城市的数据，并给出该机场出租车司机的选择方案，并对模型的合理性以及对相关因素的依赖性。我们基于问题一建立的司机最优决策模型，并结合能够找到的有关数据，对其中的一些数据进行处理以贴合模型，从而得到司机的最优决策。最后，将对模型的合理性以及对相关因素的依赖性进行分析。

5.2.2 问题二的求解

5.2.2.1 数据的来源处理及决策的确定

上海虹桥国际机场，位于中国上海长宁区，建筑面积51万平方米，共有89个机位，2017年旅客吞吐量4188.41万人次，是中国三大门负荷枢纽之一。作为一个国际化的交通枢纽，对上海虹桥机场的接续交通问题的研究就显得尤为重要，因此，我们搜集到上海虹桥机场2015年到2018年中四个季度航班到港数量统计值，一个典型日中每隔三十分钟进行采集的航运数据，以及上海市出租车收费标准、机场旅客主要目的地及车费参考价格，并针对这些数据对机场出租车司机的决策进行分析。相关数据如下所示：

表 1 2015-2019 年四个季度航班到港数量统计表

季度	历史数据			
	2015	2016	2017	2018
第一季度	932.24	973.64	997.64	1056.76
第二季度	984.19	989.72	1057.25	1100.13
第三季度	999.04	1039.12	1098.68	1139.68
第四季度	993.63	1044.51	1106.37	1036.76

表 2 上海出租车收费标准

里程数	日间 （05：00-23：00）	夜间 （23：00-05：00）
0-3 公里	14 元	18 元
3-10 公里	2.5 元/公里	3.1 元/公里
15 公里以上	3.6 元/公里	4.7 元/公里

表 3 上海浦东机场乘客主要目的地及出租车参考乘车价格

目的地	里程(KM)	里程(KM)	参考价格（元）	夜间
商圈	徐家汇（东方商厦、港汇广场、太平洋百货）	48	180	230
	莘庄	50	185	240
	陆家嘴（东方明珠、金茂大厦、正大广场）	44	165	210
	浦东八佰伴	44	165	210
	大拇指广场（联洋社区）	37	165	170
	五角场	45	170	215
	月星环球港	56	215	270
景点	静安寺	50	185	240
	上海体育场（上海体育馆）	48	180	230
	锦江乐园	46	180	220
	人民广场（城市规划馆、博物馆）	48	180	230
	豫园（城隍庙）	46	175	220
	外滩	50	185	240
	上海科技馆	39	145	180

此外，对上海虹桥机场的航班到港数量每三十分钟进行一次数据的采集，统计结果可绘制成如下折线图

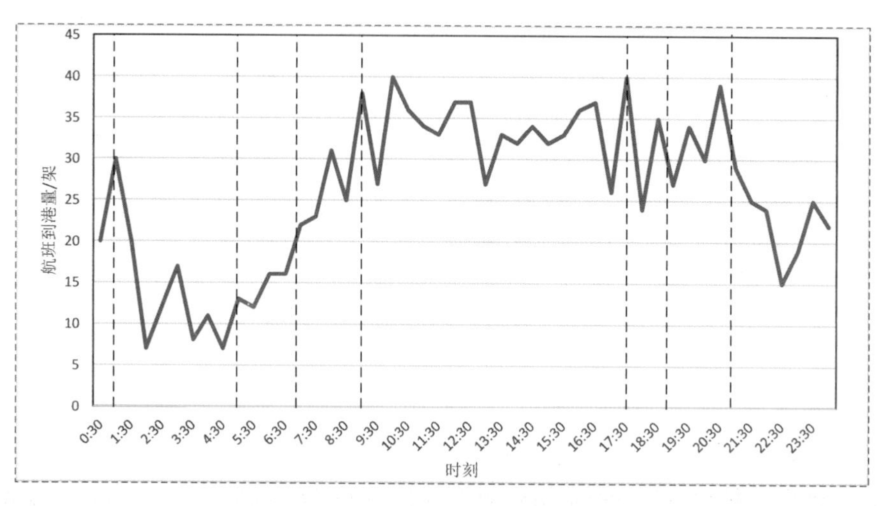

图1 30min尺度下航班到港量数据折线图

根据以上数据，我们对机场中司机的决策环境有了一个初步的了解。同时，有关数据的处理对问题二的求解也至关重要。对于最近一年即2019年的航班到港数量并没有完整的数据，因此我们采用季节指数预测法，利用表1中2015-2018年的统计数据，得到2019年四个季度到港航班数量的预测值分别为：1077.57、1143.02、1183.68、1194.33。同时，由于上海市出租车收费方式采用三阶梯收费，根据表4可以明显看出，机场乘客接续交通里程数将大于15公里，因此，可将b值选取为3.6或4.7，同时在进行综合经济效益的计算时，应注意时段对出租车收费情况的影响，具体而言，即选取合适的b值。同时，根据图1中的48个数据点，可以计算得到48个时刻航班量所占权重系数 $g_1,g_{2,}g_3\cdots g_{48}$。从图1中可看出，一天中航班量的变化规律有着较强的规律性，因此，为了简化分析与计算过程，我们将其划分为8个航段类型：早上航段、早上攀升航段、黄金航段，傍晚小低谷航段、晚间黄金航段、晚间低谷航段、凌晨小高峰航段、红眼航班航段。同时，以每个航段内的权重系数算出每个航段的半小时内乘客占比，再结合季节指数预测，可计算出某个季度、某个时段的等待时间，代入综合经济效益模型，可计算得到预估等待时间，进而得到最优决策模型中 $F(W_a,W_b)$ 与季度和航段的关系。

表 4 虹桥机场航班时刻分异结构表

航段类型	时间段	航班数量（次）	航班密度（次/小时）	航班比例(%)	每半小时乘客占比(%)
早上航段	05：00-07：00	66	33.00	5.28	1.32
早上攀升航段	07：00-09：00	117	58.50	9.36	2.34
黄金航段	09：00-17：30	574	67.53	45.92	2.70
傍晚小低谷航段	17：30-19：00	86	57.33	6.88	2.29
晚间黄金航段	19：00-21：00	132	66.00	10.56	2.64
晚间低估航段	21：00-23：00	83	41.50	6.64	1.66
凌晨小高峰航段	23：00-01：30	117	46.80	9.36	1.87
红眼航班航段	01：30-05：00	75	21.43	6.00	0.86

表5 基于上海虹桥机场数据F函数计算值与季度和航段关系表

航段类型 \ w1/w2 \ 季度	第一季度	第二季度	第三季度	第四季度
早上航段	0.94	0.97	0.88	0.93
早上攀升航段	1.07	1.11	1.13	1.09
黄金航段	1.1	1.16	1.22	1.13
傍晚小低谷航段	0.91	0.98	1.01	0.89
晚间黄金航段	1.19	1.22	1.09	1.13
晚间低谷航段	1.01	0.92	0.96	0.91
凌晨小高峰航段	1.31	1.33	1.21	1.26
红眼航班航段	0.83	0.94	0.89	0.81

<1　　1　　>1

由表5得到的计算结果，可以给出出租车司机的决策如下：

（1） 在早上航段均选择决策B。

（2） 早上攀升阶段均选择决策A。

（3） 黄金航段均选择决策A。

（4） 傍晚小低谷航段在第三季度时可偏向决策A，其余季度均选择决策B为最优。

（5） 晚间黄金航段均选择决策A。

（6） 晚间低谷航段除在第一季度可偏向决策A，其余季度选择决策B为最优。

（7） 凌晨小高峰航段均选择决策A。

（8） 红眼航班阶段均选择决策B。

5.2.2.2 模型的合理性检验

在本小节我们将对最优决策模型的合理性进行检验。在此模型的合理性，我们可以理解为决策模型的相对于无决策模型的优越性，这个优越性可通过综合经济收益值计算结果得大小来体现。我们选取虹桥机场出站旅客的14个主要目的地作为自变量，依据上述综合经济收益模型计算得到收益值，再按照决策A与决策B各占一般概率计算得到收益期望值，将依据模型计算得到的收益值与收益期望值进行对比，可得出表6所示结果：

表6 最优决策下收益与不决策期望收益对比统计表

目的地	里程(KM)	Wa	Wb	Wa/Wb（计算值）	不决策期望收益	决策模型收益增益	溢出收益比率
太平洋百货	48	182	110	1.1	146	36	0.246575342
莘庄	50	211	188	0.94	199.5	-11.5	-0.05764411
金茂大厦	44	165	180	0.91	172.5	7.5	0.043478261
浦东八佰伴	44	166	190	0.94	178	12	0.06741573
联洋社区	37	174	110	1.19	142	32	0.225352113
五角场	45	170	151	1.19	160.5	9.5	0.059190031
月星环球港	56	217	140	1.1	178.5	38.5	0.215686275
静安寺	50	185	130	1.31	157.5	27.5	0.174603175
上海体育场	48	180	121	1.1	150.5	29.5	0.196013289
锦江乐园	46	180	153	1.19	166.5	13.5	0.081081081
博物馆	48	180	182	0.83	181	1	0.005524862
豫园	46	175	160	1.1	167.5	7.5	0.044776119
外滩	50	185	130	1.1	157.5	27.5	0.174603175
上海科技馆	39	145	107	1.31	126	19	0.150793651

表中的数据结果可以看出，问题一中所建立的最优决策下除了目的地为莘庄时决策模型收益小于不决策期望收益，其余情况下，最优决策模型下的收益较不决策期望收益均体现出了经济上的优越性，证明该模型下司机经济收益增加11.6%，即证明所建模型具有合理性。

5.2.2.3 模型的对相关因素的依赖性检验

针对模型对不同因素的依赖性检验问题，本文选取四个对模型计算结果影响较大的因素：期望里程、乘客总数量、出租车数量、出租车长途计价数，并基于收集到的上海机场及上海出租车的数据，确定了8个不同的变化幅度，并根据控制变量原理分别对每个因素的每个变化幅度对最终决策函数$F(W_a,W_b)$的影响进行分析，计算出每种情况下F的变化幅度。计算结果如表7所示：

表7 四种因素不同变化幅度下F值变化幅度统计表

F变化幅度 变化幅度 变化量	-20%	-15%	-10%	-5%	0	15%	10%	15%	20%
期望里程	-0.24	-0.145	-0.075	-0.05	0	0.065	0.11	0.16	0.245
乘客总数量	-0.25	-0.14	-0.055	-0.02	0	0.015	0.06	0.15	0.26
出租车数量	0.19	0.16	0.08	0.05	0	-0.06	-0.09	-0.16	-0.37
出租车长途计价	-0.16	-0.15	-0.05	-0.035	0	0.045	0.075	0.145	0.18

根据表中数据，获得各因素条件下得到的各因素依赖性图表如下：

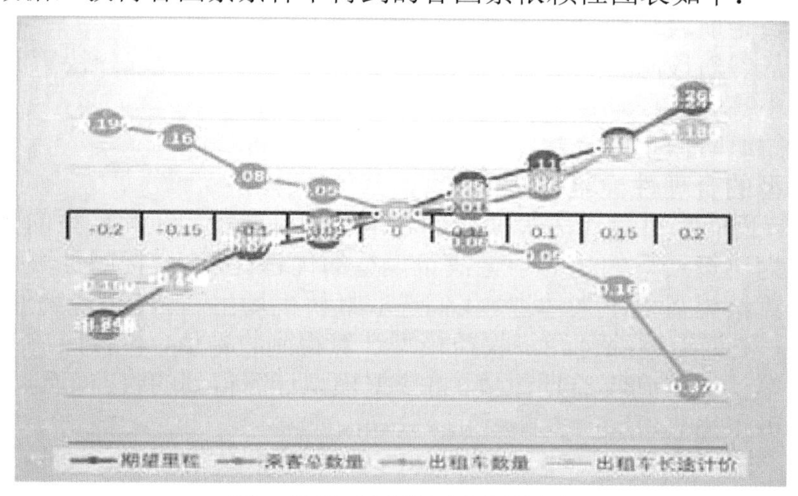

图2 模型对四种因素折线图

由此可得出结论：四种因素对决策函数F的计算结果的影响程度总体较小，出租车数量的影响最大，最大可对F值产生37%的影响，并且只有出租车数量对于F值计算结果的影响为负向影响，即出租车数量变化幅度增加导致F值减小，其余三个因素对于F值变化幅度的影响均为正向影响。

5.3 问题3的分析

出租车客流集散的客流量约占铁路客流集总量的 10% ～ 30%，占据较大的权重，在“乘车区”通常存在出租车等候或是乘客等候的情况，机场接续交通系统的疏散效率将在很大程度上影响机场的工作情况，如何合理解决乘客及出租车司机候车、候客的问题，即设置合理的“上车点”，并对乘客与出租车进行合理的安排使得总乘车效率最高成为亟待解决的问题。在本小节本文将基于排队论，分别从旅客排队上车模式和出租车排队服务系统的服务台数两个方面进行分析，由于系统综合乘车效率最优计算较为复杂，本文将从两个方面考虑达到分别最优，并认为其组成系统的效率即为最优。

5.3.1 旅客排队方式优化

通常而言，旅客排队方式可分为单队列以及多队列，即M/M/1型与M/M/c型。

（1） M/M/1排队模型

λ：乘客到达率，也即单位时间内到达乘车区进行排队等候的顾客数量。

μ：服务率，即单位时间内完成乘车服务的顾客数量。

排队等待系统达到平衡时，队长N的概率分布用p_n表示，并记为：

$$p_n = P\{N = n\}(n = 1, 2, \cdots)$$

单队列排队系统属于典型的生灭过程，由生灭求解过程可得：

$$p_0 = \frac{1}{1 + \sum_{n=0}^{\infty} \rho^n} = \left(\frac{1}{1-\rho}\right)^{-1} = 1 - \rho \tag{11}$$

其中服务强度$\rho = \dfrac{\lambda}{\mu}$，因此可得：

$$p_n = (1-\rho)\rho^n, n = 1, 2, \cdots \tag{12}$$

则排队系统的平均队长为：

$$L_i = \sum_{n=0}^{\infty} n p_n = \frac{\rho}{1-\rho} = \frac{\lambda}{\mu - \lambda} \tag{13}$$

根据大量研究表明，顾客接受服务的时间服从指数分布，且满足：

$$P\{T > t\} = e^{-(\mu-\lambda)t}, t \ge 0 \tag{14}$$

因此，平均滞留时间为：

$$W_s = \frac{1}{\mu - \lambda} \tag{15}$$

则顾客在队列中的平均等待时间为：

$$W_q = W_s - \frac{1}{\mu} = \frac{\lambda}{\mu(\mu - \lambda)} \tag{16}$$

(3) M/M/c等待制排队模型

在M/M/c排队系统中，旅客单个到达乘车区，系统中共有c个服务台，并且每个服务台的的服务时间时相互独立、互不影响的。其中：

$$p_n = \begin{cases} \dfrac{\rho^n}{n!} p_0, n = 1, 2, \cdots c \\ \dfrac{\rho^n}{c! c^{n-c}} p_0, n \ge c \end{cases} \tag{17}$$

其中，

$$p_0 = \left[\sum_{n=0}^{c-1} \frac{\rho^n}{n!} + \frac{\rho^c}{c!(1-\rho_c)}\right]^{-1} \tag{18}$$

利用Erlang等待公式可求得旅客必须等待的概率，也即旅客数量$n \ge c$时的概率，即为：

$$c(c,p)=\sum_{n=c}^{\infty}p_n=\frac{\rho^c}{c!(1-\rho_c)}p_0 \tag{19}$$

系统平稳状态下，M/M/c排队系统的平均队长L_q的计算公式为：

$$L_q=\sum_{n=c+1}^{\infty}(n-c)p_n=\frac{p_0\rho^c}{c!}\sum(n-c)\rho_c^{n-c}=\frac{p_0\rho^c\rho_c}{c!(1-\rho_c)^2} \tag{20}$$

根据经典排队理论，多个服务点相对单个服务点具有显著的优越性，因此，在实际生活中机场出租车乘车区乘车点的个数往往是多个的。鉴于此种情况，本节着重研究并联M/M/1模型，即多队列多乘车点模型与M/M/c即单队列多服务台模型的比较。

5.3.1.1 单队列多乘车点模型

单队列多乘车点模型的主要特征为：乘客到达“乘车区”是独立的，排队服务系统中共有c个乘车点，并且每个乘车点的服务时间是相互独立的。当旅客到达乘车区时，如果有空闲的乘车点，即可以马上上车无需排队等候；如果没有空闲的乘车点，所有乘客根据顺序排成一个队列等待乘车点出租车的更新。根据5.3.1节的运算公式，可得出：

（1） 排队系统中顾客为0的概率

$$p_0=\left[\sum_{n=0}^{c-1}\frac{\rho^n}{n!}+\frac{\rho^c}{c!(1-\rho_c)}\right]^{-1}$$

（2） 平均队列长以及平均队长

$$L_q=\sum_{N=c+1}^{\infty}(n-c)\frac{(cp)^c\rho}{c!(1-\rho)^2}p_0\ ,L_S=L_q+\rho$$

（3） 平均等待时间和逗留时间：

$$W_q=W_s-\frac{1}{\mu}\quad W_s=\frac{L_s}{\lambda}$$

5.3.1.2 多队列多乘车点模型

对队列多乘车点模型的特征为：顾客到达乘车区，对并列的c个队伍进行长度的判断，选取其中最短的队列并加入其中，并且中途不再换队。经典运筹学将多队列多服务点的排队系统即c个并联的M/M/1系统进行简化，简化为c个独立的M/M/1系统，利用简化后的系统计算得到：

（1） 排队系统中顾客为0的概率

$$p_0=1-\rho$$

（2） 平均队列长以及平均队长

$$L_s=\frac{\lambda}{\mu-\lambda}\ ,L_q=\frac{\lambda^2}{\mu(\mu-\lambda)}$$

（3）平均等待时间和逗留时间

$$W_s=W_q+\frac{1}{\mu}=\frac{1}{\mu-\lambda}$$

5.3.1.3 两种排队系统性能比较

根据上海机场历年数据，选择λ取0.9，μ取0.4。可得到两种排队系统的性能对比数据如表所示

表 8 基于解析法的系统性能指标对比表

模型 指标	M/M/c 型	M/M/1 型
平均队列长 Lq	1.56	2.41
平均队长 Ls	2.97	2.86
平均逗留时间 Ws	4.36 分钟	9.73 分钟
平均等待时间 Wq	1.93 分钟	8.3 分钟

分析表中数据可以得出结论，多队列多乘车点模型在实际运行过程中所耗费的总时间最少，能使乘车效率达到最高。

5.3.2 出租车排队方式模型的优化

出租车排队接待乘客的过程大致如下：出租车在机场送客后进入蓄车池等待，听候机场管理部门的安排分批次进入乘车区，在接客上车后驶离上车区，之后又由后面的出租车补位。用服务机构的繁忙率ρ'用以描述排队系统中的繁忙程度，其在两并行车道排队系统中可定义为：

$$\rho' = \frac{\lambda}{C\mu}$$

其中，C为服务台个数，在本模型中具体表现为乘车区中乘车点数目。λ为乘客的平均到达率，也即单位时间内到达排队系统的乘客数量；μ为单个乘车点的平均服务率。

在排队系统到达平稳状态时时，C个服务台并联工作，系统中乘客数量n的概率如下：

$$P_n(C) = \begin{cases} \frac{1}{n}\left(\frac{\lambda}{\mu}\right)^k P_0(C) \quad n = 1,2,3\cdots,C \\ \frac{1}{C!C^{n-C}}\left(\frac{\lambda}{\mu}\right) P_0(C) \quad n = C+1 \end{cases} \tag{21}$$

假设乘客等待时间的成本为$Z_1 = \alpha L_s$，其中α为每个乘客单位时间的等待时间成本。当时间成本最低时，乘车效率最高，排队系统将会得到优化，即：

$$\begin{cases} Z(C) \le Z(C+1) \\ Z(C-1) \le Z(C) \end{cases} \tag{22}$$

将λ与μ值分别代入，求解得当C=4时，满足式22的要求。即设立4个乘车点，乘车区规划示意图如下：

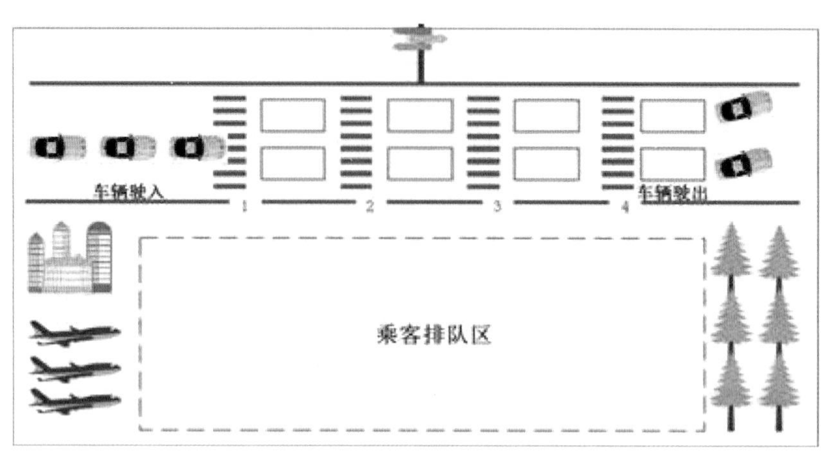

图3 乘车区排队系统示意图

5.4.1 问题四分析

机场由于客流量较大，导致排队等候的出租车在蓄车池内等候时间过长，具有相当多的时间成本，鉴于此种情况，等待排队时间较长的出租车司机街道短途订单后盈利相对长途较少，甚至出现亏本现象，因此滋生不少出租车司机拒载的情况。为了整治这种现状，促进机场交通工作正常运行，鉴于计价规则等因素机场管理部门无权修改，本小节将提出一种安排方法，在等待时间方面给予短途出租车优先权，力求平衡长、短途出租车盈利均衡，调动出租车司机接短途订单的积极性，增强机场出租车接续交通运营系统的公平性。

5.4.2 “短途优先”安排方法

对于接短途订单的出租车，机场管理部门能够提供的最有效的方案即为节省其等待时间，增强接客效率。上海虹桥机场基于这种现状，已经提出了一个有效措施：规定22公里为短途、长途的分界线，承接短途订单的出租车司机可以领取一张“短途票”，如果保证往返时间不超过1小时，回到机场后将被承诺不必排队即可承担新的业务，为司机节约了时间成本，增大了接客效率。但这种机制下仍然存在一些问题，例如，有些司机为获取短途票，一路上超速行驶，威胁到司机与乘客的乘车安全；对于略超过22公里的目的地，司机既不能获取较多的收益，也不能享受“短途票”所带来的时间成本的节约；还有些司机为骗取“短途票”，将乘客撇在半路上等等。为解决以上所述问题，基于以上安排方法，提出了一种新的优化安排方法。

5.4.3 优化后的“短途优先”安排方法

鉴于以上情况，需要确定司机的实际行车路线以杜绝先前的发票造假,合理确定司机以安全速度行驶的路程和时间上限,随着智能化的到来.这两项主要问题的逐步得到了解决方案.

首先,通过与手机软件结合的GPS轨迹识别定位系统直接取代先前由调度人员手动开发票,消灭造假行为.其次,以机场为中心,做一个“短程上限距离”为半径的圆域，利用长途单与短途单经济效益相对平衡的约束方程,确定一个最合适的半径,从而给合理车程内的出租车优先权,使出租车收益均衡。

5.4.3.1模型的建立：

设合适圆域半径为a，短途票每车每日限n次，起步价为c元，起步价公里数为d公里，超出起步价每公里为f元，为均衡出租车收益，使出租车司机在机场收益与从机场接一次长途单收入差别在10%以内。根据GPS轨迹跟踪识别，要求司机在放下短途单之后返回机场的路途中不再接受新的订单，且要求从出发到返航所花时间为行驶路程为所花时间的2倍，另外，考虑到乘客上下车时间及路况等因素，允许缓冲时间为总时间的5%。

5.4.3.2模型的求解：

记司机在机场接到n次短途单的收入为K_i，记每次跑的距离为S_i，接到长途单的收入为K_0，超出范围a即认为失去短途票资格。

目标函数：K=K0【1（+-）10%】

条件方程：

若$S_i \le a$，则记为取消短途票资格。记为长途单收入K0,K的初始值记为0.

$$\begin{cases} K_i = c + (\mathrm{S}_i - d)\mathrm{f} \\ K = K + K_i \\ t = t_1 \cdot 2(1 + 5\%) \end{cases} \tag{23}$$

5.4.3.3模型的检验：

分析上海虹桥情况，得出最优解为n=2；a=33.3公里，即短途票每车每日限2次，“短程上限距离”为33.3公里，合理单次短途时长为36分钟，实现使拥有短程票多次往返的出租车司机收益与接一次长途单收入差别在10%以内，满足相对车程5%的缓冲时间，总体使得这些出租车的收益均衡。

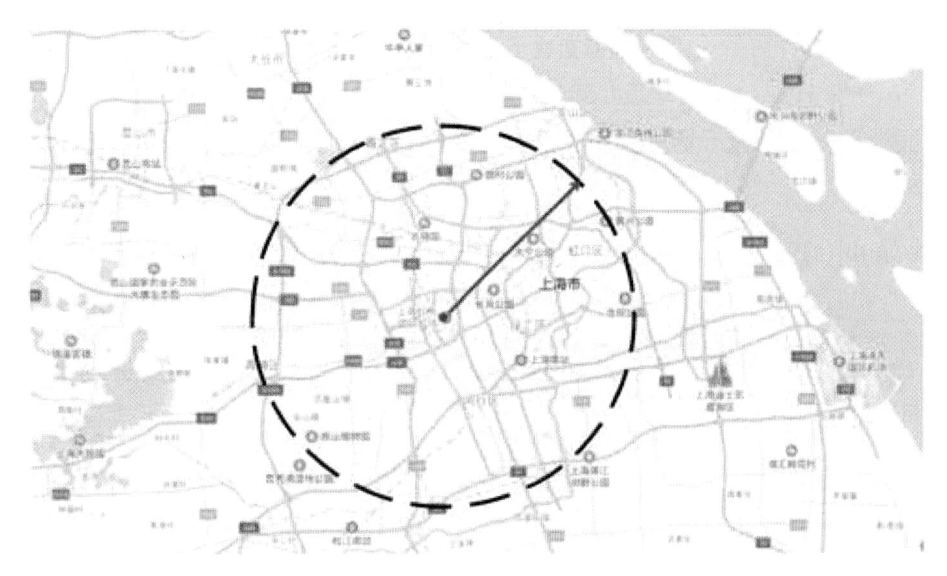

图4 优化模型“短途半径”选择示意图

六、模型评价

6.1 模型优势

1.本文在第一问中所建立的依托综合经济收益值建立的最优决策模型，将难以统一考量的等待时间成本与空载成本通过量化为经济量的方法，纳入到同一个评价体系下，解决了评价不同物理量的难题，为司机的决策提供有效的依据。
2. 本文所建立的最优决策模型针对不同城市、不同机场的情况进行的改进幅度较小，改进方式较为方便，具有较强的普适性。
3. 本文在问题三中所建立的综合排队服务系统模型充分考虑了乘客排队和出租车排队两方面问题，以两方面分别最优的组合解代替系统最优解，减轻了计算与分析过程中的负担。

6.2 模型劣势

1. 问题一中模型的建立过程中，预估等待时间与时段的关系是仅基于某机场航班某一典型日航段与航班到港量的数据进行回归分析所得，可能造成误差。
2. 由于相关具体数据获取较难，模型建立中的某些量，例如市区内平均载客收入只能根据局部数据推断全体，并采用平均值进行模型的求解

七、参考文献

[1] 基于排队论的航空枢纽陆侧旅客服务资源建模与仿真[D]. 中国矿业大学(北京)，2017。

[2] 王燕，罗秀秀. 公共蓄车场服务质量测评——以上海浦东机场出租车蓄车场为例[J]. 经济管理，2013(8):171-180.

[3] 首都机场接续运输协调保障技术研究与实现[D]. 电子科技大学，2015.

八、附录

代码部分

```
#include <iostream>

using namespace std;
double x;
double a;
double b;
double S;
double p0;
double L;
double nmt;
double ckl[48];
long long cksl[4];
int seasondays[4]={90,91,92,92};
double T;
double v;
void init()
{
    cout<<" \b 请输入当地出租车的计价规则"<<endl;
//1
    cout<<" \b 请输入起步里程（单位：千米）"<<endl;
    cin>>x;
    while(x<=0)
    {
        cout<<" \b 你输入起步里程非法，请输入正确的起步里程"<<endl;
        cin>>x;
    }

//2
    cout<<" \b 请输入起步价格（单位：元）"<<endl;
    cin>>a;
    while(a<=0)
    {
        cout<<" \b 你输入起步价格非法，请输入正确的起步价格"<<endl;
        cin>>a;
    }

//3
    cout<<" \b 请输入起步之后每公里的价格（单位：元）"<<endl;
    cin>>b;
    while(b<=0)
```

```
    {
        cout<<" \b 你输入起步价格非法，请输入正确的起步价格"<<endl;
        cin>>b;
    }

//4
    cout<<" \b 请输入出租车期望到达市路程（单位：千米）"<<endl;
    cin>>S;
    while(S<=0)
    {
        cout<<" \b 你输入路程数据非法，请输入正确的路程"<<endl;
        cin>>S;
    }

//5
    cout<<" \b 请输入油价（单位：元/升）"<<endl;
    cin>>p0;
    while(p0<=0)
    {
        cout<<" \b 你输入油价非法，请输入正确的油价"<<endl;
        cin>>p0;
    }

//6
    cout<<" \b 请输入出租车油耗（单位：升/千米）"<<endl;
    cin>>L;
    while(L<=0)
    {
        cout<<" \b 你输入的油耗数据，请输入正确的油耗"<<endl;
        cin>>L;
    }

//7
    cout<<" \b 请输入因载客而产生的怎么增加系数"<<endl;
    cin>>nmt;
    while(nmt<=0)
    {
        cout<<" \b 你输入的增加系数非法，请输入正确的增加系数"<<endl;
        cin>>L;
    }
}
void init2()
{
    cout<<" \b 请输入机场的相关信息";
```

```
    cout<<" \b 请输入机场一天 24 小时的乘车率"<<endl;
    for(int i=0;i<48;i++)
    {
        cin>>ckl[i];
    }
    cout<<" \b 请输入预测的四个季度的乘客数量"<<endl;
    for(int i=0;i<4;i++)
    {
        cin>>cksl[i];
    }
}
double ww1()
{
    return a+(S-x)*b-S*p0*L*(1+nmt);
}
double ww2()
{
    return T*v*p0*L;
}
void test()
{
    int pre=1;
    int time=-1;
    int season;
    while(pre!=0)
    {
        cout<<" \b 请输入现在几点(请输入整点)"<<endl;
        cin>>time;
        cout<<" \b 请输入现在第几季度"<<endl;
        cin>>season;
        cout<<" \b 请输入现在机场出租车的数量单位（辆）"<<endl;
        int cz;
        cin>>cz;
        cout<<" \b 请输入增益系数"<<endl;
        int zy;
        cin>>zy;
        cout<<" \b 请输入车速（单位：千米/小时）"<<endl;
        cin>>v;
        T=(cksl[season]/seasondays[season]*ckl[time])/cz*zy;
        double w1=ww1();
        double w2=ww2();
        cout<<"w1/w2="<<w1/w2<<endl;
        if(w1/w2<1)
        {
```

```
            cout<<" \b 建议前往空载市区"<<endl;
        }
        else
        {
            cout<<" \b 建议留在机场等待"<<endl;
        }
    }

}
int main()
{
    init();
    init2();
    test();
    return 0;
}


#include <iostream>
#include <cstdio>
#include <algorithm>

using namespace std;
const int Max = 1000010;

int f[Max];
int n,m,x,y;

int Find(int x)
{
    if(x==f[x]) return x;
    else return f[x]=Find(f[x]);
}

void Union(int x,int y)
{
    int a=Find(x),b=Find(y);
    if(a!=b) f[a]=b;
}

struct node
{
    int x,y,step;
};
```

```
node a[Max];

bool cmp(node a,node b)
{
    return a.step<b.step;
}

void kruskal()
{
    for(int i=1;i<=n;i++)//并查集 f 数组初始化
        f[i]=i;

    for(int i=0;i<m;i++){
        cin>>a[i].x>>a[i].y>>a[i].step;
    }
    sort(a,a+m,cmp);//按路径长度从小到大排序
    int sum=0,t=0;//sum 为最小路径总和，t 为已经添加了的路径条数
    for(int i=0;i<m;i++)
    {
        if(t==n-1) break;//最小生成树的路径有 n-1 条
        if(Find(a[i].x)!=Find(a[i].y)) {
            sum +=a[i].step;
            Union(a[i].x,a[i].y);
            t++;
        }
    }
    if(t==n-1)cout<<sum<<endl;
    else cout<<"impossible";
}

int main()
{
    cin>>n>>m;
    kruskal();
    return 0;
}
```

```
#include <iostream>

using namespace std;
double ddl;//代表 λ
```

```
double fwl;//代表 μ
double p0;
double px; //代表 ρ
double pxn;//代表 ρ 的 n 次方
double pn;
double n;
double Li=1;
double Ws;
double Wq;
double test(int n)
{
    px=ddl/fwl;
    p0=1-px;
    pxn=1;
    for(int i=0;i<n;i++);
    {
        pxn=pxn*px;
    }
    pn=(1-px)*pxn;
    return pn;
}
int main()
{
    cout<<" \bM/M/1 型"<<endl;
    cout<<" \b 请输入乘客到达率"<<endl;
    cin>>ddl;
    cout<<" \bq 请输入服务率"<<endl;
    cin>>fwl;
    cout<<" \b 请输入 n 的值"<<endl;
    cin>>n;
    for(int i=0;i<n;i++)
    {
        Li=Li+i*test(i);
    }
    Ws=1/(fwl-ddl);
    Wq=ddl/(fwl*(fwl-ddl));
    cout<<"Ws:"<<Ws<<endl;
    cout<<"Wq:"<<Wq<<endl;
    return 0;
}
```

```
#include <iostream>
```

```
using namespace std;
double ddl;//代表λ
double fwl;//代表μ
double p0;
double px; //代表ρ
double pxn;//代表ρ的 n 次方
double pn;
double n;
double Li=1;
double Lq=1;
double Ws;
double Wq;
int nn=1;
int cc=1;
int ccn=1;
int c;
double pc=0;
double test(int n)
{
    px=ddl/fwl;
    p0=1-px;
    pxn=1;
    for(int i=0; i<n; i++);
    {
        pxn=pxn*px;
    }
    for(int i=1; i<=n; i++)
    {
        nn=nn*i;
    }
    for(int i=1; i<=c; i++)
    {
        cc=cc*i;
    }
    if(n<=c)
    {
        pn=pxn/nn*p0;
    }
    else
    {
        for(int i=1; i<=n-c; i++)
        {
            ccn=ccn*i;
```

```
            }
            pn=pxn/cc/ccn*p0;
        }
        return pn;
}
int main()
{
        cout<<" \bM/M/c 型"<<endl;
        cout<<" \b 请输入乘客到达率 λ "<<endl;
        cin>>ddl;
        cout<<" \bq 请输入服务率 μ "<<endl;
        cin>>fwl;
        cout<<" \b 请输入 n 的值"<<endl;
        cin>>n;
        cout<<" \b 请输入站台个数 c"<<endl;
        cin>>c;
        for(int i=c; i<=n; i++)
        {
            pc=pc+test(i);
        }
        cout<<" \b 旅客数量 n≥c 的概率"<<pc;
        for(int i=c+1;i<=n;i++)
        {
            Lq=Lq*(i-c)*test(i);
        }
        cout<<" \b 平均队长为"<<Lq;
        return 0;
}
```

数据部分

1

表 1 2015-2019 年四个季度航班到港数量统计表

季度	历史数据			
	2015	2016	2017	2018
第一季度	932.24	973.64	997.64	1056.76
第二季度	984.19	989.72	1057.25	1100.13
第三季度	999.04	1039.12	1098.68	1139.68
第四季度	993.63	1044.51	1106.37	1036.76

2

表 2 上海出租车收费标准

里程数	日间 （05：00-23：00）	夜间 （23：00-05：00）
0-3 公里	14 元	18 元
3-10 公里	2.5 元/公里	3.1 元/公里
15 公里以上	3.6 元/公里	4.7 元/公里

3

表 3 上海浦东机场乘客主要目的地及出租车参考乘车价格

目的地	里程(KM)	里程(KM)	参考价格 （元）	夜间
商圈	徐家汇（东方商厦、港汇广场、太平洋百货）	48	180	230
	莘庄	50	185	240
	陆家嘴（东方明珠、金茂大厦、正大广场）	44	165	210
	浦东八佰伴	44	165	210
	大拇指广场（联洋社区）	37	165	170
	五角场	45	170	215
	月星环球港	56	215	270
景点	静安寺	50	185	240
	上海体育场（上海体育馆）	48	180	230
	锦江乐园	46	180	220
	人民广场（城市规划馆、博物馆）	48	180	230
	豫园（城隍庙）	46	175	220
	外滩	50	185	240
	上海科技馆	39	145	180

9.高压油管的压力控制

摘要：本文研究的是燃油系统的高压油管的压力控制问题。故设计燃油进入和喷出的工作时间控制策略内压力，进而控制管内压力，对燃油发动的工作效率的提高具有积极作用。本文主要采用微积分方程、模拟退火算法、非线性无条件规划算法、计算机仿真模拟等方法对问题进行分析和解决。

针对问题一，第一小问，首先依据质量守恒定律，进入油的总质量等于喷出油的总质量，已知了喷油时高压油管密度及进油口处密度都为恒值。其次运用积分法，对喷油速率时间图像积分，求得喷油总体积，而每次进油体积可用含未知量单向阀开启时长的变量表示，得出一元一次方程求解即可。对于喷油进油的分配方案，分两种情况进行讨论（这里假设进油与出油并不同一时刻进行）（1）先喷油后进油（2）先作预充油处理，先进油后喷油。通过讨论发现，在采取预冲油方法，使高压油管内的压力在100MPa做上下对称均匀波动，此时波动图形与横轴围成面积最小，并在此基础上得到最小波动幅度的具体时间分配，获得单向阀开启时间为0.2876ms。

第二小问，基于前一小问，增加达到稳态的时间限制，利用所给公式通过MATLAB解微分方程拟合出压力与密度关系式，然后先取得一个初始的过程所需总时间的值，采用模拟退火法，基于计算机仿真模拟方法模拟管内压力趋于确定值的过程，不断迭代压力与密度，直到高压油管内的压力达标时刻停止迭代，得到最终所求过程所需总时间的最优近似值，并依据第一小问的分配方案进行分配每个总时间的喷油进油时间，得到所求结果为：经过2 S的调整过程，单向阀开启的时长t_1=0.8090s；经过5 S的调整，开启的时长t_1=0.5650s；经过10 S的调整，开启的时长t_1=0.4958s。

针对问题二，在问题一的基础上，考虑了供油来源以及喷油嘴的针阀升程情况，依据凸轮转动曲线与柱塞上下移动距离关系和针阀的升程与油通过的小孔面积的变化关系，亦可通过模拟退化法，不断迭代密度与压力，到进油圈数与喷油时间的最小正周期时喷油进油总质量相等时，得出相应进油总质量，换算出凸轮转动时间以及曲线长度，得出凸轮边缘转动的角速度为0.027rad/ms。

针对问题三第一小问，要求在增加一个喷油嘴的情况下，调整喷油和供油策略。本题采取两种方法，喷口B和喷口C同时喷油和喷口B和喷口C错时喷油。根据绘制的假想图，喷口B和喷口C错时喷油可使压力波动幅度最小。针对问题三第二小问，要求在增加一个单向减压阀使高压油管内的压力减小，使其继续稳定，为使单向减压阀工作且不影响高压油管的压强，必须使凸轮的角速度增加，当处在两喷油嘴都停止喷油的状态下，高压油泵向高压油管供油，这时单向减压阀应当开始工作来维持高压油管的压力稳定。

关键词：模拟退火算法 微积分方程 质量守恒

作　　者：姬明月，余文杰，董苏营

指导教师：罗虎明

奖　　项：2019年全国大学生数学建模竞赛国家二等奖

1 问题重述

1.1 问题背景

燃油系统承接燃油加注与运输，它确保燃油顺利到达发动机，提供给发动机工作。高压燃油泵向高压油管输送高压燃油，保证向燃油嘴供应持续的燃油，它的作用是提高燃油压力，高压喷射，达到雾化效果。燃油进入和喷出的间歇性工作过程会导致高压油管内压力的变化，使得所喷出的燃油量出现偏差，从而影响发动机的工作效率。

1.2 任务目标

（1）问题一：给出某型号高压油管的内腔尺寸、供油入口尺寸，供油单向阀每打开一次后就要关闭 10ms，喷油器每秒工作 10 次，每 100ms 为工作一个周期，每次工作时喷油时间为 2.4ms。

①第一小问，要求在单向阀开启时间没有限制的情况下，要将高压油管内的压力尽可能稳定在 100 MPa 左右，设置单向阀每次开启的时长策略；

②第二小问，增加了时间限制，求要将高压油管内的压力从 100 MPa 增加到 150 MPa，且分别经过约 2 s、5 s 和 10 s 的调整过程达到在 150 MPa 的情况下，单向阀开启的时长调整策略。

（2）问题二：在第一小问的基础上，又考虑到高压油管的燃油来源问题，以及喷油嘴具体喷油方式问题。高压油管处的燃油来自高压油泵的柱塞腔出口，由凸轮驱动柱塞上下运动，柱塞向上运动时压缩柱塞腔内的燃油，达到一定压力阈值后，柱塞腔与高压油管连接的单向阀开启，燃油进入高压油管内。喷油由喷油嘴的针阀控制，针阀升程为 0 时，针阀关闭；针阀升程大于 0 时，针阀开启，燃油向喷孔流动，通过喷孔喷出。

要求在给出的喷油器工作次数、高压油管尺寸和初始压力已知的情况下，确定凸轮的角速度，使得高压油管内的压力尽量稳定在100 MPa左右。

（3）问题三：在问题 2 的基础上，再增加一个喷油嘴，每个喷嘴喷油规律相同，调整相应的喷油和供油策略。为了更有效地控制高压油管的压力，现计划在装一个单向减压阀。单向减压阀出口为直径为 1.4mm 的圆，打开后高压油管内的燃油可以在压力下回流到外部低压油路中，从而使得高压油管内燃油的压力减小。要求给出请给出高压油泵和减压阀的控制方案。

2 问题分析

2.1 问题一分析

在高压油管的压力变化方面，需要考虑诸多影响因素，如高压油泵的进油量、进油时间，喷油嘴的喷油量、喷油时间，喷油进油的时长分配等。

第一小问，依据物理学的质量守恒定律和假设 2,为使高压油管压力保持稳定在

100Mpa，应使进油质量量等于出油质量，求出单向阀开关控制供油时间。但是由于喷油过程与进油过程的时间点不能确定，分情况讨论喷油与进油时间分配问题，以找到使总压力变化最小的方案，从而使压力尽可能稳定在 100MPa。

第二小问，在第一小问的基础上增加难度，总过程历时增加压力调整至稳定在 150MPa 为一个动态过程。

由假设 3、4 可得，进油周期的高压油管密度用进油开始时刻的高压油管密度代替，结束后发生变化；喷油周期的高压油管密度用喷油开始时刻的高压油管密度代替，结束后发生变化。由已知条件，“燃油的压力变化量与密度变化量成正比”，“进出高压油管的流量”，由微分模型法以及非线性回归的拟合法，可得高压油管内密度与管内压力的确定关系式，即确定密度对应确定压力。

采用二分法选取单向阀开关控制供油时间，然后采用迭代法，不断迭代修改密度与管内压力，至压力无限趋近于150MPa时停止模拟过程，得到最终所求过程所需总时间，并使总时间不断接近 2s，5s，10s，再次更改单向阀开关控制供油时间使其可以稳定在150Mpa 并依据第一小问的分配方案进行分配每个总时间的喷油进油时间，得到结果。

2.2 问题二分析

在问题一的基础上，又考虑了进油入口的供油来源--高压油泵的柱塞腔，以及喷油嘴的针阀升程情况，凸轮的转动导致高压油泵的柱塞上下运动，柱塞上下运动导致高压油泵的压力发生变化，进而推动单向阀的开闭，从而控制进油情况；

由假设 2 得出，此情况下，高压油管内的压力和密度为定值，所以喷油嘴的喷油速率只取决于针阀底面所在平面截密封座的圆减去针阀底面的面积所得的圆环面积和喷油嘴喷孔的最小值决定，从而改变喷油情况；

由问题一的思想和假设2得出，当凸轮旋转一周时，高压油管内的压力恒稳定在100Mpa，从而利用连续性方程的思想，计算出凸轮旋转一周时向高压油管内注入的燃油质量，进而算出凸轮和喷油嘴工作的最小周期，从而求出凸轮的角速度，使得高压油管内的压力尽量稳定在100 MPa左右。

2.3 问题三分析

第一小问：问题分析问题三根据题目已知信息，每个喷嘴喷油规律相同，因此喷口B和喷口C喷油质量相等，为了保持高压油管内压力维持在某一个数值，根据质量守恒定律，$M_{进}=M_{出}$，高压油泵也要提供相同的燃油质量，从而可推导出凸轮的转速是问题二转速的两倍。其次，列举两种方法：喷口B和喷口C同时喷油和喷口B和喷口C错时喷油。按照相同比例绘制出假想图，比较在两种方案下图形的面积，哪种方法面积最小，就是最优策略。

第二小问：要求在增加一个单向减压阀使高压油管内的压力减小，使其继续稳定。为使单向减压阀在工作时，并且对高压油管的压强不造成太大影响，凸轮的角速度增加到原来的二倍，当处在两喷油口都处在停止喷油的状态下，高压油泵向高压油管供油，这时单向减压阀应当开始工作来维持高压油管的压力稳定。

3 符号说明

表格 1 符号说明

变量名称	含义	单位
$V_{出}$	B 口的喷油体积	mm^3
$V_{进}$	A 口的进油体积	mm^3
t_1	单向阀开启时长	ms
ρ_1	燃油进入高压油管内的密度	mg/mm^3
Q_i	A 口每次进油量	mm^3/ms
ρ	燃油密度	mg/mm^3
p	高压油罐内压力	MPa
$M_{出}$	从 B 口喷出的燃油总质量	mg
$M_{进}$	从 A 口进入的燃油总质量	mg
V_1	高压油管体积	mm^3
r	针阀底面所在平面截密封座圆的半径	mm
S_1	圆环面积	mm^2
S_2	喷口面积	mm^2
h_0	针阀升程为 0 时圆锥定点到针阀的高度	mm
$u(t)$	喷油速率随时间变化函数	mm^3/ms
S_{pt1}	方案一中，实际进油与出油过程中的管内时压力与时间的函数与压力为 100Mpa 时所围成的面积	mm^2
S_{pt2}	方案二中，实际出油过程中的管内时压力与时间的函数与压力为 100Mpa 时所围成的面积	mm^2
$P(t)$	高压管内压力随时间变化的关系式	Mpa
r	针阀底面所在平面截密封座圆的半径	mm
t_2	经过调整时间后单向阀开启时长	ms
m_1	每次从 A 口进入的燃油质量	mg

4 模型假设

1.假设进油时燃油以油蒸汽状态存在；

2.由于高压油管内压力一直稳定在某一数值，压力变化幅度较小，即可合理假设在维持高压油管内压力稳定的情况下，认为高压油管内压力不变为稳定值；

3.由于在进油周期这一小段时间内，高压油管密度变化较小，我们将做出合理假设，进油周期的高压油管密度用进油开始时刻的高压油管密度代替，结束后发生变化；

4.由于在喷油周期这一小段时间内，高压油管密度变化较小，我们将做出合理假设，喷油周期的高压油管密度用喷油开始时刻的高压油管密度代替，结束后发生变化；

5 模型建立与求解

5.1 问题一模型的建立

为了便于理解，将原题目中高压油管工作过程示意图进行了简化，

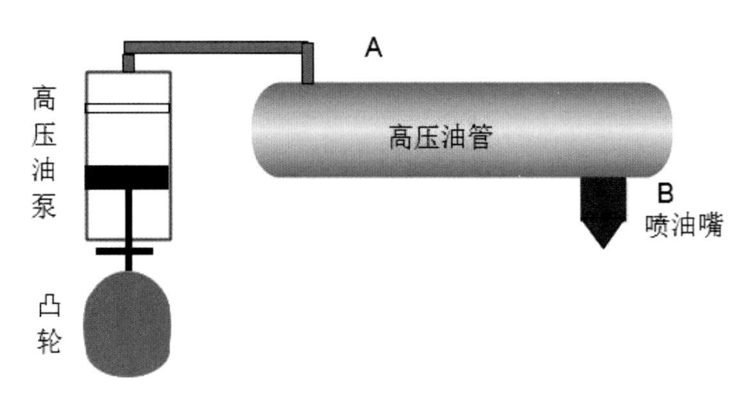

图表 1 高压油管工作过程简化图

5.1.1 问题一第一问模型准备

首先计算燃油进入高压油管内的密度$\rho_{160MPa}=0.8710\text{mg/mm}^3$，根据气体的密度特性，可知燃油进入高压油管内被汽化为气态。

问题一的第一个问题要求设置单向阀每次开启的时长，使得高压油管内的压力稳定在 100MPa 左右。为了维持动态状态下高压油管内压力的稳定，由流体力学的连续性方程，依据质量守恒定律：

$$M_{进}=M_{出} \tag{1}$$

只要使喷油嘴喷油质量等于进油口进油总质量即可。

又由假设密度在进油和出油的极短时间内保持不变，故可转化成体积变化问题。只需通过控制进油时间与次数进而控制进油总体积，即可以使高压油罐内的压力在 100 MPa 处于小幅度上下波动，即趋于稳态。

根据题目所给的信息，由题目所给的图 2 喷油速率示意图，通过积分的方式可以计

算出来 B 口的喷油总体积是一个恒定值，

$$V_{出}=\int_0^{2.4} u(t)dt=44mm^3 \tag{2}$$

一个喷油周期是 100ms，真正的喷油的时间是 2.4 ms，也就是说停止喷油的时间是 97.6 ms。因为喷油时间与进油时间的分配方案直接决定了高压油管的压力变化情况。

基于分类讨论的思想，可以提出以下方案使高压油管压力稳定在 100MPa 左右：

方案一：先让 B 口喷油，喷油结束后再打开 A 口进油。

因此高压油管内压力先从 100MPa 减小，在 2.4 ms喷油结束后，管内压力下降至一个最低值后，再从 A 口进油，之后单向阀再关闭 10ms；不断循环进油过程，直至累加时间达到一个周期 100ms。根据方案画出高压油管压力波动假想图：

（注：为简便模拟，假设忽略了单向阀关闭时间 10ms 以放大进油效果，以及放大了出油时间以仿真作用效果，便于讨论）

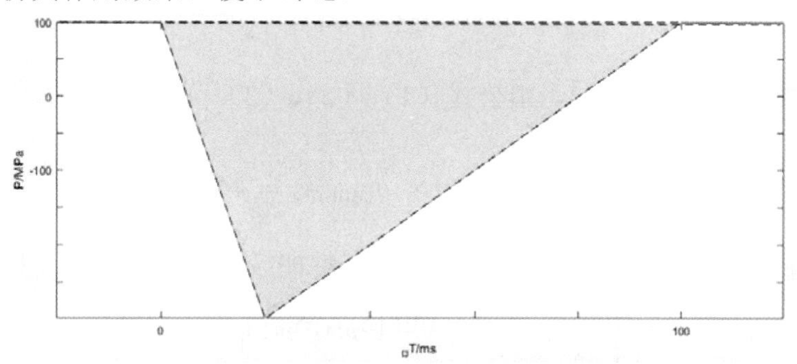

图表 2 高压油管压力波动假想图

如图所示，阴影部分面积为所探讨的总过程压力变化总量，理论上使阴影面积达到最小为最佳方案。

题目要求所求的波动幅度即压力与时间乘机的积分，即$S_{pt1}=\int_a^b |P(t)|\mathrm{d}t$ 最小。

方案二：先给高压油管做预冲油处理。

因此高压油管内压力必定高于 100 MPa，预冲油结束后，开始喷油，2.4ms 后高压油管内压力必定小于 100 MPa，从此刻再进行进油过程，循环往复直到累加时间达到一个周期。根据方案给出高压油管压力波动假想图：

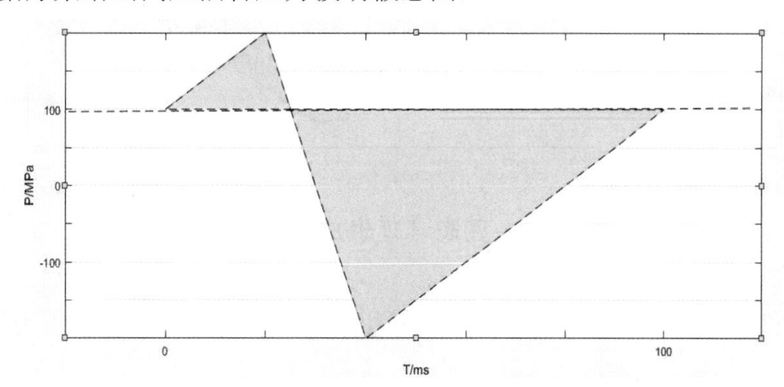

图表 3 高压油管压力波动假想图

两张假想图的横轴与纵轴按照相同比例绘制，图形大小也按照相同的比例绘制。根据图形可以直观地看出波形面积，只有波动面积的绝对值越小，高压油管内的压力越趋向稳定。

题目要求所求的波动幅度即压力与时间乘机的积分，即S_{pt2}=$\int_a^b |P(t)|\mathrm{d}t$ 最小。

5.1.2 模型建立

对于问题一，我们建立非线性规划无约束模型来表示两个方案的假想图：

$$\min\{S_{pt1},S_{pt2}\} \tag{3}$$

（1）式中$a=0$，$b=100$。对于单向阀的开启时间，我们建立如下公式：

$$\mathrm{M}_{进}=\frac{100}{t_1+10}\times t_1\times Q_i\times \rho_{100\mathrm{MPa}} \tag{4}$$

$$\mathrm{M}_{进}=\mathrm{M}_{出} \tag{5}$$

由题中已知信息$Q_{出}$=44mm^3，由公式（1）、（2）、（3）可得

$$\frac{100}{t_1+10}\times t_1\times Q_i\times \ \rho_{160\mathrm{MPa}}=44\times \ \rho_{100\mathrm{MPa}} \tag{6}$$

（4）式的前提条件是高压油罐内压力不变。整理后问题一第一问模型为：

$$\begin{cases}\min\{S_{pt1},S_{pt2}\}\\ \frac{100}{t_1+10}\times t_1\times Q_i\times \ \rho_{160\mathrm{MPa}}=44\times \ \rho_{100\mathrm{MPa}}\end{cases} \tag{7}$$

5.1.3 模型求解

由假想图观察可得，预冲油情况下，正弦函数与 t 轴所围成的面积最小，方案二是理想方案。并进一步优化找到方案二最小的情况，如下图所示：

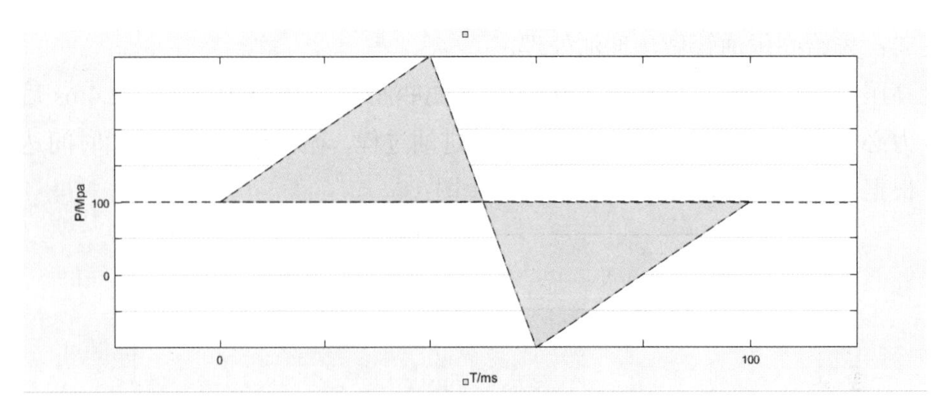

图表 4 理想方案

利用 MATLAB 计算单向阀开启时间大约为 0.2876ms，可使高压油管压力维持在 100 MPa 左右。

5.1.4 问题一第二问模型建立

题目要求将高压油管内的压力从 100 MPa 增加到 150 MPa，在压力升高的整个过程中，燃油的密度势必会发生改变，

可知当燃油密度发生变化时，由进出高压油管的流量公式$Q = CA\sqrt{\frac{2\Delta P}{\rho}}$，进出高压管的流量发生变化；燃油密度发生变化，又根据题意“燃油的压力变化量与密度变化量成正比”，压力必定发生变化。可据此求得高压油管内密度与管内压力的确定关系式，即确定密度对应确定压力。

首先由附件3的弹性模量与压力的关系，可采用微分方程模型以及非线性回归的拟合法，通过MATLAB拟合得出弹性模量与压力的函数关系$\mathrm{E} = \mathrm{f}(p)$，

$$\mathrm{E}(p) = \alpha_1 p^5 + \alpha_2 p^4 + \alpha_3 p^3 + \alpha_4 p^2 + \alpha_5 p + \alpha_6 \tag{8}$$

利用 MATLAB 计算压力的系数α_i，可得结果为：

$$\begin{cases} \alpha_1 = 1.199\mathrm{e} - 09 \\ \alpha_2 = -2.539\mathrm{e} - 07 \\ \alpha_4 = 0.008696 \\ \alpha_3 = 6.835\mathrm{e} - 05 \\ \alpha_5 = 4.914 \\ \alpha_6 = 1538 \end{cases} \tag{9}$$

同时绘制图像：

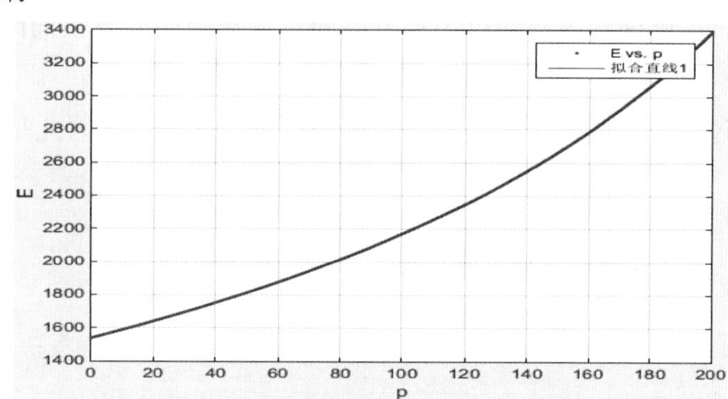

图表 5 弹性模量与压力的关系

燃油的压力与密度的关系可由公式表示：

$$\frac{\mathrm{dp}}{\mathrm{d\rho}} = \frac{E}{\rho} \tag{10}$$

联合（6）、（7）、（8）可得密度关于压力的函数关系与压力关于密度的函数关系：

$$\mathrm{P}(\rho) = \varepsilon_1 \rho^4 + \varepsilon_2 \rho^3 + \varepsilon_3 \rho^2 + \alpha_4 \rho + \alpha_5 \tag{11}$$

$$\rho(p) = \beta_1 p^4 + \beta_2 p^3 + \beta_3 p^2 + \beta_4 p + \beta_5 \tag{12}$$

利用 MATLAB 计算压力的系数ε_i、β_i可得结果为：

$$\begin{cases} \varepsilon_1 = 1.108e + 07 \\ \varepsilon_2 = -3.508e + 07 \\ \varepsilon_3 = 4.537e + 07 \\ \varepsilon_4 = -2.608e + 07 \\ \varepsilon_5 = 5.624e + 06 \end{cases} \tag{13}$$

$$\begin{cases}\beta_1 = 1.474e-12 \\ \beta_2 = -1.064e-09 \\ \beta_3 = -3.683e-07 \\ \beta_4 = 0.0004874 \\ \beta_5 = 0.8059\end{cases} \tag{14}$$

根据密度关于压力的函数关系与压力关于密度的函数关系，绘制出图像：

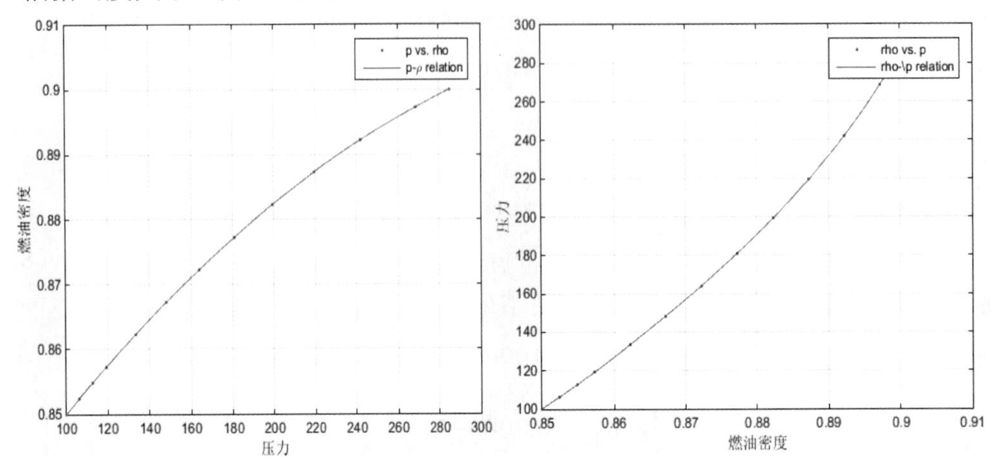

图表 6 压力与燃油密度的关系

由假设 2、3,知，分别用进油和喷油的初始时刻的压力和密度分别代替燃油进入和喷出的压力和密度，高压油管内部的压力和密度随进油和喷油的发生而改变。为了得到单向阀开启的时长，建立以下数学模型：

为了简化计算过程，由假设 1 知，采取先进油后喷油的策略。因此每次从 A 口进入的燃油质量为：

$$m_1 = Q_i \times t_1 \times \rho' \quad (i=1,2,3\ldots) \tag{15}$$

根据式（8）可以计算出密度为ρ'时的压力，记为P'

$$\Delta P = 160 - P' \tag{16}$$

$$Q = CA\sqrt{\frac{2\Delta P}{\rho'}} \tag{17}$$

Q_i为每时间段从 A 口进入的油量，ρ'燃油从 A 口进入时的密度。

再计算当燃油进入高压油管后密度：

$$\frac{m_1+m_2}{V_1} = \rho'' \tag{18}$$

m_2为高压油管内质量，ρ''是在燃油进入高压油管后新密度，根据式（11）可以计算出密度为ρ''时的压力，记为P''。

迭代过程采用计算机仿真模拟方法，采用二分法选取不同得单向阀开启时长，计算出达到 150 MPa 所需总时间的值，判断其值是否非常接近 2s，5s，10s，采用模拟退火法，结合（15）、（16）、（17）、（18）式进行求解，直到高压油管内的压力大于或等于 150 MPa 时停止迭代，，得到最终所求过程所需总时间的最优近似值，并依据第一小问的分配方案进行分配每个总时间的喷油进油时间，得到所求结果。

5.1.5 模型求解

利用 MATLAB 对上述模型进行迭代寻求满足条件不等式的最优解，结果得出后，跳出循环，得出结果：

经过 2 S 的调整过程后稳定在 150 MPa，单向阀开启的时长t_1=0.8090s；

经过 5 S 的调整过程后稳定在 150 MPa，单向阀开启的时长t_1=0.5650s；

经过 10 S 的调整过程后稳定在 150 MPa，单向阀开启的时长t_1=0.4958s。

5.1.6 对问题一第二问模型的检验与修正

在经过迭代算法求出最优解后，虽然能得到使高压油管稳定在 150 MPa 的单向阀开启时长，但根据实际情况来看，经过调整时间高压油管内压力虽然能稳定在 150 MPa，如果去掉约束条件，再进行对上述模型进行迭代，会发现迭代下去的结果一直稳定在 155 MPa 左右。也就是说，经过调整时间后，高压油管内的压力并不是一直稳定在 150 MPa，而是以一个很小的幅度在持续增长。

针对该问题，将单向阀的开启时长分为两部分，第一个部分是经过调整时间后的开启时长，第一个部分的单向阀开启时长沿用上述所求结果即可；第二个部分是在高压油管内的压力达到 150 MPa 后，单向阀开启时长。

调整结果是根据式（2）来计算出单向阀的开启时长，在到达 150 MPa 后单向阀再开启的时长在 2 S、5 S、10 S 的结果是相等的，因此t_2 =0.7517ms。

5.2 问题二模型的建立

首先绘制出油器喷嘴放大后的简单图形：

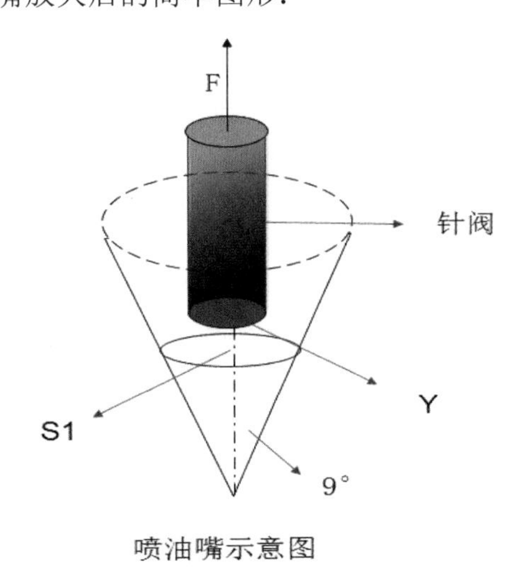

图表 7 喷油嘴示意图

根据附件 2 模拟出针阀升程与时间的图形：

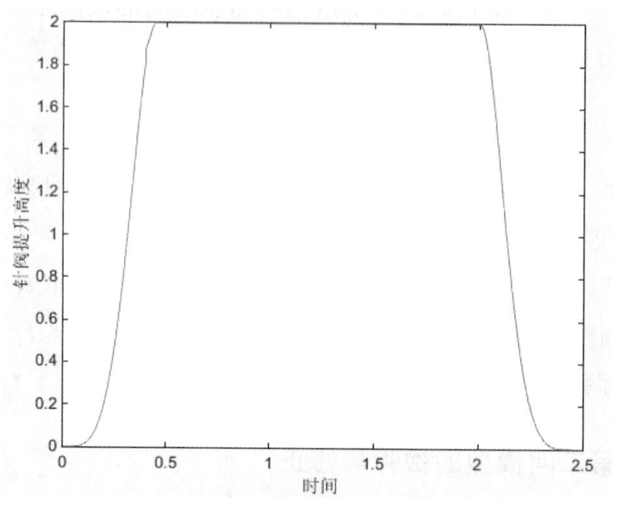

图表 8 针阀升程与时间

根据该图形，得到针阀升程与时间的关系式：

计算出针阀底面所在平面截密封座圆的半径

$$r=(h_0+\mathrm{h})\times\tan\frac{\pi}{20} \tag{19}$$

h 为针阀底面所在平面截密封座圆的圆心与圆锥顶点的连线。

由（16）式与题中所给信息可以计算出S_1圆环面积，同样可以计算出S_2喷口面积。

首先计算在喷油嘴只喷油的情况下，高压油管内密度的变化，喷油时密度为 0.8492mg/mm^3，在此问中，要重新计算流量公式$Q=CA\sqrt{\frac{2\Delta P}{\rho}}$中 A 的值，用公式表示为：

$$\mathrm{A}=\min\{S_1,S_2\} \tag{20}$$

当$S_1>S_2$时，限制燃油流量的是S_2喷口面积，当$S_1<S_2$时，限制燃油流量的是S_1圆环面积。因此，Q 与时间 t 存在一定的关系。将他们之间的关系用图表示：

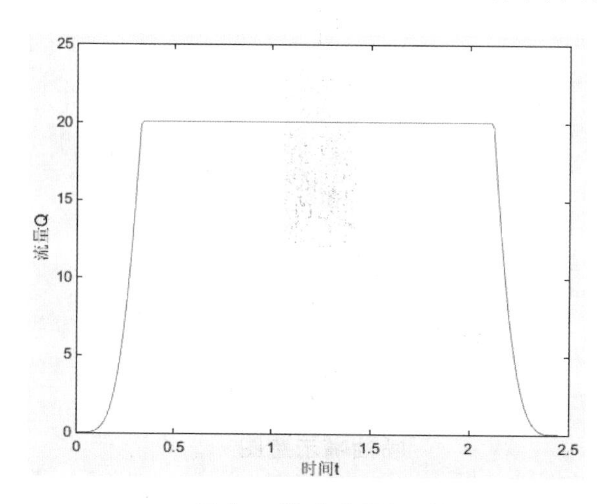

图表 9 燃油流量与时间

通过对上图的微积分可以计算出喷油的体积$V_{出}$。首先对上图的左半部分进行函数拟合，$Q=\gamma_1 t^5+\gamma_2 t^4+\gamma_3 t^3+\gamma_4 t^2+\gamma_5 t+\gamma_6$，通过求解得到参数值为：

$$\begin{cases}\gamma_1=-5.415e+04 \\ \gamma_2=4.218e+04 \\ \gamma_3=-1.06e+04 \\ \gamma_4=1158 \\ \gamma_5=-45.72 \\ \gamma_6=0.3499\end{cases} \tag{21}$$

对上图的右半部分进行函数拟合得到，$Q=\delta_1 t^5+\delta_2 t^4+\delta_3 t^3+\delta_4 t^2+\delta_5 t+\delta_6$，通过求解参数值为：

$$\begin{cases}\delta_1=-5.415e+04 \\ \delta_2=4.218e+04 \\ \delta_3=-1.06e+04 \\ \delta_4=1158 \\ \delta_5=-45.72 \\ \delta_6=0.3499\end{cases} \tag{22}$$

将喷油口喷出油的质量计算出来：

$$m_2=V_{出}\times\rho_1 \tag{23}$$

接下来建立高压油泵的数学模型。由题中所给信息，柱塞向上运动到上止点时，剩余体积的容积为 V $_{上止点}$=20mm^3，高度$h_1\frac{V_{上止点}}{\pi\times r_{柱塞腔}}=1.0186\text{mm}$，柱塞向下运动到下止点位置是剩余体积高度为$h_2=1.0186\ +7.239-2.413=5.8446mm$，因此，剩余体积$V_2=114.7585$。题目给出下止点的压力，根据（9）式可以计算得出此时柱塞在下止点位置时高压油泵里燃油的密度，记为ρ_1。

也就是说此时柱塞在下止点时燃油的质量为：

$$m_3=V_2\times\rho_1 \tag{24}$$

凸轮开始旋转，凸轮的旋转周期是下止点 → 上止点 → 下止点，高压油泵的压力达到 100Mpa 开始向高压油管里注油，以这个状态作为初始状态，将凸轮旋转一周的时间分为若干个时间段，每个时间段为 0.001ms。当经过 0.001ms 时，油泵供给 A 口的油量为$m_i(i=1,2,3\ldots)$，柱塞上升，高压油泵剩余体积改变，剩余燃油的压力改变，因此剩余燃油的密度也会随之改变。循环此过程，直到$\sum_{i=1}m_i$与喷油口喷出油的质量值出现一个最小公倍数质量时，出现此最小公倍数所耗用的时间，就是使高压油管的压力稳定在 100 MPa 时凸轮的旋转周期。

角速度公式为：

$$\omega=\frac{T}{2\pi} \tag{25}$$

5.3 模型求解

首先计算出喷油的体积$V_{出}=38.1714mm^3$，因此$m_2=32.4457$mg,根据题意计算凸轮旋转一周质量减少$\sum_{i=1} m_i$=75.4037mg。利用 MATLAB 计算$\sum_{i=1} m_i$与喷油口喷出油的质量值出现一个最小公倍数质量，可得该质量为2.44655×10^7mg。易知，喷油口想要喷出该质量需要 754030 个周期，即时间为 75403000ms，凸轮应旋转 324450 圈才可以注入该质量的油。所以角速度为 0.027rad/ms。

5.4 问题三模型建立

问题三要求在问题二的基础上，再增加一个喷油嘴，每个喷嘴喷油规律相同，调整喷油和供油策略。每个喷嘴喷油规律相同，喷口 B 和喷口 C 喷油质量相等，在问题二的计算基础上，此时喷口的质量为$2m_2$，为了保持高压油管内压力维持在某一个数值，根据质量守恒定律，$M_{进}$=$M_{出}$，高压油泵也要提供相同的燃油质量，从而可推导出凸轮的转速是问题二转速的两倍。

绘制出不同喷油和供油策略下的假想图：

（1）喷口 B 和喷口 C 同时喷油假想图：

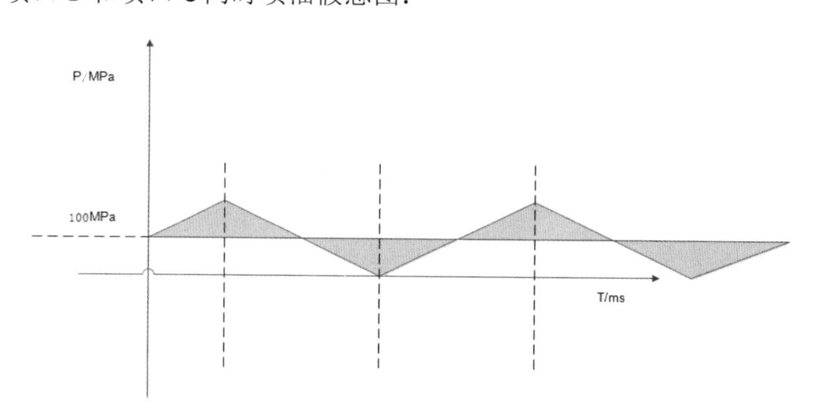

图表 10 喷口 B 和喷口 C 同时喷油假想图

（2）喷口 B 和喷口 C 错时喷油假想图：

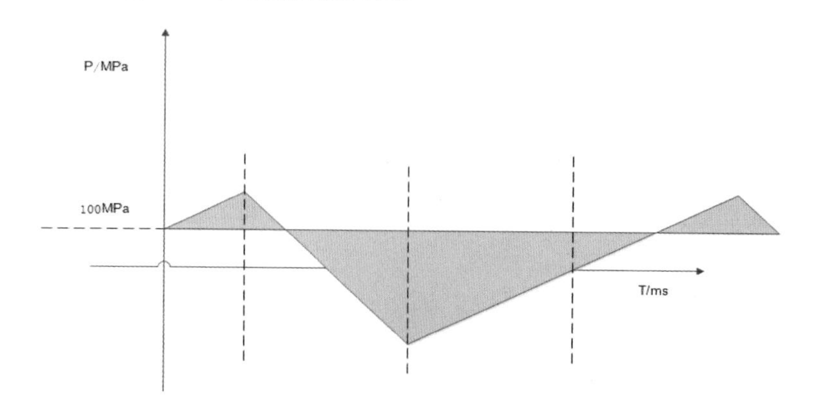

图表 11 喷口 B 和喷口 C 错时喷油假想图

两张假想图根据相同比例绘制，因此可以直观地看出，喷口 B 和喷口 C 错时喷油使高压油管内的压力在某一定值上下波动幅度小。

针对第二小问，由于缺乏数据，在这里给出一个具体的思路：要求在增加一个单向减压阀使高压油管内的压力减小，使其继续稳定。为使单向减压阀在工作时，并且对高压油管的压强不造成太大影响，凸轮的角速度增加到原来的二倍，当处在两喷油口都处在停止喷油的状态下，高压油泵向高压油管供油，这时单向减压阀应当开始工作来维持高压油管的压力稳定。

6 模型评价

6.1 模型优点分析

1.在模型假设时,对模型进行了合理假设，使得在计算机上模拟时得到简化，更易验证出模型的正确性。

6.2 模型缺点分析

1.在模型二中，将进油/喷油周期的高压油管密度用进油/喷油开始时刻的高压油管密度代替，结束后发生变化，由于在进油/喷油过程中高压油管的密度也会有轻微的变化，在模型二进行计算机模拟求解时，会有轻微误差

2.在模型三中，要重新计算流量公式$Q=CA\sqrt{\frac{2\Delta P}{\rho}}$，$\Delta P=100\ MPa-0\ MPa$，此时假设了与喷油嘴连接的空间内部压力为 0 MPa，在实际运用中，与喷油嘴连接的空间内部压力为肯定大于 0 MPa，因此该假设会使被喷油的质量偏大，造成误差。

参考文献

[1]周丽[1]，黄素珍[2]. 基于模拟退火的混合遗传算法研究[J]. 计算机应用研究, 2005(9).

附录:

1.test.m

```
%用来计算出压强和密度的关系
%并且模拟了第一题第二小问
data=xlsread('abc.xlsx');
x = data(:,1);%压力
y = data(:,2); %E
[fitresult, gof] = createFit(x,y);
rhospan = [0.85 0.9];
p0 = 100;
[rho,p] = ode15s(@(rho,x) (fitresult.p1*x^5 + fitresult.p2*x^4 +
fitresult.p3*x^3 +...
    fitresult.p4*x^2 + fitresult.p5*x^1 + fitresult.p6)/rho, rhospan, p0);
figure;plot(rho,p);xlabel('燃油密度\rho');ylabel('压力p')%拟合图像

[fitr2, ~] = createFit1(p, rho);%拟合关系
[fitr3, ~] = createFit2(rho, p);%拟合关系
```

2.test2.m

```
%通过修改t的值，从而观察sumt的值，sumt即为通过多少时间可以达到
endp = 100;
C=0.85;
A=pi*0.7*0.7;
t = 0.809;
sumt = 0;
count = 0;
while endp < 150
```

```
    endrho=(fitr2.p1*endp^4 + fitr2.p2*endp^3+fitr2.p3*endp^2+fitr2.p4*endp +
fitr2.p5);
    Q = C*A*sqrt(2*(160-endp)/endrho);
    endrho = (Q*t*endrho+500*pi*5*5*endrho)/(500*pi*5*5);
    endp = (fitr3.p1*endrho^4 +
fitr3.p2*endrho^3+fitr3.p3*endrho^2+fitr3.p4*endrho + fitr3.p5);
    sumt = sumt+t;
    sumt = sumt+10;
    if(sumt > count * 100+50)
        count = count + 1;
        endrho = (500*pi*5*5*endrho - 44 * endrho)/(500*pi*5*5);
        endp = (fitr3.p1*endrho^4 +
fitr3.p2*endrho^3+fitr3.p3*endrho^2+fitr3.p4*endrho + fitr3.p5);
    end
    sprintf('%f\n', endp)
end
```

3.test3.m

```
%第二问漏油
data=xlsread('b.xlsx');%把文件附件2-针阀运动曲线数据读入（已改名为b）
t = data(:,1);
h = data(:,2);
r = (7.8922 + h) * 0.1584;%h和r的关系式
A = min(pi*r.*r - pi * 1.25^2, pi*0.7^2);%计算流过的
C = 0.85;
Q = C * A * sqrt(2 * 100 / 0.85);
plot(t, Q);
xlabel('时间t');
ylabel('流量Q');
sumans = 0;
t1=data(1:38, 1);
h1=data(1:38, 2);
r = (7.8922 + h1) * 0.1584;
A = min(pi*r.*r - pi * 1.25^2, pi*0.7^2);
Q1 = C * A * sqrt(2 * 100 / 0.85);
[xData, yData] = prepareCurveData(t1,Q1);
ft = fittype( 'poly5' );
[fitresult11, ~] = fit( xData, yData, ft);%拟合出t和q的关系
syms x;
y = fitresult11.p1*x^5 + fitresult11.p2*x^4 + fitresult11.p3*x^3 +
fitresult11.p4*x^2 + fitresult11.p5*x + fitresult11.p6;
sumans = sumans + int(y,x,0, 0.37);%积分
```

```
t2=data(39:209, 1);
h2=data(39:209, 2);
r = (7.8922 + h2) * 0.1584;
A = min(pi*r.*r - pi * 1.25^2, pi*0.7^2);
Q2 = C * A * sqrt(2 * 100 / 0.85);
syms x;
y = 20.0711+ 0 * x;
plot(t2, Q2);
sumans = sumans + int(y,x,0.38, 2.06);%积分

t3=data(209:246, 1);
h3=data(209:246, 2);
r = (7.8922 + h3) * 0.1584;
A = min(pi*r.*r - pi * 1.25^2, pi*0.7^2);
Q3 = C * A * sqrt(2 * 100 / 0.85);
[xData, yData] = prepareCurveData(t3,Q3);%拟合出t和q的关系
ft = fittype( 'poly5' );
[fitresult33, gof] = fit( xData, yData, ft);
syms x;
y = fitresult33.p1*x^5 + fitresult33.p2*x^4 + fitresult33.p3*x^3 +
fitresult33.p4*x^2 + fitresult33.p5*x + fitresult33.p6;
sumans = sumans + int(y,x,2.08, 2.45);%积分
```

4.test4.m

```
%运行过test1才可以运行test4，test1用来生成压强和密度的拟合

w = 0.1;
data = xlsread('a.xlsx');
the = data(:,1);
rh = data(:,2);
[xData, yData] = prepareCurveData(the,rh);
ft = fittype( 'poly8' );
[fitresult4, ~] = fit( xData, yData, ft);%拟合出极角与极径的关系图
```

5.test5.m

```
%模拟气缸压气的过程
th = pi - 2.63;
tendp = 100;
sumt = th/w;
h = 5.5365;
tendrh = 0.85;
s = 6.25*pi;
```

```
m = 92.4025;
C = 0.85;
A = pi * 0.7 * 0.7;
while tendp >= 100
    Q = C * A * sqrt(2*(tendp - 100)/tendrh);

    tm = Q * 0.0001 * tendrh;
    m = m - tm;
    h0 = fitresult4.p1*th^8 + fitresult4.p2*th^7 + fitresult4.p3*th^6 +
fitresult4.p4*th^5 ...
        + fitresult4.p5*th^4 + fitresult4.p6*th^3 + fitresult4.p7*th^2 +
fitresult4.p8*th + fitresult4.p9;
    th = th + 0.0001*w;
    h1 = fitresult4.p1*th^8 + fitresult4.p2*th^7 + fitresult4.p3*th^6 +
fitresult4.p4*th^5 ...
        + fitresult4.p5*th^4 + fitresult4.p6*th^3 + fitresult4.p7*th^2 +
fitresult4.p8*th + fitresult4.p9;
    ttt = h1 - h0;
    tendrh = m/(s*h);
    tendp =
fitr3.p1*tendrh^4+fitr3.p2*tendrh^3+fitr3.p3*tendrh^2+fitr3.p4*tendrh+fitr3
.p5;
    h = h  - h1 + h0;
    sumt = sumt + 0.0001;
end
```

6. paint.m

```
%画个图
data=xlsread('a.xlsx');
x = data(:,1);
y = data(:,2);
plot(x, y);
```

7. createFit2.m

```
function [fitrt, gof] = createFit2(p, rho)
[x, y] = prepareCurveData(p, rho);
ft = fittype( 'poly4' );
[fitrt, gof] = fit( x, y, ft );
figure( 'Name', 'rho-\p relation' );
h = plot( fitrt, x, y);
legend( h, 'rho vs. p', 'rho-\p relation', 'Location', 'NorthEast',
```

```
'Interpreter', 'none' );
xlabel( '燃油密度', 'Interpreter', 'none' );
ylabel( '压力', 'Interpreter', 'none' );
grid on
```

createFit1.m

```
function [fiti, gof] = createFit1(rho, p)
[x, y] = prepareCurveData( rho, p );
ft = fittype( 'poly4' );
[fiti, gof] = fit( x, y, ft );
figure( 'Name', 'p-\rho relation' );
h = plot( fiti, x, y);
legend( h, 'p vs. rho', 'p-\rho relation', 'Location', 'NorthEast',
'Interpreter', 'none' );
xlabel( '压力', 'Interpreter', 'none' );
ylabel( '燃油密度', 'Interpreter', 'none' );
grid on
```

createFit.m

```
function [fitrt, gof] = createFit(p, E)
[x, y] = prepareCurveData( p, E);
ft = fittype( 'poly5' );
[fitrt, gof] = fit( x, y, ft );
figure( 'Name', '拟合直线1' );
h = plot( fitrt, x, y);
legend( h, 'E vs. p', '拟合直线1', 'Location', 'NorthEast', 'Interpreter',
'none' );
xlabel( 'p', 'Interpreter', 'none' );
ylabel( 'E', 'Interpreter', 'none' );
grid on
```

2018年高教社杯全国大学生数学建模竞赛题目

（请先阅读“全国大学生数学建模竞赛论文格式规范”）

A题　高温作业专用服装设计

在高温环境下工作时，人们需要穿着专用服装以避免灼伤。专用服装通常由三层织物材料构成，记为I、II、III层，其中I层与外界环境接触，III层与皮肤之间还存在空隙，将此空隙记为IV层。

为设计专用服装，将体内温度控制在37ºC的假人放置在实验室的高温环境中，测量假人皮肤外侧的温度。为了降低研发成本、缩短研发周期，请你们利用数学模型来确定假人皮肤外侧的温度变化情况，并解决以下问题：

（1）专用服装材料的某些参数值由附件1给出，对环境温度为75ºC、II层厚度为6 mm、IV层厚度为5 mm、工作时间为90分钟的情形开展实验，测量得到假人皮肤外侧的温度（见附件2）。建立数学模型，计算温度分布，并生成温度分布的Excel文件（文件名为problem1.xlsx）。

（2）当环境温度为65ºC、IV层的厚度为5.5 mm时，确定II层的最优厚度，确保工作60分钟时，假人皮肤外侧温度不超过47ºC，且超过44ºC的时间不超过5分钟。

（3）当环境温度为80ºC时，确定II层和IV层的最优厚度，确保工作30分钟时，假人皮肤外侧温度不超过47ºC，且超过44ºC的时间不超过5分钟。

附件1. 专用服装材料的参数值

附件2. 假人皮肤外侧的测量温度

2018国赛链接

2018年高教社杯全国大学生数学建模竞赛题目

问题B　智能RGV的动态调度策略

图1是一个智能加工系统的示意图，由8台计算机数控机床（Computer Number Controller，CNC）、1辆轨道式自动引导车（Rail Guide Vehicle，RGV）、1条RGV直线轨道、1条上料传送带、1条下料传送带等附属设备组成。RGV是一种无人驾驶、能在固定轨道上自由运行的智能车。它根据指令能自动控制移动方向和距离，并自带一个机械手臂、两只机械手爪和物料清洗槽，能够完成上下料及清洗物料等作业任务（参见附件1）。

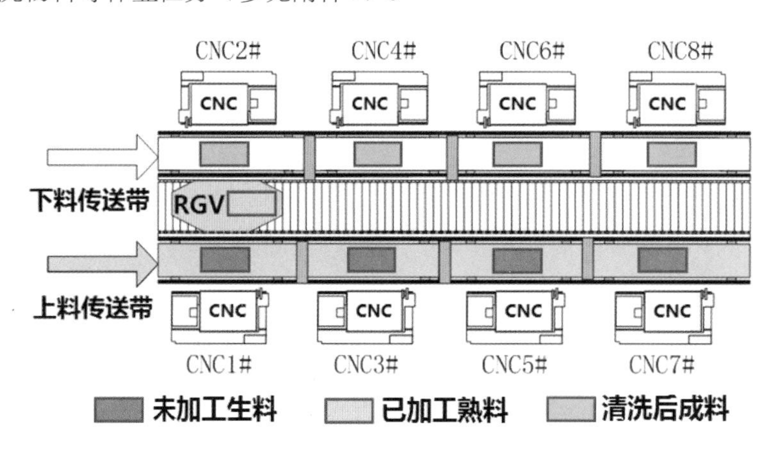

图1　智能加工系统示意图

针对下面的三种具体情况：

（1）一道工序的物料加工作业情况，每台CNC安装同样的刀具，物料可以在任一台CNC上加工完成；

（2）两道工序的物料加工作业情况，每个物料的第一和第二道工序分别由两台不同的CNC依次加工完成；

（3）CNC在加工过程中可能发生故障（据统计：故障的发生概率约为1%）的情况，每次故障排除（人工处理，未完成的物料报废）时间介于10~20分钟之间，故障排除后即刻加入作业序列。要求分别考虑一道工序和两道工序的物料加工作业情况。

请你们团队完成下列两项任务：

任务1：对一般问题进行研究，给出RGV动态调度模型和相应的求解算法；

任务2：利用表1中系统作业参数的3组数据分别检验模型的实用性和算法的有效性，给出RGV的调度策略和系统的作业效率，并将具体的结果分别填入附件2的EXCEL表中。

表1：智能加工系统作业参数的3组数据表　　　　时间单位：秒

系统作业参数	第1组	第2组	第3组
RGV移动1个单位所需时间	20	23	18
RGV移动2个单位所需时间	33	41	32
RGV移动3个单位所需时间	46	59	46
CNC加工完成一个一道工序的物料所需时间	560	580	545
CNC加工完成一个两道工序物料的第一道工序所需时间	400	280	455
CNC加工完成一个两道工序物料的第二道工序所需时间	378	500	182
RGV为CNC1#，3#，5#，7#一次上下料所需时间	28	30	27
RGV为CNC2#，4#，6#，8#一次上下料所需时间	31	35	32
RGV完成一个物料的清洗作业所需时间	25	30	25

注：每班次连续作业8小时。

附件1：智能加工系统的组成与作业流程

附件2：模型验证结果的EXCEL表（完整电子表作为附件放在支撑材料中提交）

附件1：智能加工系统的组成与作业流程

1. 系统的场景及实物图说明

在附图1中，中间设备是自带清洗槽和机械手的轨道式自动引导车RGV，清洗槽每次只能清洗1个物料，机械手臂前端有2个手爪，通过旋转可以先后各抓取1个物料，完成上下料作业。两边排列的是CNC，每台CNC前方各安装有一段物料传送带。右侧为上料传送带，负责为CNC输送生料（未加工的物料）；左边为下料传送带，负责将成料（加工并清洗完成的物料）送出系统。其他为保证系统正常运行的辅助设备。

附图1：RGV—CNC车间布局图

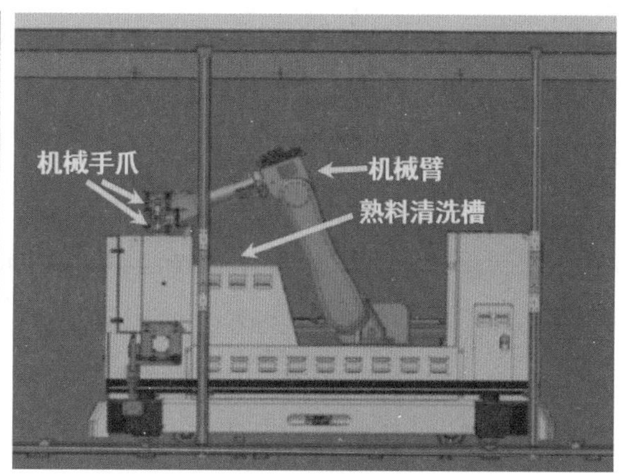

附图2：带机械手臂和清洗槽的RGV实物图

附图2是RGV的实物图，包括车体、机械臂、机械手爪和物料清洗槽等。

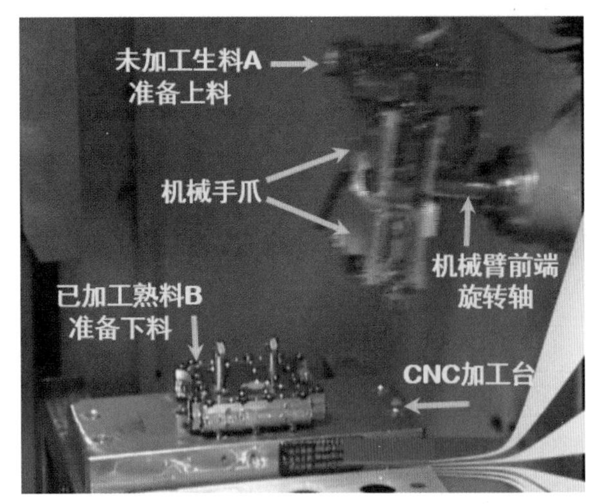

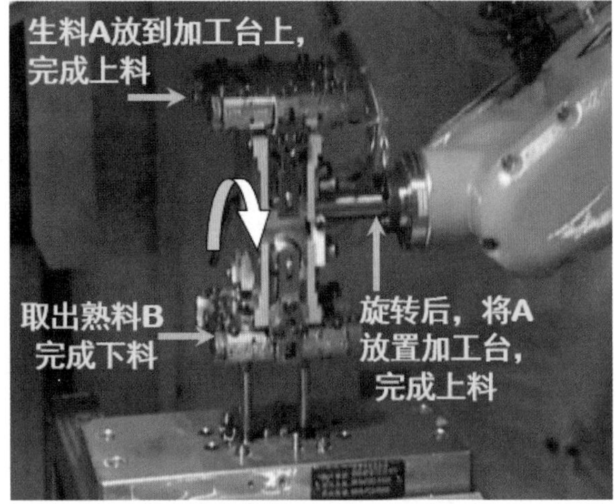

附图3：RGV机械手臂前端的2个手爪实物图

在附图3左图中，机械臂前端上方手爪抓有1个生料A，CNC加工台上有1个熟料B。RGV机械臂移动到CNC加工台上方，机械臂下方空置的手爪准备抓取熟料B，在抓取了熟料B后即完成下料作业。

在附图3右图中，RGV机械臂下方手爪已抓取了CNC加工台上的熟料B抬高手臂，并旋转手爪，将生料A对准加工位置，安放到CNC加工台上，即完成上料作业。

2. 系统的构成及说明

智能加工系统由8台CNC、1台带机械手和清洗槽的RGV、1条RGV直线轨道、1条上料传送带和1条下料传送带等附属设备构成。

（1）CNC：在上料传送带和下料传送带的两侧各安装4台CNC，等距排列，每台CNC同一时间只能安装1种刀具加工1个物料。

如果物料的加工过程需要两道工序，则需要有不同的CNC安装不同的刀具分别加工完成，在加工过程中不能更换刀具。第一和第二道工序需要在不同的CNC上依次加工完成，完成时间也不同，每台CNC只能完成其中的一道工序。

（2）RGV：RGV带有智能控制功能，能够接收和发送指令信号。根据指令能在直线轨道上移动和停止等待，可连续移动1个单位（两台相邻CNC间的距离）、2个单位（三台相邻CNC间的距离）和3个单位（四台相邻CNC间的距离）。RGV同一时间只能执行移动、停止等待、上下料和清洗作业中的一项。

（3）上料传送带：上料传送带由4段组成，在奇数编号CNC1#、3#、5#、7#前各有1段。由系统传感器控制，只能向一个方向传动，既能连动，也能独立运动。

（4）下料传送带：下料传送带由4段组成，在偶数编号CNC2#、4#、6#、8#前各有1段。由传感器控制，只能向同一个方向传动，既能连动，也能独立运动。

3. 系统的作业流程

（1）智能加工系统通电启动后，RGV在CNC1#和CNC2#正中间的初始位置，所有CNC都处于空闲状态。

（2）在工作正常情况下，如果某CNC处于空闲状态，则向RGV发出上料需求信号；否则，

CNC处于加工作业状态，在加工作业完成即刻向RGV发出需求信号。

（3）RGV在收到某CNC的需求信号后，它会自行确定该CNC的上下料作业次序，并依次按顺序为其上下料作业。根据需求指令，RGV运行至需要作业的某CNC处，同时上料传送带将生料送到该CNC正前方，供RGV上料作业。

RGV为偶数编号CNC一次上下料所需时间要大于为奇数编号CNC一次上下料所需时间。

（4）在RGV为某CNC完成一次上下料作业后，就会转动机械臂，将一只机械手上的熟料移动到清洗槽上方，进行清洗作业（只清洗加工完成的熟料）。

具体过程：首先用另一只机械手抓取出清洗槽中的成料、转动手爪、放入熟料到清洗槽中，然后转动机械臂，将成料放到下料传送带上送出系统。这个作业过程所需要的时间称为RGV清洗作业时间，并且在这个过程中RGV不能移动。

熟料在清洗槽中的实际清洗时间是很短的，远小于机械手将成料放到下料传送带上的时间。

（5）RGV在完成一项作业任务后，立即判别执行下一个作业指令。此时，如果没有接到其他的作业指令，则RGV就在原地等待直到下一个作业指令。

某CNC完成一个物料的加工作业任务后，即刻向RGV发出需求信号。如果RGV没能即刻到达为其上下料，该CNC就会出现等待。

（6）系统周而复始地重复（3）至（5），直到系统停止作业，RGV回到初始位置。

2019高教社杯全国大学生数学建模竞赛题目

（请先阅读“全国大学生数学建模竞赛论文格式规范”）

A题　高压油管的压力控制

燃油进入和喷出高压油管是许多燃油发动机工作的基础，图1给出了某高压燃油系统的工作原理，燃油经过高压油泵从A处进入高压油管，再由喷口B喷出。燃油进入和喷出的间歇性工作过程会导致高压油管内压力的变化，使得所喷出的燃油量出现偏差，从而影响发动机的工作效率。

图1　高压油管示意图

问题1. 某型号高压油管的内腔长度为500 mm，内直径为10 mm，供油入口A处小孔的直径为1.4 mm，通过单向阀开关控制供油时间的长短，单向阀每打开一次后就要关闭10 ms。喷油器每秒工作10次，每次工作时喷油时间为2.4 ms，喷油器工作时从喷油嘴B处向外喷油的速率如图2所示。高压油泵在入口A处提供的压力恒为160 MPa，高压油管内的初始压力为100 MPa。如果要将高压油管内的压力尽可能稳定在100 MPa左右，如何设置单向阀每次开启的时长？如果要将高压油管内的压力从100 MPa增加到150 MPa，且分别经过约2 s、5 s和10 s的调整过程后稳定在150 MPa，单向阀开启的时长应如何调整？

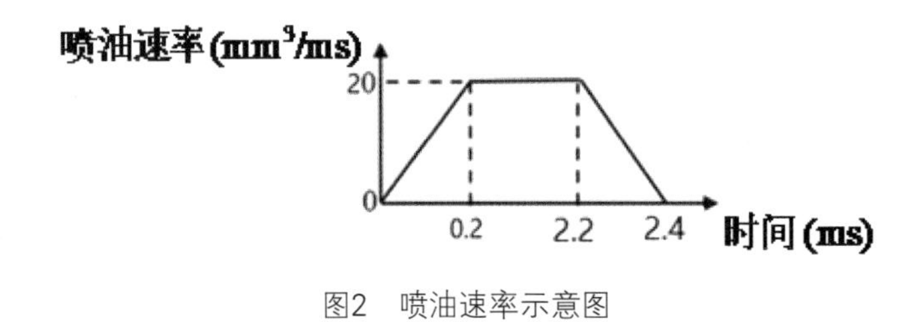

图2　喷油速率示意图

问题2. 在实际工作过程中，高压油管A处的燃油来自高压油泵的柱塞腔出口，喷油由喷油嘴的针阀控制。高压油泵柱塞的压油过程如图3所示，凸轮驱动柱塞上下运动，凸轮边缘曲线与角度的关系见附件1。柱塞向上运动时压缩柱塞腔内的燃油，当柱塞腔内的压力大于高压油管内的压力时，柱塞腔与高压油管连接的单向阀开启，燃油进入高压油管内。柱塞腔内直径为5 mm，柱塞运动到上止点位置时，柱塞腔残余容积为20 mm^3。柱塞运动到下止点时，低压燃油会充满柱塞腔（包括残余容积），低压燃油的压力为0.5 MPa。喷油器喷嘴结构如图4所示，针阀直径为2.5 mm、密封座是半角为9°的圆锥，最下端喷孔的直径为1.4 mm。针阀升程为0时，针阀关闭；针阀升程大于0时，针阀开启，燃油向喷孔流动，通过喷孔喷出。在一个喷油周期内针阀升程与时间的关系由附件2给出。在问题1中给出的喷油器工作次数、高压油管尺寸和初始压力下，确定凸轮的角速度，使得高压油管内的压力尽量稳定在100 MPa左右。

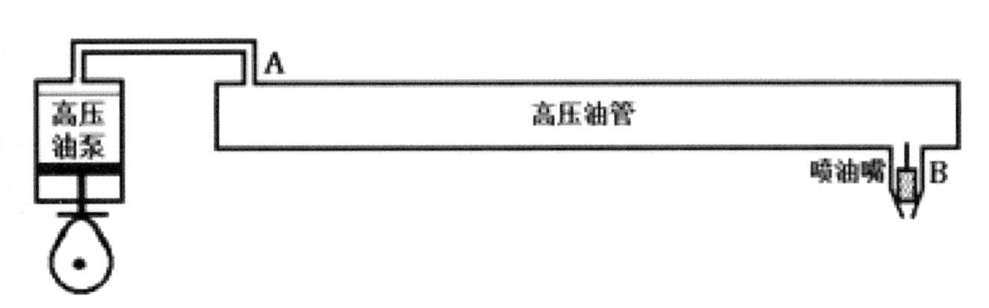

图3　高压油管实际工作过程示意图

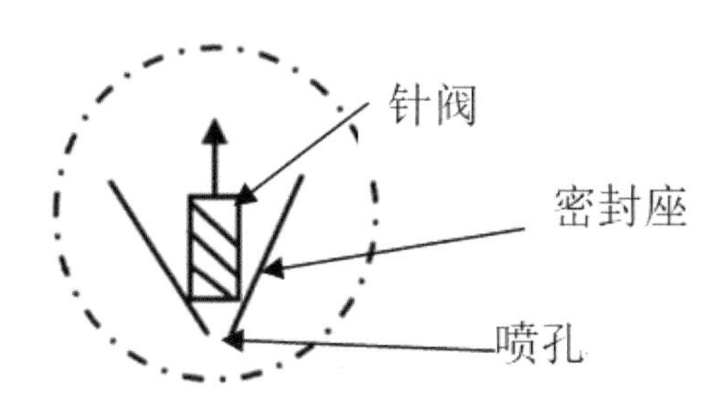

图4　喷油器喷嘴放大后的示意图

问题3. 在问题2的基础上，再增加一个喷油嘴，每个喷嘴喷油规律相同，喷油和供油策略应如何调整？为了更有效地控制高压油管的压力，现计划在D处安装一个单向减压阀（图5）。单向减压阀出口为直径为1.4 mm的圆，打开后高压油管内的燃油可以在压力下回流到外部低压油路中，从而使得高压油管内燃油的压力减小。请给出高压油泵和减压阀的控制方案。

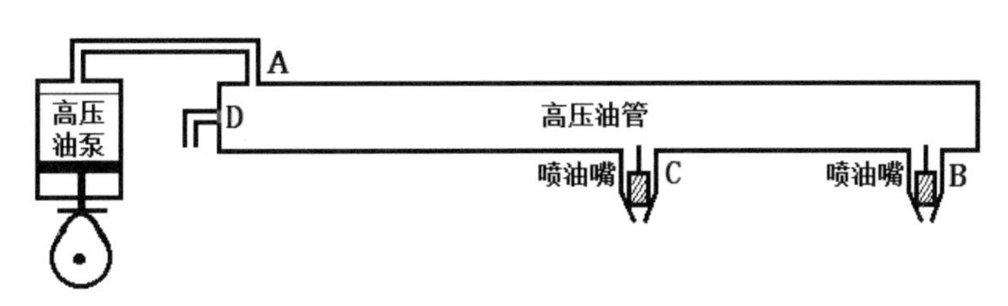

图5　具有减压阀和两个喷油嘴时高压油管示意图

注1. 燃油的压力变化量与密度变化量成正比，比例系数为$\frac{E}{\rho}$，其中ρ为燃油的密度，当压力为100 MPa时，燃油的密度为0.850 mg/mm^3。E为弹性模量，其与压力的关系见附件3。

注2. 进出高压油管的流量为$Q=CA\sqrt{\frac{2\Delta P}{\rho}}$，其中$Q$为单位时间流过小孔的燃油量（$mm^3/ms$），$C$=0.85为流量系数，$A$为小孔的面积（$mm^2$），ΔP为小孔两边的压力差（MPa），$\rho$为高压侧燃油的密度（$mg/mm^3$）。

附件1：凸轮边缘曲线

附件2：针阀运动曲线

附件3：弹性模量与压力的关系

2019国赛链接

2019高教社杯全国大学生数学建模竞赛题目

（请先阅读“全国大学生数学建模竞赛论文格式规范”）

B题　“同心协力”策略研究

“同心协力”（又称“同心鼓”）是一项团队协作能力拓展项目。该项目的道具是一面牛皮双面鼓，鼓身中间固定多根绳子，绳子在鼓身上的固定点沿圆周呈均匀分布，每根绳子长度相同。团队成员每人牵拉一根绳子，使鼓面保持水平。项目开始时，球从鼓面中心上方竖直落下，队员同心协力将球颠起，使其有节奏地在鼓面上跳动。颠球过程中，队员只能抓握绳子的末端，不能接触鼓或绳子的其他位置。

图片来源：https://yjs.syu.edu.cn/_mediafile/yjs/2017/10/26/32yuesec78.png

项目所用排球的质量为270 g。鼓面直径为40 cm，鼓身高度为22 cm，鼓的质量为3.6 kg。队员人数不少于8人，队员之间的最小距离不得小于60 cm。项目开始时，球从鼓面中心上方40 cm处竖直落下，球被颠起的高度应离开鼓面40 cm以上，如果低于40cm，则项目停止。项目的目标是使得连续颠球的次数尽可能多。

试建立数学模型解决以下问题：

1. 在理想状态下，每个人都可以精确控制用力方向、时机和力度，试讨论这种情形下团队的最佳协作策略，并给出该策略下的颠球高度。

2. 在现实情形中，队员发力时机和力度不可能做到精确控制，存在一定误差，于是鼓面可能出现倾斜。试建立模型描述队员的发力时机和力度与某一特定时刻的鼓面倾斜角度的关系。设队员人数为8，绳长为1.7 m，鼓面初始时刻是水平静止的，初始位置较绳子水平时下降11 cm，表1中给出了队员们的不同发力时机和力度，求0.1 s时鼓面的倾斜角度。

表1 发力时机（单位：s）和用力大小（单位：N）取值

序号	用力参数	1	2	3	4	5	6	7	8	鼓面倾角（度）
1	发力时机	0	0	0	0	0	0	0	0	
	用力大小	90	80	80	80	80	80	80	80	
2	发力时机	0	0	0	0	0	0	0	0	
	用力大小	90	90	80	80	80	80	80	80	
3	发力时机	0	0	0	0	0	0	0	0	
	用力大小	90	80	80	90	80	80	80	80	
4	发力时机	–0.1	0	0	0	0	0	0	0	
	用力大小	80	80	80	80	80	80	80	80	
5	发力时机	–0.1	–0.1	0	0	0	0	0	0	
	用力大小	80	80	80	80	80	80	80	80	
6	发力时机	–0.1	0	0	–0.1	0	0	0	0	
	用力大小	80	80	80	80	80	80	80	80	
7	发力时机	–0.1	0	0	0	0	0	0	0	
	用力大小	90	80	80	80	80	80	80	80	
8	发力时机	0	–0.1	0	0	–0.1	0	0	0	
	用力大小	90	80	80	90	80	80	80	80	
9	发力时机	0	0	0	0	–0.1	0	0	–0.1	
	用力大小	90	80	80	90	80	80	80	80	

3. 在现实情形中，根据问题2的模型，你们在问题1中给出的策略是否需要调整？如果需要，如何调整？

4. 当鼓面发生倾斜时，球跳动方向不再竖直，于是需要队员调整拉绳策略。假设人数为10，绳长为2 m，球的反弹高度为60 cm，相对于竖直方向产生1度的倾斜角度，且倾斜方向在水平面的投影指向某两位队员之间，与这两位队员的夹角之比为1∶2。为了将球调整为竖直状态弹跳，请给出在可精确控制条件下所有队员的发力时机及力度，并分析在现实情形中这种调整策略的实施效果。

2019高教社杯全国大学生数学建模竞赛题目

（请先阅读“全国大学生数学建模竞赛论文格式规范”）

问题C　机场的出租车问题

大多数乘客下飞机后要去市区（或周边）的目的地，出租车是主要的交通工具之一。国内多数机场都是将送客（出发）与接客（到达）通道分开的。送客到机场的出租车司机都将会面临两个选择：

（A）前往到达区排队等待载客返回市区。出租车必须到指定的“蓄车池”排队等候，依“先来后到”排队进场载客，等待时间长短取决于排队出租车和乘客的数量多少，需要付出一定的时间成本。

（B）直接放空返回市区拉客。出租车司机会付出空载费用和可能损失潜在的载客收益。

在某时间段抵达的航班数量和“蓄车池”里已有的车辆数是司机可观测到的确定信息。通常司机的决策与其个人的经验判断有关，比如在某个季节与某时间段抵达航班的多少和可能乘客数量的多寡等。如果乘客在下飞机后想“打车”，就要到指定的“乘车区”排队，按先后顺序乘车。机场出租车管理人员负责“分批定量”放行出租车进入“乘车区”，同时安排一定数量的乘客上车。在实际中，还有很多影响出租车司机决策的确定和不确定因素，其关联关系各异，影响效果也不尽相同。

请你们团队结合实际情况，建立数学模型研究下列问题：

（1）分析研究与出租车司机决策相关因素的影响机理，综合考虑机场乘客数量的变化规律和出租车司机的收益，建立出租车司机选择决策模型，并给出司机的选择策略。

（2）收集国内某一机场及其所在城市出租车的相关数据，给出该机场出租车司机的选择方案，并分析模型的合理性和对相关因素的依赖性。

（3）在某些时候，经常会出现出租车排队载客和乘客排队乘车的情况。某机场“乘车区”现有两条并行车道，管理部门应如何设置“上车点”，并合理安排出租车和乘客，在保证车辆和乘客安全的条件下，使得总的乘车效率最高。

（4）机场的出租车载客收益与载客的行驶里程有关，乘客的目的地有远有近，出租车司机不能选择乘客和拒载，但允许出租车多次往返载客。管理部门拟对某些短途载客再次返回的出租车给予一定的“优先权”，使得这些出租车的收益尽量均衡，试给出一个可行的“优先”安排方案。

2018 MCM
Problem A: Multi-hop HF Radio Propagation

Background: On high frequencies (HF, defined to be 3 – 30 mHz), radio waves can travel long distances (from one point on the earth's surface to another distant point on the earth's surface) by multiple reflections off the ionosphere and off the earth. For frequencies below the *maximum usable frequency* (MUF), HF radio waves from a ground source reflect off the ionosphere back to the earth, where they may reflect again back to the ionosphere, where they may reflect again back to the earth, and so on, travelling further with each successive hop. Among other factors, the characteristics of the reflecting surface determine the strength of the reflected wave and how far the signal will ultimately travel while maintaining useful signal integrity. Also, the MUF varies with the season, time of day, and solar conditions. Frequencies above the MUF are not reflected/refracted, but pass through the ionosphere into space. In this problem, the focus is particularly on reflections off the ocean surface. It has been found empirically that reflections off a turbulent ocean are attenuated more than reflections off a calm ocean. Ocean turbulence will affect the electromagnetic gradient of seawater, altering the local permittivity and permeability of the ocean, and changing the height and angle of the reflection surface. A turbulent ocean is one in which wave heights, shapes, and frequencies change rapidly, and the direction of wave travel may also change.

Problem:

Part I: Develop a mathematical model for this signal reflection off the ocean. For a 100-watt HF constant-carrier signal, below the MUF, from a point source on land, determine the strength of the first reflection off a turbulent ocean and compare it with the strength of a first reflection off a calm ocean. (Note that this means that there has been one reflection of this signal off the ionosphere.) If additional reflections (2 through n) take place off calm oceans, what is the maximum number of hops the signal can take before its strength falls below a usable signal-to-noise ratio (SNR) threshold of 10 dB?

Part II: How do your findings from **Part I** compare with HF reflections off mountainous or rugged terrain versus smooth terrain?

Part III: A ship travelling across the ocean will use HF for communications and to receive weather and traffic reports. How does your model change to accommodate a shipboard receiver moving on a turbulent ocean? How long can the ship remain in communication using the same multi-hop path?

Part IV: Prepare a short (1 to 2 pages) synopsis of your results suitable for publication as a short note in *IEEE Communications Magazine.*

Your submission should consist of:

- One-page Summary Sheet,
- Two-page synopsis,
- Your solution of no more than 20 pages, for a maximum of 23 pages with your summary and synopsis.
- Note: Reference list and any appendices do not count toward the 23-page limit and should appear after your completed solution.

2018 MCM
Problem B: How Many Languages?

Background: There are currently about 6,900 languages spoken on Earth. About half the world's population claim one of the following ten languages (in order of most speakers) as a native language: Mandarin (incl. Standard Chinese), Spanish, English, Hindi, Arabic, Bengali, Portuguese, Russian, Punjabi, and Japanese. However, much of the world's population also speaks a second language. When considering total numbers of speakers of a particular language (native speakers plus second or third, etc. language speakers), the languages and their order change from the native language list provided. The total number of speakers of a language may increase or decrease over time because of a variety of influences to include, but not limited to, the language(s) used and/or promoted by the government in a country, the language(s) used in schools, social pressures, migration and assimilation of cultural groups, and immigration and emigration with countries that speak other languages. Moreover, in our globalized, interconnected world there are additional factors that allow languages that are geographically distant to interact. These factors include international business relations, increased global tourism, the use of electronic communication and social media, and the use of technology to assist in quick and easy language translation.

Native Language Rank	Native Language	Family	Native Speakers	Second (or 3rd, etc) Language Speakers	Second Language Rank	Total
1	Mandarin Chinese (incl. Standard Chinese)	Sino-Tibetan, Sinitic	897 million	193 million	4	1.09 billion
2	Spanish	Indo-European, Romance	436 million	91 million	8	527 million
3	English	Indo-European, Germanic	371 million	611 million	1	983 million
4	Hindustani (Hindi/Urdu)	Indo-European, Indo-Aryan	329 million	215 million	2	544 million
5	Arabic	Afro-Asiatic, Semitic	290 million (2017)	132 million	6	422 million
6	Bengali	Indo-European, Indo-Aryan	242 million	19 million in Bangladesh (2011)	13	261 million
7	Portuguese	Indo-European, Romance	218 million	11 million	15	229 million
8	Russian	Indo-European, Slavic	153 million	113 million (2010)	7	267 million
9	Punjabi	Indo-European, Indo-Aryan	148 million	?	?	148 million
10	Japanese	Japonic	128 million	1 million (2010)	19	129 million

Retrieved from https://en.wikipedia.org/wiki/List_of_languages_by_total_number_of_speakers on January 17, 2018.

Problem: A large multinational service company, with offices in New York City in the United States and Shanghai in China, is continuing to expand to become truly international. This company is investigating opening additional international offices and desires to have the employees of each office speak both in English and one or more additional languages. The Chief Operating Officer of the company has hired your team to investigate trends of global languages and location options for new offices.

Part I:

A. Consider the influences and factors described in the background paragraph above, as well as other factors your group may identify. Based on projected trends, and some or all of these influences and factors, model the distribution of various language speakers over time.

B. Use your model to predict what will happen to the numbers of native speakers and total language speakers in the next 50 years. Do you predict that any of the languages in the current top-ten lists (either native speakers or total speakers) will be replaced by another language? Explain.

C. Given the global population and human migration patterns predicted for the next 50 years, do the geographic distributions of these languages change over this same period of time? If so, describe the change.

Part II:

A. Based on your modeling from **Part I**, and assuming your client company wants to open six new international offices, where might you locate these offices and what languages would be spoken in the offices? Would your recommendations be different in the short term versus the long term? Explain your choices.

B. Considering the changing nature of global communications, and in an effort to save your client company resources, might you suggest that the company open less than six international offices? Indicate what additional information you would need and describe how you would analyze this option in order to advise your client.

Part III:

Write a 1-2 page memo to the Chief Operating Officer of the service company summarizing your results and recommendations.

Note: In your analysis, ignore unpredictable or high-impact, low probability events such as asteroid collisions that would cause a catastrophic jump in evolutionary trends over time, and possibly render all languages extinct.

Your submission should consist of:

- One-page Summary Sheet,
- Two-page memo,
- Your solution of no more than 20 pages, for a maximum of 23 pages with your summary and memo.
- Note: Reference list and any appendices do not count toward the 23-page limit and should appear after your completed solution.

Attachments:
List of Languages by Total Numbers of Speakers

References:

Lane, J. (2017). *The 10 Most Spoken Languages in the World.* Babbel Magazine. Retrieved from https://www.babbel.com/en/magazine/the-10-most-spoken-languages-in-the-world

Noack, R. and Gamio, L. (April 23, 2015). *The World's Languages in 7 Maps and Charts.* The Washington Post. Retrieved from https://www.washingtonpost.com/news/worldviews/wp/2015/04/23/the-worlds-languages-in-7-maps-and-charts/?utm_term=.a993dc2a15cb

List of Languages by Total Numbers of Speakers
https://en.wikipedia.org/wiki/List_of_languages_by_total_number_of_speakers

List of Languages by Total Numbers of Speakers. Retrieved from https://en.wikipedia.org/wiki/List_of_languages_by_total_number_of_speakers **on January 17, 2018.**

Ethnologue (2017 20th edition)

The following 26 languages are listed as having 50 million or more total speakers in the 2017 edition of *Ethnologue*, a language reference published by SIL International based in the United States[2] (although *Ethnologue* also lists more than only these 26 languages as having 50 million or more total speakers, e.g., the Wikipedia page for the Tagalog language reports 70+ million speakers by as early as 2000 and 73+ million speakers by 2013: 28 million L1 speakers as of 2007 and 45 million L2 speakers as of 2013; these are largely based on *Ethnologue* reports and would, e.g., unless further updated, rank it as the language with the 26th most L1 speakers, the 13th most L2 speakers, and the 23rd most speakers in total). Speaker totals are generally not reliable, as they add together estimates from different dates and (usually uncited) sources; language information is not collected on most national censuses.

Rank	Language	Family	L1 speakers	L1 Rank	L2 speakers	L2 Rank	Total
1	**Mandarin Chinese**(incl. Standard Chinese)	Sino-Tibetan, Sinitic	897 million	1	193 million	4	**1.09 billion**
2	**English**	Indo-European, Germanic	371 million	3	611 million	1	**983 million**
3	**Hindustani (Hindi/Urdu)**[Note 1]	Indo-European, Indo-Aryan	329 million	4	215 million	2	**544 million**
4	**Spanish**	Indo-European, Romance	436 million	2	91 million	8	**527 million**
5	**Arabic**	Afro-Asiatic, Semitic	290 million (2017)	5	132 million	6	**422 million**[5][6]

Rank	Language	Family	L1 speakers	L1 Rank	L2 speakers	L2 Rank	Total
6	**Malay** (incl. Indonesian and Malaysia n)	Austronesian, Malayo-Polynesian	77 million (2007)	15	204 million	3	**281 million**[7]
7	**Russian**	Indo-European, Slavic	153 million	8	113 million (2010)	7	**267 million**
8	**Bengali**	Indo-European, Indo-Aryan	242 million	6	19 million in Bangladesh (2011)	13	**261 million**
9	**Portuguese**	Indo-European, Romance	218 million	7	11 million	15	**229 million**
10	**French**	Indo-European, Romance	76 million	17	153 million	5	**229 million**
11	**Hausa**	Afro-Asiatic, Chadic	85 million	11	65 million	10	**150 million**[8]
12	**Punjabi**	Indo-European, Indo-Aryan	148 million[9]	9	?	?	**148 million**
13	**Japanese**	Japonic	128 million	10	1 million (2010)[10]	19	**129 million**

Rank	Language	Family	L1 speakers	L1 Rank	L2 speakers	L2 Rank	Total
14	**German**	Indo-European, Germanic	76 million	18	52 million	12	**129 million**
15	**Persian**	Indo-European, Iranian	60 million (2009)	25	61 million[11]	11	**121 million**[11]
16	**Swahili**	Niger–Congo language, Coastal Tanzanian, Bantu	16 million	26	91 million	8	**107 million**
17	**Telugu**	Dravidian	80 million (2011)	13	12 million in India (2011)	14	**92 million**
18	**Javanese**	Austronesian, Malayo-Polynesian	84 million (2000)	12	?	?	**84 million**
19	**Wu Chinese** (incl. Shanghainese)	Sino-Tibetan, Chinese	80 million (2013)	14	?	?	**80 million**
20	**Korean**	Koreanic	77 million (2008–2010)	16	?	?	**77 million**
21	**Tamil**	Dravidian	67 million (2001)	23	8 million in India	16	**75 million**

Rank	Language	Family	L1 speakers	L1 Rank	L2 speakers	L2 Rank	Total
22	**Marathi**	Indo-European, Indo-Aryan	71 million (2001)	20	3 million in India	17	**74 million**
23	**Yue Chinese** (incl. Cantonese)	Sino-Tibetan, Chinese	72 million	19	?	?	**72 million**
24	**Turkish**	Turkic, Oghuz	71 million	21	<1 million	20	**71 million**
25	**Vietnamese**	Austroasiatic, Viet–Muong	68 million	22	?	?	**68 million**
26	**Italian**	Indo-European, Romance	63 million	24	3 million	17	**66 million**

Notes

1. Refers to *Modern Standard Hindi* and *Modern Standard Urdu*. Modern Hindi and Urdu are mutually intelligible and are considered by linguists to be dialects of the same language; the two distinct registers are the outcome of nationalist tendencies.[3] The Census of India defines Hindi on a loose and broad basis. In addition to Standard Hindi, it incorporates a set of other Indo-Aryan languages written in Devanagari script including Awadhi, Bhojpuri, Haryanvi, Dhundhari etc. under Hindi group which have more than 422 million native speakers as on 2001.[4] However, the census also acknowledges Standard Hindi, the above mentioned languages and others as separate mother tongues of Hindi language and provides individual figures for all these languages.[4]

References

1. *Crystal, David (March 2008). "Two thousand million?". English Today. doi:10.1017/S0266078408000023.*

2. *"Summary by language size". Ethnologue. Retrieved 2016-04-06.*

3. Abdul Jamil Khan (2006). Urdu/Hindi: an artificial divide. Algora. p. 290. ISBN 978-0-87586-437-2.

4. Abstract of speakers' strength of languages and mother tongues – 2000, Census of India, 2001

5. "Världens 100 största språk 2010" (The World's 100 Largest Languages in 2010), in Nationalencyklopedin

6. *"World Arabic Language Day | United Nations Educational, Scientific and Cultural Organization". www.unesco.org. Retrieved 2017-07-14.*

7. Indonesia 258 million (World Bank, 2015); Malaysia 19.4 million Bumiputera (Dept of Statistics, Malaysia, 2016); Brunei 0.43 million (World Bank, 2015); Singapore 0.5 million (University of Hawaii 2012); Thailand 3 million (University of Hawaii, 2012)

8. *"Hausa speakers in Nigeria now 120m– Communique - Vanguard News". vanguardngr.com. Retrieved 2017-04-26.*

9. Lahnda/Western Punjabi 116.6 million Pakistan (c. 2014). Eastern Punjabi: 28.2 million India (2001), other countries: 1.1 million. *Ethnologue* 19.

10. *"Japanese". Ethnologue. Retrieved 2016-03-07.*

11. Windfuhr, Gernot: *The Aryan Languages*, Routledge 2009,

2018 MCM
Problem C: Energy Production

Background: Energy production and usage are a major portion of any economy. In the United States, many aspects of energy policy are decentralized to the state level. Additionally, the varying geographies and industries of different states affect energy usage and production. In 1970, 12 western states in the U.S. formed the Western Interstate Energy Compact (WIEC), whose mission focused on fostering cooperation between these states for the development and management of nuclear energy technologies. An interstate compact is a contractual arrangement made between two or more states in which these states agree on a specific policy issue and either adopt a set of standards or cooperate with one another on a particular regional or national matter.

Problem: Along the U.S. border with Mexico, there are four states – California (CA), Arizona (AZ), New Mexico (NM), and Texas (TX) – that wish to form a realistic new energy compact focused on increased usage of cleaner, renewable energy sources. Your team has been asked by the four governors of these states to perform data analysis and modeling to inform their development of a set of goals for their interstate energy compact.

The attached data file "ProblemCData.xlsx" provides in the first worksheet ("**seseds**") 50 years of data in 605 variables on each of these four states' energy production and consumption, along with some demographic and economic information. The 605 variable names used in this dataset are defined in the second worksheet ("**msncodes**").

Part I:
A. Using the data provided, create an energy profile for each of the four states.

B. Develop a model to characterize how the energy profile of each of the four states has evolved from 1960 – 2009. Analyze and interpret the results of your model to address the four states' usage of cleaner, renewable energy sources in a way that is easily understood by the governors and helps them to understand the similarities and difference between the four states. Include in your discussion possible influential factors of the similarities and differences (e.g. geography, industry, population, and climate).

C. Determine which of the four states appeared to have the "best" profile for use of cleaner, renewable energy in 2009. Explain your criteria and choice.

D. Based on the historical evolution of energy use in these states, and your understanding of the differences between the state profiles you established, predict the energy profile of each state, as you have defined it, for 2025 and 2050 in the absence of any policy changes by each governor's office.

Part II:

A. Based on your comparison between the four states, your criteria for "best" profile, and your predictions, determine renewable energy usage targets for 2025 and 2050 and state them as goals for this new four-state energy compact.

B. Identify and discuss at least three actions the four states might take to meet their energy compact goals.

Part III:

Prepare a one-page memo to the group of Governors summarizing the state profiles as of 2009, your predictions with regard to energy usage absent any policy changes, and your recommended goals for the energy compact to adopt.

Your submission should consist of:

- One-page Summary Sheet,
- One-page memo,
- Your solution of no more than 20 pages, for a maximum of 22 pages with your summary and memo.
- Note: Reference list and any appendices do not count toward the 22-page limit and should appear after your completed solution.

Attachments:

ProblemCData.xlsx

Includes two worksheets **seseds** and **msncodes**.

References:

State Energy Data System (SEDS) Complete Dataset through 2009 (All 50 states)
https://catalog.data.gov/dataset/state-energy-data-system-seds-complete-dataset-through-2009#sec-dates

2018 ICM
Problem D: Out of Gas and Driving on E (for electric, not empty)

For both environmental and economic reasons, there is global interest in reducing the use of fossil fuels, including gasoline for cars. Whether motivated by the environment or by the economics, consumers are starting to migrate to electric vehicles. Several countries are seeing early signs of the potential for rapid growth in the adoption of electric vehicles. In the US and other countries, the release of the more affordable all-electric Tesla Model 3 has resulted in record numbers of pre-orders and long wait lists (https://www.wired.com/story/tesla-model-3-delivery-timeline/). To further accelerate the switch to electric vehicles, some countries, including China, have announced that they will ban gasoline and diesel cars in the coming years (http://money.cnn.com/2017/09/11/news/china-gas-electric-car-ban/index.html).

Eventually, when a ban goes into effect, there needs to be a sufficient number of vehicle charging stations in all the right places so that people can use their vehicles for their daily business, as well as make occasional long-distance trips. The migration from gasoline and diesel cars to electric vehicles, however, is not simple and can't happen overnight. In a fantasy world, we would wake up one day with every gas vehicle replaced by an electric one, and every gas station replaced with a charging station. In reality, there are limited resources, and it will take time for consumers to make the switch. In fact, the location and convenience of charging stations is critical as early adopters and eventually mainstream consumers volunteer to switch (http://www.govtech.com/fs/Building-Out-Electric-Vehicle-Infrastructure-Where-Are-the-Best-Locations-for-Charging-Stations.html).

As nations plan this transition, they need to consider the final network of charging stations (the number of stations, where they will be located, the number of chargers at the stations, and the differences in the needs of rural areas, suburban areas, and urban areas), as well as the growth and evolution of the network of charging stations over time. For example, what should the network look like when electric vehicles represent 10% of all cars, 30% of all cars, 50% of all cars, and 90% of all cars?

As nations seek to develop policies that promote the migration towards electric vehicles, they will need to design a plan that works best for their individual country. Before they can begin, they would like your team's help in determining the final architecture of the charging network to support the full adoption of all-electric vehicles. Additionally, they would like you to identify the key factors that will be important as they plan their timeline for an eventual ban or dramatic reduction of gasoline and diesel vehicles.

To help your team manage the scope of this problem, we ask that you focus only on personal passenger vehicles (i.e. cars, vans, and light trucks used for passengers). At the end of your report, you may briefly comment on the relevance of your findings and conclusions on commercial vehicles to include heavy trucks and busses.

Your tasks are the following:

Task 1: Explore the current and growing network of Tesla charging stations in the United States. Tesla currently offers two types of charging stations: (1) destination charging designed for charging for several hours at a time or even overnight (https://www.tesla.com/destination-charging); and (2) supercharging designed for longer road trips to provide up to 170 miles of range in as little as 30 minutes of charging (https://www.tesla.com/supercharger). These stations are in addition to at-home charging used by many Tesla owners who have a personal garage or a driveway with power. Is Tesla on track to allow a complete switch to all-electric in the US? If everyone switched to all-electric personal passenger vehicles in the US, how many charging stations would be needed, and how should they be distributed between urban, suburban, and rural areas?

Task 2: Select one of the following nations (South Korea, Ireland, or Uruguay).

2a. Determine the optimal number, placement, and distribution of charging stations if your country could migrate all their personal passenger vehicles to all-electric vehicles instantaneously (no transition time required). What are the key factors that shaped the development of your plan?

2b. While these countries have already started installing chargers, you get to start with a clean slate. Present a proposal for evolving the charging network of your chosen country from zero chargers to a full electric-vehicle system. How do you propose the country invest in chargers? Should the country build all city-based chargers first, or all rural chargers, or a mix of both? Will you build the chargers first and hope people buy the cars, or will you build chargers in response to car purchases? What are the key factors that shaped your proposed charging station plan?

2c. Based on your growth plan, what is the timeline you propose for the full evolution to electric vehicles in your country? To get started, you may wish to consider how long it will take for there to be 10% electric vehicles, 30% electric vehicles, 50% electric vehicles, or 100% electric vehicles on your selected country's roads. What are the key factors that shaped your proposed growth plan timeline?

Task 3: Now consider countries with very different geographies, population density distributions, and wealth distributions, such as Australia, China, Indonesia, Saudi Arabia, and Singapore. Would your proposed plan for growing and evolving the network of chargers still apply to each of these countries? What are the key factors that trigger the selection of different approaches to growing the network? Discuss the feasibility of creating a classification system that would help a nation determine the general growth model they should follow in order for them to successfully migrate away from gasoline and diesel vehicles to all electric cars.

Task 4: The technological world continues to change and is impacting transportation options such as car-share and ride-share services, self-driving cars, rapid battery-swap stations for electric cars, and even flying cars and a Hyperloop. Comment on how these technologies might impact your analyses of the increasing use of electric vehicles.

Task 5: Prepare a one-page handout written for the leaders of a wide range of countries who are attending an international energy summit. The handout should identify the key factors the leaders should consider as they return to their home country to develop a national plan to migrate personal transportation towards all-electric cars and set a gas vehicle-ban date.

Your submission should consist of:

- One-page Summary Sheet,
- One-page handout,
- Your solution of no more than 20 pages, for a maximum of 22 pages with your summary and handout.
- Note: Reference list and any appendices do not count toward the 22-page limit and should appear after your completed solution.

2018 ICM
Problem E: How does climate change influence regional instability?

The effects of Climate Change, to include increased droughts, shrinking glaciers, changing animal and plant ranges, and sea level rise, are already being realized and vary from region to region. The Intergovernmental Panel on Climate Change suggests that the net damage costs of climate change are likely to be significant. Many of these effects will alter the way humans live, and may have the potential to cause the weakening and breakdown of social and governmental structures. Consequently, destabilized governments could result in fragile states.

A fragile state is one where the state government is not able to, or chooses not to, provide the basic essentials to its people. For the purpose of this problem "state" refers to a sovereign state or country. Being a fragile state increases the vulnerability of a country's population to the impact of such climate shocks as natural disasters, decreasing arable land, unpredictable weather, and increasing temperatures. Non-sustainable environmental practices, migration, and resource shortages, which are common in developing states, may further aggravate states with weak governance (Schwartz and Randall, 2003; Theisen, Gleditsch, and Buhaug, 2013). Arguably, drought in both Syria and Yemen further exacerbated already fragile states. Environmental stress alone does not necessarily trigger violent conflict, but evidence suggests that it enables violent conflict when it combines with weak governance and social fragmentation. This confluence can enhance a spiral of violence, typically along latent ethnic and political divisions (Krakowka, Heimel, and Galgano 2012).

Your tasks are the following:

Task 1: Develop a model that determines a country's fragility and simultaneously measures the impact of climate change. Your model should identify when a state is fragile, vulnerable, or stable. It should also identify how climate change increases fragility through direct means or indirectly as it influences other factors and indicators.

Task 2: Select one of the top 10 most fragile states as determined by the Fragile State Index (http://fundforpeace.org/fsi/data/) and determine how climate change may have increased fragility of that country. Use your model to show in what way(s) the state may be less fragile without these effects.

Task 3: Use your model on another state not in the top 10 list to measure its fragility, and see in what way and when climate change may push it to become more fragile. Identify any definitive indicators. How do you define a tipping point and predict when a country may reach it?

Task 4: Use your model to show which state driven interventions could mitigate the risk of climate change and prevent a country from becoming a fragile state. Explain the effect of human intervention and predict the total cost of intervention for this country.

Task 5: Will your model work on smaller "states" (such as cities) or larger "states" (such as continents)? If not, how would you modify your model?

Your submission should consist of:

- One-page Summary Sheet,
- Your solution of no more than 20 pages, for a maximum of 21 pages with your summary.
- Note: Reference list and any appendices do not count toward the 21-page limit and should appear after your completed solution.

References:

Krakowka, A.R., Heimel, N., and Galgano, F. "Modeling Environmenal Security in Sub-Sharan Africa – ProQuest." The Geographical Bulletin, 2012, 53 (1): 21-38.

Schwartz, P. and Randall, D. "An Abrupt Climate Change Scenario and Its Implications for United States National Security", October 2003. http://eesc.columbia.edu/courses/v1003/readings/Pentagon.pdf

Theisen, O.M., Gleditsch, N.P., and Buhaug, H. "Is climate change a driver of armed conflict?" *Climate Change*, April 2013, V117 (3), 613-625.

Helpful Links:

Fragile States Index: http://fundforpeace.org/fsi/

The World Bank: http://www.worldbank.org/en/topic/fragilityconflictviolence/brief/harmonized-list-of-fragile-situations

2018 ICM
Problem F: Cost of Privacy

Pervasiveness of, and reliance on, electronic communication and social media have become widespread. One result is that some people seem willing to share private information (PI) about their personal interactions, relationships, purchases, beliefs, health, and movements, while others hold their privacy in these areas as very important and valuable. There are also significant differences in privacy choices across various domains. For example, some people are quick to give away the protection of their purchasing information for a quick price reduction, but at the same time are unlikely to share information about their disease conditions or health risks. Similarly, some populations or subgroups may be less willing to give up particular types of personal information if they perceive it posing a personal or community risk. The risk may involve loss of safety, money, valuable items, intellectual property (IP), or the person's electronic identity. Other risks include professional embarrassment, loss of a position or job, social loss (friendships), social stigmatization, or marginalization. While a government employee who has voiced political dissent against the government might be willing to pay to keep their social media data private, a young college student may feel no pressure to restrict their posting of political opinion or social information. It seems that individual choices on PI protection and internet and system security in cyber space can create risks and rewards in elements of freedom, privacy, convenience, social standing, financial benefits, and medical treatment.

Is private information (PI) similar to private personal property (PP) and intellectual property (IP)? Once lawfully obtained, can PI be sold or given to others who then have the right or ownership of the information? As detailed information and meta-data of human activity becomes more and more valuable to society, specifically in the areas of medical research, disease spread, disaster relief, businesses (e.g. marketing, insurance, and income), records of personal behaviors, statements of beliefs, and physical movement, these data and detailed information may become a valuable and quantifiable commodity. Trading in one's own private data comes with a set of risks and benefits that may differ by the domain of information (e.g. purchasing, social media, medical) and by subgroup (e.g. citizenship, professional profile, age).

Can we quantify the cost of privacy of electronic communications and transactions across society? That is, what is the monetary value of keeping PI protected, or how much would it cost for others to have or use PI? Should the government regulate this information or is it better left to privacy industry or the individual? Are these information and privacy issues merely personal decisions that individuals must evaluate to make their own choices and provide their own protection?

There are several things to consider when evaluating the cost of privacy. First, is data sharing a public good? For example, Center for Disease Control may use the data to trace the spread of disease in order to prevent further outbreak. Other examples include managing at risk populations, such as children under 16, people at risk of suicide, and the elderly. Moreover, consider groups of extremists who seek to hide their activities. Should their data be trackable by the government for national security concerns? Consider a person's browser, phone system, and internet feed with their personalized advertisements; how much is this customization worth?

Overall, when evaluating cost of privacy we need to consider all of these tradeoffs. What is the potential gain from keeping data private and what is lost by doing so?

As a policy analysis team for a national decision maker, your team's tasks are:

Task 1: Develop a price point for protecting one's privacy and PI in various applications. To evaluate this, you may want to categorize individuals into subgroups with reasonably similar levels of risk or into related domains of the data. What are the set of parameters and measures that would need to be considered to accurately model risk to account for both 1) characteristics of the individuals, and 2) characteristics of the specific domain of information?

Task 2: Given the set of parameters and measures from Task 1, model for cost of privacy across at least three domains (social media, financial transactions, and health/medical records). In your base model consider how the tradeoffs and risks of keeping data protected affect your model. You may consider giving some of the tradeoffs and risks more weight than others as well as stratifying weights by subgroup or category. Consider how different basic elements of the data (e.g. name, date of birth, gender, social security or citizenship number) contribute to your model. Are some of these elements worth more than others? For example, what is the value of a name alone compared with value of a name with the person's picture attached? Your model should design a pricing structure for PI.

Task 3: Not long ago, people had no knowledge about which agencies had purchased their PI, how much their PI was worth, or how PI was being used. New proposals are being put forth which would turn PI into a commodity. With the pricing structure you generated in Task 2, establish a pricing system for individuals, groups, and entire nations. With data becoming a commodity subject to market fluctuations, is it appropriate to consider forces of supply and demand for PI? Assuming people have control to sell to their own data, how does this change the model?

Task 4: What are the assumptions and constraints of your model? Assumptions and constraints should address issues such as government regulations (e.g. price regulations, specific data protections such as certain records that may not be subject to the economic system) and cultural and political issues. Based on your model and the political and cultural issues, consider if information privacy should be made a basic human right when thinking about policy recommendations. Consider introducing a dynamic element to your model by introducing the variations over time in human decision-making given changing personal beliefs about the worth of their own data (e.g. personal data such as name, address, picture), transaction data (e.g. on-line purchases, search history), and social media data (e.g. posts, pictures).

Task 5: Are there generational differences in perceptions of the risk-to-benefit ratio of PI and data privacy? As generations age, how does this change the model? How is PI different or similar to PP and IP?

Task 6: What are the ways to account for the fact that human data is highly linked and often each individual's behaviors are highly correlated with others? Data on one person can provide information about others whom they are socially, professionally, economically, or demographically connected. Therefore, personal decisions to share one's own data can affect countless others. Are there good ways to capture the network effects of data sharing? Does that effect the price system for individuals, subgroups, and entire communities and nations? If communities have shared privacy risks, is it the responsibility of the communities to protect citizens' PI?

Task 7: Consider the effects of a massive data breach where millions of people's PI are stolen and sold on the dark web, sold as part of an identity theft ring, or used as ransom. How does such a PI loss or cascade event impact your model? Now that you have a pricing system that quantifies the value of data per individual or loss type, are agencies that are to blame for the data breach responsible to pay individuals directly for misuse or loss of PI?

Task 8: Write a two-page policy memo to the decision maker on the utility, results, and recommendations based your policy modeling on this issue. Be sure to specify what types of PI are included in your recommendations.

Your submission should consist of:

- One-page Summary Sheet,
- Two-page memo,
- Your solution of no more than 20 pages, for a maximum of 23 pages with your summary and memo.
- Note: Reference list and any appendices do not count toward the 23-page limit and should appear after your completed solution.

2018−MCM−Problem−DATA

2019 MCM
Problem A: Game of Ecology

In the fictional television series *Game of Thrones*, based on the series of epic fantasy novels *A Song of Ice and Fire*[1], three dragons are raised by Daenerys Targaryen, the "Mother of Dragons." When hatched, the dragons are small, roughly 10 kg, and after a year grow to roughly 30-40 kg. They continue to grow throughout their life depending on the conditions and amount of food available to them.

For the purposes of this problem, consider these three fictional dragons are living today. Assume that the basic biology of dragons described above is accurate. You will need to make some additional assumptions about dragons that might include, for example, that dragons are able to fly great distances, breath fire, and resist tremendous trauma. As you address the problem requirements, it should be clear how your assumptions are related to the physical constraints of the functions, size, diet, changes, or other characteristics associated with the animals.

Your team is assigned to analyze dragon characteristics, behavior, habits, diet, and interaction with their environment. To do so, you will have to consider many questions. At a minimum, address the following: What is the ecological impact and requirements of the dragons? What are the energy expenditures of the dragons, and what are their caloric intake requirements? How much area is required to support the three dragons? How large a community is necessary to support a dragon for varying levels of assistance that can be provided to the dragons? Be clear about what factors you are considering when addressing these questions.

As with other animals that migrate, dragons might travel to different regions of the world with very different climates. How important are the climate conditions to your analysis? For example, would moving a dragon between an arid region, a warm temperate region, and an arctic region make a big difference in the resources required to maintain and grow a dragon?

Once your dragon analysis is complete, draft a two-page letter to the author of *A Song of Ice and Fire*, George R.R. Martin, to provide guidance about how to maintain the realistic ecological underpinning of the story, especially with respect to the movement of dragons from arid regions to temperate regions and to arctic regions.

While your dragon analysis does not directly apply to a real physical situation, the mathematical modeling itself makes use of many realistic features used in modeling a situation. Aside from the modeling activities themselves, describe and discuss a situation outside of the realm of fictional dragons that your modeling efforts might help inform and provide insight?

Your submission should consist of:

- One-page Summary Sheet,
- Two-page letter,
- Your solution of no more than 20 pages, for a maximum of 23 pages with your summary and letter.
- Note: Reference list and any appendices do not count toward the 23-page limit and should appear after your completed solution.

NOTE: You should not make use of unauthorized images and materials whose use is restricted by copyright laws. Please be careful in how you use and cite the sources for your ideas and the materials used in your report.

Reference

1. Penguin Random House (2018). *A Song of Ice and Fire Series*. Retrieved from https://www.penguinrandomhouse.com/series/SOO/a-song-of-ice-and-fire/.

2019 MCM
Problem B: Send in the Drones: Developing an Aerial Disaster Relief Response System

Background: In 2017, the worst hurricane to ever hit the United States territory of Puerto Rico (see Attachment 1) left the island with severe damage and caused over 2900 fatalities. The combined destructive power of the hurricane's storm surge and wave action produced extensive damage to buildings, homes, and roads, particularly along the east and southeast coast of Puerto Rico. The storm, with its fierce winds and heavy rain, knocked down 80 percent of Puerto Rico's utility poles and all transmission lines, resulting in loss of power to essentially all of the island's 3.4 million residents. In addition, the storm damaged or destroyed the majority of the island's cellular communication networks. The electrical power and cell service outages lasted for months across much of the island, and longer in some locations. Widespread flooding blocked and damaged many highways and roads across the island, making it nearly impossible for emergency services ground vehicles to plan and navigate their routes. The full extent of the damage in Puerto Rico remained unclear for some time; dozens of areas were isolated and without communication. Demands for medical supplies, lifesaving equipment, and treatment strained health-care clinics, hospital emergency rooms, and *non-governmental organizations' (NGOs)* relief operations. Demand for medical care continued to surge for some time as the chronically ill turned to hospitals and temporary shelters for care.

Problem: Non-governmental organizations (NGOs) are often challenged to provide adequate and timely response during or after natural disasters, such as the hurricane that struck the United States territory of Puerto Rico in 2017. One NGO in particular – HELP, Inc. - is attempting to improve its response capabilities by designing a transportable disaster response system called "DroneGo." DroneGo will use rotor wing *drones* to deliver pre-packaged medical supplies and provide high-resolution aerial video reconnaissance. Selected drones should be able to perform these two missions – medical supply delivery and video reconnaissance – simultaneously or separately, depending on relief conditions and scheduling. HELP, Inc. has identified various candidate rotor wing drones that it would like your team to consider for possible use in designing its *DroneGo fleet* (see Attachments 2, 3).

DroneGo's pre-packaged medical supplies, called *medical packages*, are meant to augment, not replace, the supplies provided by local medical assistance organizations on-site within the country affected by the disaster. HELP, Inc. is planning on three different medical packages referred to as MED1, MED2, and MED3. Drones will carry these medical packages within *drone cargo bays* for delivery to selected locations (see Attachments 4, 5). Depending on the specific drone being used to transport medical supplies, it may be possible that multiple medical packages can be transported in a single drone cargo bay. Note that drones must land on the ground to offload medical supplies from the drone cargo bays. The video capability of the drones will provide high-resolution video of damaged and serviceable transportation road networks to HELP, Inc.'s command and control center for ground-based route planning.

HELP, Inc. will use International Standards Organization (ISO) standard dry *cargo containers* to quickly transport a complete DroneGo disaster response system to a particular disaster area. The individual shipping containers for all drones in the DroneGo fleet, along with all required medical packages, must fit within a maximum of three of the ISO cargo containers to be delivered to a single location, or up to three different locations if three cargo containers are used in the disaster area. Each shipping container's contents should be packed in order to minimize any need for buffer materials for unused space. Table 1 shows the dimensions of an ISO standard dry cargo container.

Table 1. Standard ISO Container Dimensions

	Exterior			Interior			Door Opening	
	Length	Width	Height	Length	Width	Height	Width	Height
20' Standard Dry Container	20'	8'	8'6"	19'3"	7'8"	7' 10"	7'8"	7'5"

HELP, Inc. is asking your team to use the 2017 situation in Puerto Rico to design a DroneGo disaster response system that will fit within the containers noted while meeting the anticipated medical supply demands during a potential similar future disaster scenario. It is possible that the demand requirements of this scenario may exceed the capabilities of the drone fleet your team identifies. If this occurs, HELP, Inc. wants to clearly understand any tradeoffs that it must make for implementing solutions to address these shortcomings.

Part 1. Develop a DroneGo disaster response system to support the Puerto Rico hurricane disaster scenario.

Consider the background information, the requirements identified in the problem statement, and the information provided in the problem attachments to address the following.

A. Recommend a drone fleet and set of medical packages for the HELP, Inc. DroneGo disaster response system that will meet the requirements of the Puerto Rico hurricane scenario. Design the associated packing configuration for each of up to three ISO cargo containers to transport the system to Puerto Rico.

B. Identify the best location or locations on Puerto Rico to position one, two, or three cargo containers of the DroneGo disaster response system to be able to conduct both medical supply delivery and video reconnaissance of road networks.

C. For each type of drone included in the DroneGo fleet:

a. Provide the *drone payload packing configurations* (i.e. the medical packages packed into the drone cargo bay), delivery routes and schedule to meet the identified emergency medical package requirements of the Puerto Rico hurricane scenario.

b. Provide a drone flight plan that will enable the DroneGo fleet to use onboard video cameras to assess the major highways and roads in support of the Help, Inc. mission.

Part 2. Memo

Write a 1–2 page memo to the Chief Operating Officer (CEO) of HELP, Inc. summarizing your modeling results, conclusions, and recommendations so that she can share with her Board of Directors.

Your MCM team submission should consist of:

- One-page Summary Sheet,
- One- to Two-page memo to the HELP, Inc. CEO
- Your solution of no more than 20 pages, for a maximum of 23 pages with your summary and memo.
- Note: Reference list and any appendices do not count toward the 23-page limit and should appear after your completed solution.

Attachments:

1. Map of Puerto Rico
2. Potential Candidate Drones for DroneGo Fleet Consideration (with Drone payload capability)
3. Drone Cargo Bay Packing Configuration/Dimensions by Type
4. Anticipated Medical Package Demand
5. Emergency Medical Package Configuration/Dimensions

Attachment 1: Map of Puerto Rico

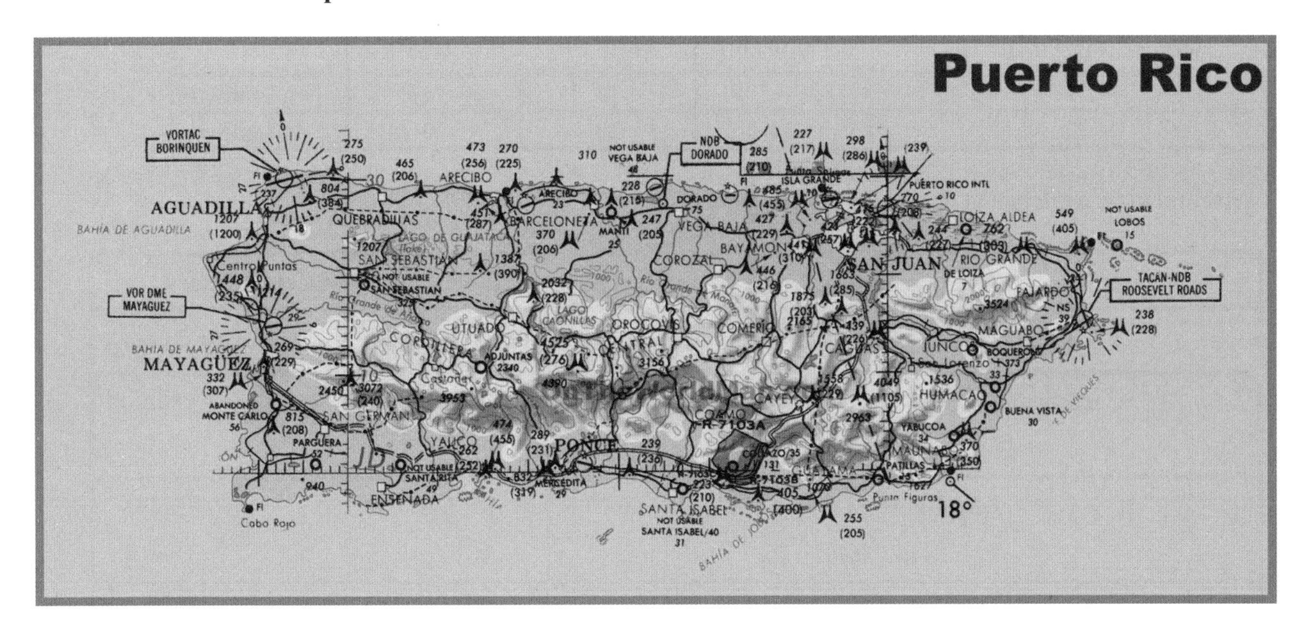

Attachment 2: Potential Candidate Drones for DroneGo Fleet Consideration (with Drone *Payload Capability)*

	Shipping Container Dimensions			Performance Characteristics/Capabilities			Configurations Capabilities		
Drone	Length (in.)	Width (in.)	Height (in.)	Max Payload Capability (lbs.)	Speed (km/h)	Flight Time No Cargo (min)	Video Capable	Medical Package Capable	Drone Cargo Bay Type*
A	45	45	25	3.5	40	35	Y	Y	1
B	30	30	22	8	79	40	Y	Y	1
C	60	50	30	14	64	35	Y	Y	2
D	25	20	25	11	60	18	Y	Y	1
E	25	20	27	15	60	15	Y	Y	2
F	40	40	25	22	79	24	N	Y	2
G	32	32	17	20	64	16	Y	Y	2
H Tethered	65	75	41	N/A	N/A	Indefinite	N	N	N/A

*Note that cargo bays are affixed to the drone and that drone must be on the ground to offload cargo. See Attachment 3 for Drone Cargo Bay Type Configuration/Dimensions.

Attachment 3: Drone Cargo Bay Packing Configuration/Dimensions by Type

Drone Cargo Bay Type	Length (in)	Width (in)	Height (in)	
1	8	10	14	Top Loaded
2	24	20	20	Top Loaded

Attachment 4: Anticipated Medical Package Demand

Delivery Location			Emergency Medical Packages **		
Location Name	Latitude	Longitude	Requirement	Quantity	Frequency
Caribbean Medical Center Jajardo	18.33	-65.65	MED 1	1	Daily
			MED 3	1	Daily
Hospital HIMA San Pablo	18.22	-66.03	MED 1	2	Daily
			MED 3	1	Daily
Hospital Pavia Santurce San Juan	18.44	-66.07	MED 1	1	Daily
			MED 2	1	Daily
Puerto Rico Children's Hospital Bayamon	18.40	-66.16	MED 1	2	Daily
			MED 2	1	Daily
			MED 3	2	Daily
Hospital Pavia Arecibo Arecibo	18.47	-66.73	MED 1	1	Daily

**See Attachment 5 for Emergency Medical Packages 1, 2, and 3 Configurations/Dimensions.

Attachment 5: Emergency Medical Package Configuration/Dimensions

Emergency Medical Package Configuration		
Package ID	Weight (lbs.)	Package Dimensions (in.) (L × W × H)
MED 1	2	14 × 7 × 5
MED 2	2	5 × 8 × 5
MED 3	3	12 × 7 × 4

Glossary:

Cargo Container (Shipping Container): a large rectangular container with doors on the ends for loading and packing, and made of material suitable for shipping, storing, and handling in many weather and climate conditions.

Drone (Unmanned Aerial Vehicle, UAV): a flying robot that can be remotely controlled or fly autonomously through software-controlled flight plans in their embedded systems that work in conjunction with onboard sensors and GPS.

Drone Cargo Bay: For rotor wing drones, this is an externally carried "box" used to transport materials. For this problem, the drones under consideration have one of two types (sizes) of cargo bays. Note that each drone must land for the medical packages to be unloaded from the bay at its destination.

Drone Fleet: a set of drones for a particular mission or purpose. For this problem, the total set of drones by type (A to H) and Payload Capability (Visual and Medical) needed to meet the requirements of HELP, Inc.

Drone Payload Packing Configuration: how the drone payload bays are packed. For this problem, how the medical packages being transported by a drone are packed inside the drone cargo bay.

Medical Package: a predetermined set of medical supplies packed in a single container. For this problem, there are three Medical Package Configurations (MED1, MED2, MED3) available for transport by a drone from a deployed cargo container location to the demand location.

Non-governmental Organization (NGO): Usually non-profit and sometimes international organization independent of government and governmental organizations that is active in humanitarian, educational, healthcare, social, public policy, human rights, environmental and other areas in attempts to affect change.

Payload Capability: the carrying capacity of an aircraft or launch vehicle, usually measured in terms of weight. For this problem, the capability/capacity of the drone to carry medical packages.

2019 MCM
Problem C: The Opioid Crisis

Background: The United States is experiencing a national crisis regarding the use of *synthetic* and *non-synthetic opioids,* either for the treatment and management of pain (legal, prescription use) or for recreational purposes (illegal, non-prescription use). Federal organizations such as the Centers for Disease Control (CDC) are struggling to "save lives and prevent negative health effects of this epidemic, such as opioid use disorder, hepatitis, and HIV infections, and neonatal abstinence syndrome."[1] Simply enforcing existing laws is a complex challenge for the Federal Bureau of Investigation (FBI), and the U.S. Drug Enforcement Administration (DEA), among others.

There are implications for important sectors of the U.S. economy as well. For example, if the opioid crisis spreads to all cross-sections of the U.S. population (including the college-educated and those with advanced degrees), businesses requiring precision labor skills, high technology component assembly, and sensitive trust or security relationships with clients and customers might have difficulty filling these positions. Further, if the percentage of people with opioid addiction increases within the elderly, health care costs and assisted living facility staffing will also be affected.

The DEA/National Forensic Laboratory Information System (NFLIS), as part of the Drug Enforcement Administration's (DEA) Office of Diversion Control, publishes a data-heavy annual report addressing "drug identification results and associated information from drug cases analyzed by federal, state, and local forensic laboratories." The database within NFLIS includes data from crime laboratories that handle over 88% of the nation's estimated 1.2 million annual state and local drug cases. For this problem, we focus on the individual counties located in five (5) U.S. states: Ohio, Kentucky, West Virginia, Virginia, and Tennessee. In the U.S., a *county* is the next lower level of government below each state that has taxation authority.

Supplied with this problem description are several data sets for your use. The first file (MCM_NFLIS_Data.xlsx) contains drug identification counts in years 2010-2017 for narcotic analgesics (synthetic opioids) and *heroin* in each of the counties from these five states as reported to the DEA by crime laboratories throughout each state. A drug identification occurs when evidence is submitted to crime laboratories by law enforcement agencies as part of a criminal investigation and the laboratory's forensic scientists test the evidence. Typically, when law enforcement organizations submit these samples, they provide location data (county) with their incident reports. When evidence is submitted to a crime laboratory and this location data is not provided, the crime laboratory uses the location of the city/county/state investigating law enforcement organization that submitted the case. For the purposes of this problem, you may assume that the county location data are correct as provided.

The additional seven (7) files are zipped folders containing extracts from the U.S. Census Bureau that represent a common set of *socio-economic factors* collected for the counties of these five states during each of the years 2010-2016 (ACS_xx_5YR_DP02.zip). (Note: The same data were not available for 2017.)

A code sheet is present with each data set that defines each of the variables noted. While you may use other resources for research and background information, THE DATA SETS PROVIDED CONTAIN THE ONLY DATA YOU SHOULD USE FOR THIS PROBLEM.

---

[1] Centers for Disease Control website, (https://www.cdc.gov/features/confronting-opioids/index.html), accessed 4 September 2018.

Problem:

Part 1. Using the NFLIS data provided, build a mathematical model to describe the spread and characteristics of the reported synthetic opioid and heroin incidents (cases) in and between the five states and their counties over time. Using your model, identify any possible locations where specific opioid use might have started in each of the five states.

If the patterns and characteristics your team identified continue, are there any specific concerns the U.S. government should have? At what drug identification threshold levels do these occur? Where and when does your model predict they will occur in the future?

Part 2. Using the U.S. Census socio-economic data provided, address the following questions:

There are a good number of competing hypotheses that have been offered as explanations as to how opioid use got to its current level, who is using/abusing it, what contributes to the growth in opioid use and addiction, and why opioid use persists despite its known dangers. Is use or trends-in-use somehow associated with any of the U.S. Census socio-economic data provided? If so, modify your model from **Part 1** to include any important factors from this data set.

Part 3. Finally, using a combination of your **Part 1** and **Part 2** results, identify a possible strategy for countering the opioid crisis. Use your model(s) to test the effectiveness of this strategy; identifying any significant parameter bounds that success (or failure) is dependent upon.

In addition to your main report, include a 1-2 page memo to the Chief Administrator, DEA/NFLIS Database summarizing any significant insights or results you identified during this modeling effort.

Your submission should consist of:

- One-page Summary Sheet,
- One- to Two-page memo,
- Your solution of no more than 20 pages, for a maximum of 23 pages with your summary and memo.
- Note: Reference list and any appendices do not count toward the 23-page limit and should appear after your completed solution.

Attachments:

2019_MCMProblemC_DATA.zip - Includes seven zip folders and the NFLIS_Data file.

ACS_10_5YR_DP02.zip
ACS_11_5YR_DP02.zip
ACS_12_5YR_DP02.zip
ACS_13_5YR_DP02.zip
ACS_14_5YR_DP02.zip
ACS_15_5YR_DP02.zip
ACS_16_5YR_DP02.zip
MCM_NFLIS_Data.xlsx

Glossary:

analgesic – pain relieving medication

county – (in the U.S.) an administrative or political subdivision of a state; a region having specific boundaries and some level of governmental authority.

heroin – an illegal, euphoria producing, highly addictive analgesic drug processed from morphine (a naturally occurring substance extracted from the seed pods of certain varieties of poppy plants).

non-synthetic opioids – a class of drugs made from extracting chemicals in opium leaves, e.g. morphine, codeine, heroin.

opioids – pain relieving drugs that are often highly addictive

socio-economic factors – factors within a society that describe the relationship between social and economic status and class such as education, income, occupation, and employment.

synthetic opioid – man-made opioids

2019–MCM–Problem C–DATA

2019 ICM
Problem D: Time to leave the Louvre

The increasing number of terror attacks in France[1] requires a review of the emergency evacuation plans at many popular destinations. Your ICM team is helping to design evacuation plans at the Louvre in Paris, France. In general, the goal of evacuation is to have all occupants leave the building as quickly and safely as possible. Upon notification of a required evacuation, individuals egress to and through an optimal exit in order to empty the building as quickly as possible.

The Louvre is one of the world's largest and most visited art museum, receiving more than 8.1 million visitors in 2017[2]. The number of guests in the museum varies throughout the day and year, which provides challenges in planning for regular movement within the museum. The diversity of visitors -- speaking a variety of languages, groups traveling together, and disabled visitors -- makes evacuation in an emergency even more challenging.

The Louvre has five floors, two of which are underground.

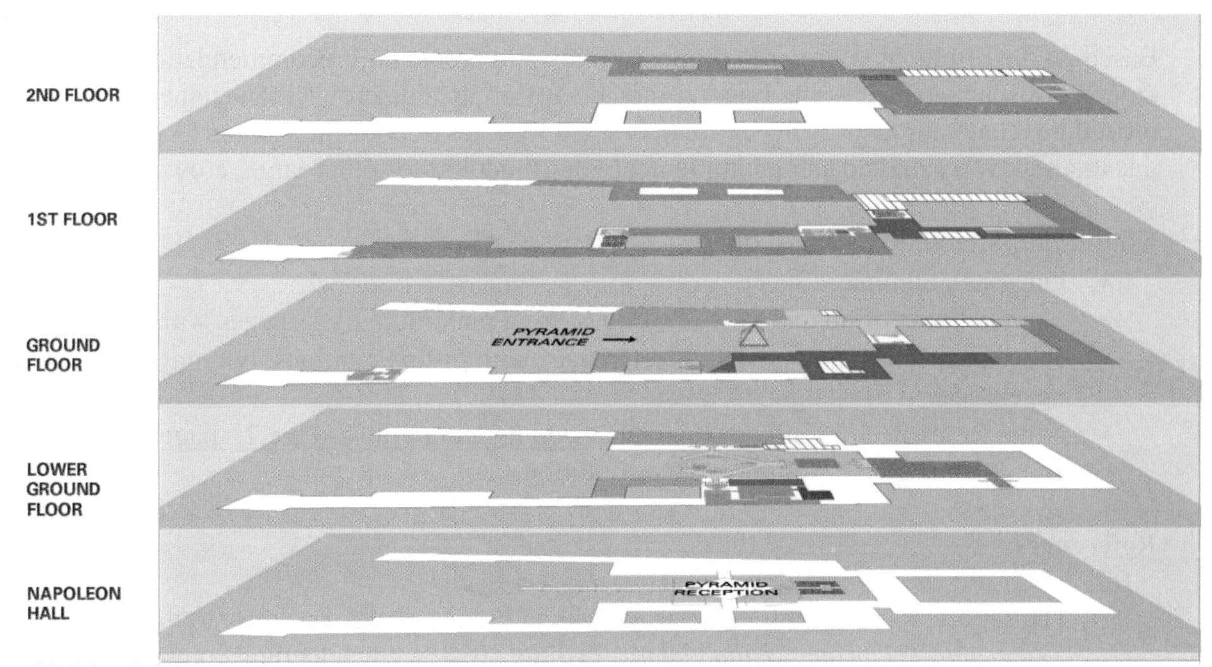

Figure 1: Floor plan of Louvre[3]

The 380,000 exhibits located on these five floors cover approximately 72,735 square meters, with building wings as long as 480 meters or 5 city blocks[3]. The pyramid entrance is the main and most used public entrance to the museum. However, there are also three other entrances usually reserved for groups and individuals with museum memberships: the Passage Richelieu entrance, the Carrousel du Louvre entrance, and the Portes Des Lions entrance. The Louvre has an online application, "Affluences" (https://www.affluences.com/louvre.php), that provides real-time updates on the estimated waiting time at each of these entrances to help facilitate entry to the museum. Your team might consider how technology, to include apps such as Affluences, or others could be used to facilitate your evacuation plan.

Only *emergency personnel* and museum officials know the actual number of total available exit points (service doors, employee entrances, VIP entrances, emergency exits, and old secret entrances built by the monarchy, etc.). While public awareness of these exit points could provide additional strength to an evacuation plan, their use would simultaneously cause security concerns due to the lower or limited security postures at these exits compared with level of security at the four main entrances. Thus, when creating your model, your team should consider carefully when and how any additional exits might be utilized.

Your supervisor wants your ICM team to develop an emergency evacuation model that allows the museum leaders to explore a range of options to evacuate visitors from the museum, while also allowing emergency personnel to enter the building as quickly as possible. It is important to identify potential *bottlenecks* that may limit movement towards the exits. The museum emergency planners are especially interested in an adaptable model that can be designed to address a broad set of considerations and various types of potential threats. Each threat has the potential to alter or remove segments of possible routes to safety that may be essential in a single optimized route. Once developed, validate your model(s) and discuss how the Louvre would implement it.

Based on the results of your work, propose policy and procedural recommendations for emergency management of the Louvre. Include any applicable crowd management and control procedures that your team believes are necessary for the safety of the visitors. Additionally, discuss how you could adapt and implement your model(s) for other large, crowded structures.

Your submission should consist of:

- One-page Summary Sheet,
- Your solution of no more than 20 pages, for a maximum of 21 pages with your summary.
- Judges expect a complete list of references with in-text citations, but may not consider appendices in the judging process.
- Note: Reference list and any appendices do not count toward the 21-page limit and should appear after your completed solution.

References:

[1] Reporters, Telegraph. "Terror Attacks in France: From Toulouse to the Louvre." *The Telegraph, Telegraph Media Group*, 24 June 2018, www.telegraph.co.uk/news/0/terror-attacks-france-toulouse-louvre/.

[2] "8.1 Million Visitors to the Louvre in 2017." *Louvre Press Release*, 25 Jan. 2018, presse.louvre.fr/8-1-million-visitors-to-the-louvre-in-2017/.

[3] "Interactive Floor Plans." *Louvre - Interactive Floor Plans | Louvre Museum | Paris*, 30 June 2016, www.louvre.fr/en/plan.

[4] "Pyramid" Project Launch – The Musée du Louvre is improving visitor reception (2014-2016)." *Louvre Press Kit*, 18 Sept. 2014, www.louvre.fr/sites/default/files/dp_pyramide%2028102014_en.pdf.

[5] "The 'Pyramid' Project - Improving Visitor Reception (2014-2016)." *Louvre Press Release,* 6 July 2016, presse.louvre.fr/the-pyramid-project/.

Glossary:

Bottlenecks – places where movement is dramatically slowed or even stopped.

Emergency personnel – people who help in an emergency, such as guards, fire fighters, medics, ambulance crews, doctors, and police.

2019 ICM
Problem E: What is the Cost of Environmental Degradation?

Economic theory often disregards the impact of its decisions on the *biosphere* or assumes unlimited resources or capacity for its needs. There is a flaw in this viewpoint, and the environment is now facing the consequences. The biosphere provides many natural processes to maintain a healthy and sustainable environment for human life, which are known as *ecosystem services*. Examples include turning waste into food, water filtration, growing food, pollinating plants, and converting carbon dioxide into oxygen. However, whenever humans alter the ecosystem, we potentially limit or remove ecosystem services. The impact of local small-scale changes in land use, such as building a few roads, sewers, bridges, houses, or factories may seem negligible. Add to these small projects, large-scale projects such as building or relocating a large corporate headquarters, building a pipeline across the country, or expanding or altering waterways for extended commercial use. Now think about the impact of many of these projects across a region, country, and the world. While individually these activities may seem inconsequential to the total ability of the biosphere's functioning potential, cumulatively they are directly impacting the *biodiversity* and causing *environmental degradation*.

Traditionally, most land use projects do not consider the impact of, or account for changes to, ecosystem services. The economic costs to *mitigate* negative results of land use changes: polluted rivers, poor air quality, hazardous waste sites, poorly treated waste water, climate changes, etc., are often not included in the plan. Is it possible to put a value on the environmental cost of land use development projects? How would environmental degradation be accounted for in these project costs? Once ecosystem services are accounted for in the cost-benefit ratio of a project, then the true and comprehensive *valuation* of the project can be determined and assessed.

Your ICM team has been hired to create an ecological services valuation model to understand the true economic costs of land use projects when ecosystem services are considered. Use your model to perform a cost benefit analysis of land use development projects of varying sizes, from small community-based projects to large national projects. Evaluate the effectiveness of your model based on your analyses and model design. What are the implications of your modeling on land use project planners and managers? How might your model need to change over time?

Your submission should consist of:

- One-page Summary Sheet,
- Your solution of no more than 20 pages, for a maximum of 21 pages with your summary.
- Judges expect a complete list of references with in-text citations, but may not consider appendices in the judging process.
- Note: Reference list and any appendices do not count toward the 21-page limit and should appear after your completed solution.

References:

Chee, Y., 2004. An ecological perspective on the valuation of ecosystem services. Biological Conservation 120, 549-565.

Costanza, R., d'Arge, R., de Groot, R., Farber, S., Grasso, M., Hannon, B., Limburg, K., Naeem, S., O'Neill, R.V., Paruelo, J., Raskin, R.G., Sutton, P., van den Belt, M., 1997. The value of the world's ecosystem services and natural capital. Nature 387, 253– 260.

Gómez-Baggethuna, E., de Groot, R., Lomas, P., Montesa, C., 1 April 2010. The history of ecosystem services in economic theory and practice: From early notions to markets and payment schemes. Ecological Economics 69 (6), 1209-1218.

Norgaard, R., 1 April 2010. Ecosystem services: From eye-opening metaphor to complexity blinder. Ecological Economics 69 (6), 1219-1227.

Richmond, A., Kaufmann R., Myneni, R., 2007, Valuing ecosystem services: A shadow price for net primary production. Ecological Economics 64, 454-462.

Yang, Q., Liu, G., Casazza, M., Campbell, E., Giannettia, B., Brown, M., December 2018. Development of a new framework for non-monetary accounting on ecosystem services valuation. Ecosystem Services 34A, 37-54.

Data sources:

US based data: https://www.data.gov/ecosystems/

Satellite data: https://www.ncdc.noaa.gov/data-access/satellite-data/satellite-data-access-datasets

Glossary:

Biodiversity - refers to the variety of life in an ecosystem; all of the living organisms within a given area.

Biosphere - the part of the Earth that is occupied by living organisms and generally includes the interaction between these organisms and their physical environment.

Ecosystem - a subset of the biosphere that primarily focuses on the interaction between living things and their physical environment.

Ecosystem Services – the many benefits and assets that humans receive freely from our natural environment and a fully functioning ecosystem.

Environmental Degradation – the deterioration or compromise of the natural environment through consumption of assets either by natural processes or human activities.

Mitigate – to make less severe, painful, or impactful.

Valuation - refers to the estimating or determining the current worth of something.

2019 ICM
Problem F: Universal, Decentralized, Digital Currency: Is it possible?

Digital currency can be used like traditional currencies to buy and sell goods, except that it is digital and has no physical representation. Digital currency enables its users to make transactions instantaneously and without any concern for national borders. *Cryptocurrency* is a subset of digital currency with unique features of privacy, decentralization, security and encryption. Cryptocurrencies have exploded in popularity in various parts of the world; moving from an *underground cult* interest to a globally accepted phenomenon. Bitcoin and Ethereum, both cryptocurrencies, have grown in value, while investors are projecting rapid growth for other cryptocurrencies such as Dogecoin or Ripple. In addition to digital and cryptocurrencies, there are also new digital methods for financial transactions that enable users to instantaneously exchange money with nothing more than an email address or a thumbprint. Peer-to-peer payment systems offered by companies like PayPal, Stripe, Venmo, Zelle, Apple Pay, Square Cash, and Google Pay offer virtual movement of money across the globe in seconds without ever having to verify the transaction through a bank or currency exchange. Digital transactions outpace cash and check transactions because they are not delayed by banking policies, national borders, citizenship, debts, or other social-economic factors. These new currency systems decentralize financial transactions, leaving many to consider a world where traditional banking may become obsolete.

Concerns about security of cryptocurrencies worry both citizens and economic analysts. These concerns have constrained its growth in some communities. On the other hand, much of the popularity of cryptocurrency is due to its departure from traditional overly-restrictive security and debt measures that rely on oversight by large banks and governments. These oversight institutions are often expensive, deeply bureaucratic, and sometimes corrupt. Some experts believe that a universal, decentralized, digital currency with internal security like *blockchain* can make markets more efficient by eliminating barriers to the flow of money. This is particularly important in countries where the majority of citizens do not have bank accounts and are unable to invest in regional or global financial markets. Some governments, however, view the lack of regulation around these currencies and their *anonymity* as too risky because of how easily they can be used in *illicit* transactions, such as tax sheltering or purchasing illegal merchandise. Others feel that a secure digital currency offers a more convenient and safer form of financial exchange. For instance, a universally accepted currency would enable truly global financial markets and would protect individual assets against regional inflation *fluctuations* and artificial manipulation of currency by regional governments. If alternative digital systems become more established, there will be many questions about how digital currency will affect current banking systems and *nation-based currencies*.

Your policy modeling team has been employed by the International Currency Marketing (ICM) Alliance to help them identify the viability and effects of a global decentralized digital financial market. ICM Alliance has asked you to construct a model that adequately represents this type of financial system, being sure to identify key factors that would limit or facilitate its growth, access, security, and stability at both the individual, national, and global levels. This requires you to consider the different needs of countries and their willingness to work with this new financial marketplace and modify their current banking and *monetary* models. It may or may not require them to abandon their own currency, so that adds a level of complexity to the market model. You are not to choose an existing digital currency, but discuss the strategies for adoption, and problems in implementation of, a general digital currency. You should also include the mechanisms for oversight of such a global digital currency. The ICM Alliance has asked you to extend your analysis to consider the long-term effects of such a system on the current banking industry; the local, regional, and world economy; and international relations between countries.

ICM requests a report of your modeling and analysis, and a separate one-page policy recommendation for national leaders, who hold mixed opinions about this effort. The policy recommendation should offer rationale for the parameters and dynamics included in your model and reflect the insights you gained from your modeling. Your polices might address, for example, growth, reach, access, security, and stability of the system.

Your team's submission should consist of:

- One-page Summary Sheet,
- One-page policy recommendation for national leaders,
- Your solution of no more than 20 pages, for a maximum of 22 pages with your summary and policy recommendation.
- Judges expect a complete list of references with in-text citations, but may not consider appendices in the judging process.
- Note: Reference list and any appendices do not count toward the 22-page limit and should appear after your completed solution.

References:

Paul Krugman, "O Canada: A neglected nation gets its Nobel". *Slate*, Oct 19, 1999. https://slate.com/business/1999/10/o-canada.html

Stephanie Lo and J. Christina Wang, "Bitcoin as Money?" *Current Policy Perspectives*, Federal Reserve Bank of Boston, 2014. https://www.bostonfed.org/publications/current-policy-perspectives/2014/bitcoin-as-money.aspx or https://www.bostonfed.org/-/media/Documents/Workingpapers/PDF/cpp1404.pdf

Glossary:

Anonymity – the state of being unnamed or unidentified; the state of being anonymous.

Blockchain – the record keeping technology that can document transactions between two parties in a verifiable and permanent way; a digital database containing information that can be shared and simultaneously used across a large publicly accessible and decentralized network.

Cryptocurrency – a digital or virtual currency that uses cryptography (protecting information through the use of codes) for security.

Digital Currency – [digital money, electronic money, electronic currency] is a type of currency in digital (electronic) versus physical (coins, paper) form.

Illicit – illegal or dishonest.

Fluctuations – variations or oscillations; rises and falls. .

Monetary – relating to money or finances, or to the mechanisms by which money is supplied to and circulates in the economy.

Nation-based currencies – [national currencies] a system of money issued by a central bank and in common use within a particular nation or group of nations; examples are United States dollar (USD), Chinese renminbi (RMB or CNY), European Euro (EUR), British pound sterling (GBP), and Japanese yen (JPY).

Underground cult – hidden or mysterious group of people sharing an excessive devotion toward a particular person, belief, or thing.

2020 MCM
Problem A: Moving North

Global ocean temperatures affect the quality of **habitats** for certain ocean-dwelling species. When temperature changes are too great for their continued thriving, these species move to seek other habitats better suited to their present and future living and reproductive success. One example of this is seen in the lobster population of Maine, USA that is slowly migrating north to Canada where the lower ocean temperatures provide a more suitable habitat. This geographic population shift can significantly disrupt the livelihood of companies who depend on the stability of ocean-dwelling species.

Your team has been hired as consultants by a Scottish North Atlantic **fishery** management consortium. The consortium wants to gain a better understanding of issues related to the potential migration of Scottish herring and mackerel from their current habitats near Scotland if and when global ocean temperatures increase. These two fish species represent a significant economic contribution to the Scottish fishing industry. Changes in population locations of herring and mackerel could make it economically impractical for smaller Scotland-based fishing companies, who use fishihing vessels without on-board refrigeration, to harvest and deliver·fresh fish to markets in Scotland fishing ports.

Requirements

1. Build a mathematical model to identify the most likely locations for these two fish species over the next 50 years, assuming that water temperatures are going to change enough to cause the populations to move.

2. Based upon how rapidly the ocean water temperature change occurs, use your model to predict best case, worst case, and most likely elapsed time(s) until these populations will be too far away for small fishing companies to harvest if the small fishing companies continue to operate out of their current locations.

3. In light of your predictive analysis, should these small fishing companies make changes to their operations?

   a. If yes, use your model to identify and assess practical and economically attractive strategies for small fishing companies. Your strategies should consider, but not be limited to, realistic options that include:

      -Relocating some or all of a fishing company's assets from a current location in a Scottish port to closer to where both fish populations are moving;

      -Using some proportion of small fishing vessels capable of operating without land- based support for a period of time while still ensuring the freshness and high quality of the catch.

      -other options that your team may identify and model.

   b. If your team rejects the need for any changes, justify reasons for your rejection based on your modeling results as they relate to the assumptions your team has made.

4. Use your model to address how your proposal is affected if some proportion of the fishery moves into the **territorial waters (sea)** of another country.

5. In addition to your technical report, prepare a one- to two-page article for *Hook Line and Sinker* magazine to help fishermen understand the seriousness of the problem and how your proposed solution(s)

will improve their future business prospects.

Your submission should consist of:

- One-page Summary Sheet
- Table of Contents
- One- to Two-page Article
- Your solution of no more than 20 pages, for a maximum of 24 pages with your summary, table of contents, and article.

Note: Reference List and any appendices do not count toward the page limit and should appear after your completed solution. You should not make use of unauthorized images and materials whose use is restricted by copyright laws. Ensure you cite the sources for your ideas and the materials used in your report.

Glossary

Fishery: The collection of fish of a given species and the area that they inhabit.

Habitat: The type of environment in which an organism or group normally lives or occurs.

Small Fishing Company: A company engaged in commercial fishing with limited or very limited financial resources to invest in new equipment/vessels.

Territorial Waters (sea):"as defined by the 1982 United Nations Convention on the Law of the Sea, is a belt of coastal waters extending at most 12 nautical miles (22.2 km; 13.8 mi) from the baseline (usually the mean low-water mark) of a coastal state. The territorial sea is regarded as the sovereign territory of the state, although foreign ships (military and civilian) are allowed innocent passage through it, or transit passage for straits; this sovereignty also extends to the airspace over and seabed below." [Territorial Waters. (n.d.). In Wikipedia. Retrieved January 28, 2020, from https://en.wikipedia.org/wiki/Territorial_ waters.]

2020 MCM Weekend 2
Problem B: The Longest Lasting Sandcastle(s)

Wherever there are recreational sandy ocean beaches in the world, there seem to be children (and adults) creating sandcastles on the seashore. Using tools, toys, and imagination, beach goers create sandcastles that range from simple mounds of sand to complicated replicas of actual castles with walls, towers, moats, and other features that mimic real castles. In all these, one typically forms an initial foundation consisting of a single, nondescript mound of wetted sand, and then proceeds to cut and shape this base into a recognizable 3-dimensional geometric shape upon which to build the more castle-defining features.

Inevitably, the inflow of ocean waves coupled with rising tides erodes sandcastles. It appears, however, that not all sandcastles react the same way to waves and tides, even if built roughly the same size and at roughly the same distance from the water on the same beach. Consequently, one wonders if there exists a best 3-dimensional geometric shape to use for a sandcastle foundation.

Requirements

1. Construct a mathematical model to identify the best 3-dimensional geometric shape to use as a sandcastle foundation that will last the longest period of time on a seashore that experiences waves and tides under the following conditions:
- built at roughly the same distance from the water on the same beach, and
- built using the same type of sand, roughly the same amount of sand, and the same water-to-sand proportion.

2. Using your model, determine an optimal sand-to-water mixture proportion for the castle foundation, assuming you use no other additives or materials (e.g. plastic or wooden supports, stones, etc.).

3. Adjust your model as needed to determine how the best 3-dimensional sandcastle foundation you identified in requirement 1 is affected by rain, and whether it remains the best 3-dimensional geometric shape to be used as a castle foundation when it is raining.

4. What other strategies, if any, might you use to make your sandcastle last longer?

5. Finally, write an informative, one- to two-page article describing your model and its results for publication in the vacation magazine: *Fun in the Sun,* whose readers are mainly non-technical.

Your submission should consist of:
- One-page Summary Sheet
- Table of Contents
- One- to Two-page Article
- Your solution of no more than 20 pages, for a maximum of 24 pages with your summary, table of contents, and article.

Note: Reference List and any appendices do not count toward the page limit and should appear after your completed solution. You should not make use of unauthorized images and materials whose use is restricted by copyright laws. Ensure you cite the sources for your ideas and the materials used in your report.

2020 MCM Weekend 2
Problem C: A Wealth of Data

In the online marketplace it created, Amazon provides customers with an opportunity to rate and review purchases. Individual ratings - called "**star ratings**" – allow purchasers to express their level of satisfaction with a product using a scale of 1 (low rated, low satisfaction) to 5 (highly rated, high satisfaction). Additionally, customers can submit text-based messages – called "**reviews**" – that express further opinions and information about the product. Other customers can submit ratings on these reviews as being helpful or not – called a "**helpfulness rating**" – towards assisting their own product purchasing decision. Companies use these data to gain insights into the markets in which they participate, the timing of that participation, and the potential success of product design feature choices.

Sunshine Company is planning to introduce and sell three new products in the online marketplace: a microwave oven, a baby **pacifier**, and a hair dryer. They have hired your team as consultants to identify key patterns, relationships, measures, and parameters in past customer-supplied ratings and reviews associated with other competing products to 1) inform their online sales strategy and 2) identify potentially important design features that would enhance product desirability. Sunshine Company has used data to inform sales strategies in the past, but they have not previously used this particular combination and type of data. Of particular interest to Sunshine Company are time-based patterns in these data, and whether they interact in ways that will help the company craft successful products.

To assist you, Sunshine's data center has provided you with three data files for this project: **hair_dryer.tsv**, **microwave.tsv**, and **pacifier.tsv**. These data represent customer-supplied ratings and reviews for microwave ovens, baby pacifiers, and hair dryers sold in the Amazon marketplace over the time period(s) indicated in the data. A glossary of data label definitions is provided as well. THE DATA FILES PROVIDED CONTAIN THE ONLY DATA YOU SHOULD USE FOR THIS PROBLEM.

Requirements

1. Analyze the three product data sets provided to identify, describe, and support with mathematical evidence, meaningful quantitative and/or qualitative patterns, relationships, measures, and parameters within and between star ratings, reviews, and helpfulness ratings that will help Sunshine Company succeed in their three new online marketplace product offerings.

2. Use your analysis to address the following specific questions and requests from the Sunshine Company Marketing Director:
   a. Identify data measures based on ratings and reviews that are most informative for Sunshine Company to track, once their three products are placed on sale in the online marketplace.
   b. Identify and discuss time-based measures and patterns within each data set that might suggest that a product's reputation is increasing or decreasing in the online marketplace.
   c. Determine combinations of text-based measure(s) and ratings-based measures that best indicate a potentially successful or failing product.

d. Do specific star ratings incite more reviews? For example, are customers more likely to write some type of review after seeing a series of low star ratings?
e. Are specific quality descriptors of text-based reviews such as 'enthusiastic', 'disappointed', and others, strongly associated with rating levels?

3. Write a one- to two-page letter to the Marketing Director of Sunshine Company summarizing your team's analysis and results. Include specific justification(s) for the result that your team most confidently recommends to the Marketing Director.

Your submission should consist of:

- One-page Summary Sheet
- Table of Contents
- One- to Two-page Letter
- Your solution of no more than 20 pages, for a maximum of 24 pages with your summary sheet, table of contents, and two-page letter.

Note: Reference List and any appendices do not count toward the page limit and should appear after your completed solution. You should not make use of unauthorized images and materials whose use is restricted by copyright laws. Ensure you cite the sources for your ideas and the materials used in your report.

Glossary

Helpfulness Rating: an indication of how valuable a particular product review is when making a decision whether or not to purchase that product.

Pacifier: a rubber or plastic soothing device, often nipple shaped, given to a baby to suck or bite on.

Review: a written evaluation of a product.

Star Rating: a score given in a system that allows people to rate a product with a number of stars.

Attachments: The Problem Datasets

Problem_C_Data.zip
The three data sets provided contain product user ratings and reviews extracted from the Amazon Customer Reviews Dataset thru Amazon Simple Storage Service (Amazon S3).

hair_dryer.tsv
microwave.tsv
pacifier.tsv

Data Set Definitions: Each row represents data partitioned into the following columns.

- marketplace (string): 2 letter country code of the marketplace where the review was written.
- customer_id (string): Random identifier that can be used to aggregate reviews written by a single author.
- review_id (string): The unique ID of the review.
- product_id (string): The unique Product ID the review pertains to.
- product_parent (string): Random identifier that can be used to aggregate reviews for the same product.
- product_title (string): Title of the product.
- product_category (string): The major consumer category for the product.
- star_rating (int): The 1-5 star rating of the review.
- helpful_votes (int): Number of helpful votes.
- total_votes (int): Number of total votes the review received.
- vine (string): Customers are invited to become Amazon Vine Voices based on the trust that they have earned in the Amazon community for writing accurate and insightful reviews. Amazon provides Amazon Vine members with free copies of products that have been submitted to the program by vendors. Amazon doesn't influence the opinions of Amazon Vine members, nor do they modify or edit reviews.
- verified_purchase (string): A "Y" indicates Amazon verified that the person writing the review purchased the product at Amazon and didn't receive the product at a deep discount.
- review_headline (string): The title of the review.
- review_body (string): The review text.
- review_date (bigint): The date the review was written.

2020 ICM Weekend 1
Problem D: Teaming Strategies

As societies become more interconnected, the set of challenges they face have become increasingly complex. We rely on interdisciplinary teams of people with diverse expertise and varied perspectives to address many of the most challenging problems. Our conceptual understanding of team success has advanced significantly over the past 50+ years allowing for better scientific, creative, or physical teams to address these complex issues. Researchers have reported on best strategies for assembling teams, optimal interactions among teammates, and ideal leadership styles. Strong teams across all sectors and domains are able to perform complex tasks unattainable through either individual efforts or a sequence of additive contributions of teammates.

One of the most informative settings to explore team processes is in competitive team sports. Team sports must conform to strict rules that may include, but are not limited to, the number of players, their roles, allowable contact between players, their location and movement, points earned, and consequences of violations. Team success is much more than the sum of the abilities of individual players. Rather, it is based on many other factors that involve how well the teammates play together. Such factors may include whether the team has a diversity of skills (one person may be fast, while another is precise), how well the team balances between individual versus collective performance (star players may help leverage the skills of all their teammates), and the team's ability to effectively coordinate over time (as one player steals the ball from an opponent, another player is poised for offense).

In light of your modeling skills, the coach of the Huskies, your home soccer (known in Europe and other places as football) team, has asked your company, **I**ntrepid **C**hampion **M**odeling (ICM), to help understand the team's dynamics. In particular, the coach has asked you to explore how the complex interactions among the players on the field impacts their success. The goal is not only to examine the interactions that lead directly to a score, but to explore team dynamics throughout the game and over the entire season, to help identify specific strategies that can improve teamwork next season. The coach has asked ICM to quantify and formalize the structural and dynamical features that have been successful (and unsuccessful) for the team. The Huskies have provided data[1] detailing information from last season, including all 38 games they played against their 19 opponents (they played each opposing team twice). Overall, the data covers 23,429 passes between 366 players (30 Huskies players, and 336 players from opposing teams), and 59,271 game events.

To respond to the Huskie coach's requests, your team from ICM should use the provided data to address the following:

- Create a network for the ball passing between players, where each player is a node and each pass constitutes a link between players. Use your passing network to identify network patterns, such as **dyadic** and **triadic configurations** and team formations. Also consider other structural indicators and network properties across the games. You should

[1] This data set was processed from a much larger dataset covering nearly 2000 matches from five European national soccer competitions, as well as the 2018 World Cup [1].

explore multiple scales such as, but not limited to, micro (pairwise) to macro (all players) when looking at interactions, and time such as short (minute-to-minute) to long (entire game or entire season).

- Identify performance indicators that reflect successful teamwork (in addition to points or wins) such as diversity in the types of plays, coordination among players or distribution of contributions. You also may consider other team level processes, such as adaptability, flexibility, tempo, or flow. It may be important to clarify whether strategies are universally effective or dependent on opponents' counter-strategies. Use the performance indicators and team level processes that you have identified to create a model that captures structural, configurational, and dynamical aspects of teamwork.

- Use the insights gained from your teamwork model to inform the coach about what kinds of structural strategies have been effective for the Huskies. Advise the coach on what changes the network analysis indicates that they should make next season to improve team success.

- Your analysis of the Huskies has allowed you to consider group dynamics in a controlled setting of a team sport. Understanding the complex set of factors that make some groups perform better than others is critical for how societies develop and innovate. As our societies increasingly solve problems involving teams, can you generalize your findings to say something about how to design more effective teams? What other aspects of teamwork would need to be captured to develop generalized models of team performance?

Your submission should consist of:

- One-page Summary Sheet
- Table of Contents
- Your solution of no more than 20 pages, for a maximum of 22 pages with your summary and table of contents.

Note: Reference List and any appendices do not count toward the page limit and should appear after your completed solution. You should not make use of unauthorized images and materials whose use is restricted by copyright laws. Ensure you cite the sources for your ideas and the materials used in your report.

Attachment

2020_Problem_D_DATA.zip

fullevents.csv
matches.csv
passingevents.csv
README.txt

Glossary

Dyadic Configurations: relationships involving pairs of players.

Triadic Configurations: relationships involving groups of three players.

Cited Reference

[1] Pappalardo, L., Cintia, P., Rossi, A. *et al. A public data set of spatio-temporal match events in soccer competitions*. *Sci Data* 6, 236 (2019).

Optional Resources

Research in football (soccer) networks has led to many articles that discuss related topics. A few articles are listed below. You are not required to use any of these sample articles in your solution, nor is it a comprehensive list. We encourage teams to utilize any journal article that supports their approach to the problem.

Buldú, J.M., Busquets, J., Echegoyen, I. *et al.* (2019). Defining a historic football team: Using Network Science to analyze Guardiola's F.C. Barcelona. *Sci Rep*, 9, 13602.

Cintia, P., Giannotti, F., Pappalardo, L., Pedreschi, D., & Malvaldi, M. (2015). The harsh rule of the goals: Data-driven performance indicators for football teams. *2015 IEEE International Conference on Data Science and Advanced Analytics (DSAA)*, 1-10, 7344823.

Duch J., Waitzman J.S., Amaral L.A.N. (2010). Quantifying the performance of individual players in a team activity. *PLoS ONE*, 5: e10937.

GÜRSAKAL, N., YILMAZ, F., ÇOBANOĞLU, H., ÇAĞLIYOR, S. (2018). Network Motifs in Football. *Turkish Journal of Sport and Exercise*, 20 (3), 263-272.

2020−Problem D−DATA

2020 ICM Weekend 1
Problem E: Drowning in Plastic

Since the 1950s, the manufacturing of plastics has grown exponentially because of its variety of uses, such as food packaging, consumer products, medical devices, and construction. While there are significant benefits, the negative implications associated with increased production of plastics are concerning. Plastic products do not readily break down, are difficult to dispose of, and only about 9% of plastics are recycled[1]. Effects can be seen by the approximately 4-12 million tons of **plastic waste** that enter the oceans each year[1,2]. Plastic waste has severe environmental consequences and it is predicted that if our current trends continue, the oceans will be filled with more plastic than fish by 2050[2]. The effect on marine life has been studied[3], but the effects on human health are not yet completely understood[4]. The rise of **single-use** and **disposable plastic products** results in entire industries dedicated to creating plastic waste. It also suggests that the amount of time the product is useful is significantly shorter than the time it takes to properly **mitigate** the plastic waste. Consequently, to solve the plastic waste problem, we need to slow down the flow of plastic production and improve how we manage plastic waste.

Your team has been hired by the **I**nternational **C**ouncil of Plastic Waste **M**anagement (ICM) to address this escalating environmental crisis. You must develop a plan to significantly reduce, if not eliminate, single-use and disposable plastic product waste.

- Develop a model to estimate the maximum levels of single-use or disposable plastic product waste that can safely be mitigated without further environmental damage. You may need to consider, among many factors, the source of this waste, the extent of the current waste problem, and the availability of resources to process the waste.

- Discuss to what extent plastic waste can be reduced to reach an environmentally safe level. This may involve considering factors impacting the levels of plastic waste to include, but not limited to, sources and uses of single-use or disposable plastics, the availability of alternatives to plastics, the impact on the lives of citizens, or policies of cities, regions, countries, and continents to decrease single-use or disposable plastic and the effectiveness of such policies. These can vary between regions, so considering regional-specific constraints may make some policies more effective than others.

- Using your model and discussion, set a target for the minimal achievable level of global waste of single-use or disposable plastic products and discuss the impacts for achieving such levels. You may consider ways in which human life is altered, the environmental impacts, or the effects on the multi-trillion-dollar plastic industry.

- While this is a global problem, the causes and effects are not equally distributed across nations or regions. Discuss the equity issues that arise from the global crisis and your intended solutions. How do you suggest ICM address these issues?

- Write a two-page memo to the ICM describing a realistic global target minimum achievable level of global single-use or disposable plastic product waste, a timeline to reach this level, and any circumstances that may accelerate or hinder the achievement of your target and timeline.

Your submission should consist of:

- One-page Summary Sheet
- Table of Contents
- Two-page Memo
- Your solution of no more than 20 pages, for a maximum of 24 pages with your summary, table of contents, and two-page memo.

Note: Reference List and any appendices do not count toward the page limit and should appear after your completed solution. You should not make use of unauthorized images and materials whose use is restricted by copyright laws. Ensure you cite the sources for your ideas and the materials used in your report.

Glossary

Disposable Plastic Products: plastic materials or products that are not recyclable and become trash.

Mitigate: To make less severe, to moderate, to alleviate.

Plastic Waste: plastic objects that have not been recycled properly or cannot be recycled; debris made of plastic.

Single-Use Plastic Products: products made of plastic intended for one time use before being discarded.

Cited References

[1] Geyer, R., Jambeck, J. R., & Law, K. L. (2017). Production, use, and fate of all plastics ever made. *Science Advances*, 3(7), e1700782.

[2] Jambeck, J. R., Geyer, R., Wilcox, C., Siegler, T. R., Perryman, M., Andrady, A., … & Law, K. L. (2015). Plastic waste inputs from land into the ocean. Science, 347(6223), 768-771.

[3] Li, W. C., Tse, H. F., & Fok, L. (2016). Plastic waste in the marine environment: A review of sources, occurrence and effects. *Science of the Total Environment*, 566, 333-349.

[4] Galloway T.S. (2015) Micro- and Nano-plastics and Human Health. In: Bergmann M., Gutow L., Klages M. (eds) *Marine Anthropogenic Litter*.

2020 ICM Weekend 2
Problem F: The Place I Called Home…

Researchers have identified several island nations, such as The Maldives, Tuvalu, Kiribati, and The Marshall Islands, as being at risk of completely disappearing due to rising sea levels. What happens, or what should happen, to an island's population when its nation's land disappears? Not only do these **environmentally displaced persons** (EDPs) need to relocate, but there is also risk of losing a unique culture, language, and way of life. In this problem, we ask you to look more closely at this issue, in terms of both the need to relocate people and the protection of culture. There are many considerations and questions to address, to include: Where will these EDPs go? What countries will take them? Given various nations' disproportionate contributions to the green-house gasses both historically and currently that have accelerated climate change linked to the rising seas, should the worst offenders have a higher obligation to address these issues? And, who gets a say in deciding where these nationless EDPs make a new home – the individuals, an intergovernmental organization like the United Nations (UN), or the individual governments of the states absorbing these persons? A more detailed explanation of these issues is given in the Issue Paper beginning on page 3.

As a result of a recent UN ruling that opened the door to the theoretical recognition of EDPs as refugees, the **I**nternational **C**limate **M**igration **F**oundation (ICM-F) has hired you to advise the UN by developing a model and using it to analyze this multifaceted issue of when, why, and how the UN should step into a role of addressing the increasing challenge of EDPs. The ICM-F plans to brief the UN on guidance for how the UN should generate a systemized response for EDPs, especially in consideration of the desire to preserve **cultural heritage**. Your assignment is to develop a model (or set of models) and use your model(s) to provide the analysis to support this briefing. The ICM-F is especially interested in understanding the scope of the issue of EDPs. For example, how many people are currently at risk of becoming EDPs[1]; what is the value of the cultures of at-risk nations; how are those answers likely to change over time? Furthermore, how should the world respond with an international policy that specifically focuses on protecting the rights of persons whose nations have disappeared in the face of climate change while also aiming to preserve culture? Based on your analysis, what recommendations can you offer on this matter, and what are the implications of accepting or rejecting your recommendations?

This problem is extremely complex. We understand that your submission will not be able to fully consider all of the aspects described in the Issue Paper beginning on page 3. However, considering the aspects that you address, synthesize your work into a cohesive answer to the ICM-F as they advise the UN. At a minimum, your team's paper should include:

- An analysis of the scope of the issue in terms of both the number of people at risk and the risk of loss of culture;
- Proposed policies to address EDPs in terms of both human rights (being able to resettle and participate fully in life in their new home) and cultural preservation;
- A description of the development of a model used to measure the potential impact of proposed policies;

---

[1] There are multiple estimates for the current and predicted number of climate refugees in the existing literature, but they are vastly different. Therefore, you need to support your conclusions with analysis based on your own model(s), either building off of existing analysis or with a new and independent analysis.

- An explanation of how your model was used to design and/or improve your proposed policies;
- An explanation, backed by your analysis, of the importance of implementing your proposed policies.

The ICM-F consists of interdisciplinary judges including mathematicians, climate scientists, and experts in refugee migration to review your work. Therefore, your paper should be written for a scientifically literate yet diverse audience.

Your submission should consist of:
- One-page Summary Sheet
- Table of Contents
- Your solution of no more than 20 pages, for a maximum of 22 pages with your summary and table of contents.

NOTE: Reference List and any appendices do not count toward the page limit and should appear after your completed solution. You should not make use of unauthorized images and materials whose use is restricted by copyright laws. Ensure you cite the sources for your ideas and the materials used in your report.

Glossary

Environmentally displaced persons (EDPs): people who must relocate as their homeland becomes uninhabitable due to climate change events

Cultural heritage: the ways of living of a group or society passed through generations to include customs, practices, art, and values.

ICM Problem F
Issue Paper

As noted in the problem statement, several island nations are at risk of completely disappearing due to rising sea levels.[1] The issue is quite complex. It is not simply a matter of identifying how to move a certain number of people around the globe – it is also about recognizing that these people are human beings who have rights and who are the last living representatives of their unique culture. In this Issue Paper, we highlight three of the essential ideas that frame this problem: relocation decisions as they relate to human rights, nation-state responsibility, and individual choice; the tension between assimilation and accommodation as part of resettlement and cultural preservation; and time factors such as the rate of the nation disappearing, the timing of these losses aligning with a global rise in nationalism, and the difficulty in making sound predictions about the size of this issue.

Relocation Decisions: Human Rights, Nation-State Responsibility, and Individual Choice

Considering the relocation issue, you might think that such EDPs would have similar rights as other UN-recognized refugees, but the United Nations High Commission on Refugees (UNHCR) and the widely adopted 1967 protocol has historically only afforded rights to those who are displaced due to politically related security issues, such as ethnic or religious persecution. However, in a very recent ruling, the UN has acknowledged this issue and recognized that some EDPs might qualify as refugees.[2] Although a ruling has now been made, there is not yet a vision on how the international community should respond as these situations increase in magnitude and frequency.[3]

Rights awarded to these refugees include right to work, freedom of movement, and protection by host governments. Additionally, the UNHCR, in collaboration with other aid organizations, work to provide aid and assistance to refugees until they are resettled in another country, become naturalized by their host state, or repatriate to their country of origin. Now, with this new ruling, the former inhabitants of the disappeared nation may be eligible for some of those rights or aid, but there is no hope of repatriation as the land itself is gone.

Even if EDPs are eligible for rights somewhere else, it is not clear where this new home would be or who would be responsible for making that decision. There are individual and international considerations related to whether the selection of a new long-term residence is made by individuals or if the choices are made or swayed by immigration policies developed by nations in isolation or as part of a cooperative effort coordinated by the United Nations. Possible migration policies could consider the financial ability of the new nation to absorb these new individuals, but there is also discussion of setting up burden-sharing based on nations' relative contributions (pollution) to the environmental conditions that is leading to the loss of these nations. In other words, the international community may press nations with high pollution records to contribute more to the resettlement of EDPs in some equitable manner.

Resettlement and Cultural Preservation: Assimilation versus Accommodation

In terms of the cultural preservation issues, the nations that are most at risk are arguably some of the most culturally distinct in the world with languages, music, art, dances, social norms, and ways of life that can be different from island to island even within the same island chain. As a result, the loss of one of these nations could represent a significant cultural loss. While the displaced inhabitants may be able to preserve some aspects of their culture, some are geographically specific. For example, traditional ocean fishing techniques used in The Marshall

Islands are unlikely to continue to be practiced by families who settle in the Alps. As another example, perhaps the language could be preserved, but this would require host nations to be more accommodating and less strict on the assimilation requirements of these special new residents who may be trying to preserve their culture in a new land. For example, France current requires refugees who resettle there to learn French, but if there were international pressure, perhaps France would waive this requirement for groups of EDPs who are trying to preserve a lost culture.

This leads to a tension between accommodation and assimilation as other nations volunteer to absorb the populations of the former nations. It is important to note that it is the lack of a UN protocol for dealing with EDPs that forces other nations to *volunteer* to settle and naturalize those affected. In fact, the loss of a nation falls into the no-man's land between several UN charges – the care of refugees (UNHCR), the protection of world culture (United Nations Educational, Scientific, and Cultural Organization (UNESCO)), and emergency aid response (United Nations Office for the Coordination of Humanitarian Affairs (UNOCHA)). And while the residents of a handful of small island nations might be absorbed relatively easily by volunteer nations, the fact is that climate change has been ushering a literal wave of more frequent and more intense environmental disasters. Imagine a major tsunami taking out a nuclear power plant and causing enough other significant damage that a more heavily inhabited nation may become uninhabitable; or a place being hit by so many repeated severe storms that rebuilding was deemed unwise; or a place where climate change is making it impossible for a nation that was formerly flush with crops to provide for its people. At what point should the UN step in, and in what role?

Time Factors: Raging Waves, Rising Seas, and Rising Nationalism

If a nation is wiped out as a result of a rapid catastrophic event, such as a tsunami or hurricane, then there is no time to prepare, even if the country knew they were at risk of such an event. When a nation is sinking as a result of slowly rising sea levels, then there are issues about how a migration could be coordinated and planned, or even how the loss could be mitigated through land-preserving measures taken by the at-risk nation with or without international support. It is not clear how the timescale of the loss would impact, or should impact, the ultimate decisions that need to be made concerning the resettlement of a population, the protection of their human rights, and the preservation of their culture.

Additionally, as the urgency to address this issue is literally rising with the sea level, the world is also experiencing a rise of nationalism, so the global response today may be very different than it would have been at other periods in history where globalism may have been more in favor than nationalism. If policies, or a lack of policies, end up pushing EDPs towards a subset of welcoming nations, then those countries may get overwhelmed and become less welcoming in response. Therefore, the changing global political climate may also be an important factor to consider.

Lastly, all of these challenges make the size of this problem extremely difficult to predict. Credible studies have predicted anywhere from 140 million to one billion EDPs by 2050.[4,5]

Summary:

In summary, as a nation disappears, it is not clear if an international cooperative and coordinated effort should be adopted to address the loss of homes, the need to resettle, and the preservation of culture. This issue is complex, and no model or report would be able to adequately address every aspect in detail, but excellent reports need to be aware of these different aspects and how they are interrelated. There is the aspect of human rights, which are now recognized in theory, but have never been applied in practice. There is the balance of individual choice versus policy-driven migration. Another aspect is defining equitable burden sharing which could be driven by the capacity for nations to absorb new residents and/or obligations due to contributions to climate change; specifically, whether the nations with the largest contributions to climate change have any ethical obligations to take on a higher burden in assisting climate refugees. Yet another aspect is a balance between assimilation and accommodation, as new residents preserve their culture and/or blend into their new home. Some nations may disappear slowly, such as sinking under rising sea levels or loss of the ability to produce food, while other nations may be wiped out in a catastrophic disaster; and the immediate needs and ability to plan for the long-term needs in these situations are different. Furthermore, the situation is evolving over time as climate change advances and as we see a global rise in nationalism. Lastly, all of this complexity has made it difficult to even measure the problem or predict how quickly it will escalate.

Cited References

Note that these are provided as citations to support claims in the Issue Paper. We have already pulled the important ideas from these resources for you, so although your team may use these sources, access to these is not required. Instead your team is encouraged to look for other sources to support your claims.

[1] Letman, J. (2018, November 19). Rising seas give island nation a stark choice: relocate or elevate. National Geographic. Retrieved from https://www.nationalgeographic.com/environment/2018/11/rising-seas-force-marshall-islands-relocate-elevate-artificial-islands/.

[2] Young, M. (2019, December 9). Climate Refugees Refused UN Protection & Denied Rights Under International Law. Retrieved from http://www.ipsnews.net/2019/12/climate-refugees-refused-un-protection-denied-rights-international-law/.

[3] Su, Y. (2020, January 29). UN ruling on climate refugees could be gamechanger for climate action. Retrieved from https://www.climatechangenews.com/2020/01/29/un-ruling-climate-refugees-gamechanger-climate-action/.

[4] The World Bank Report. (2018, March 19). Climate Change Could Force Over 140-Million to Migration Within Countries by 2050. Retrieved from https://www.worldbank.org/en/news/press-release/2018/03/19/climate-change-could-force-over-140-million-to-migrate-within-countries-by-2050-world-bank-report.

[5] Kamal, B. (2017, August 21). Climate Migrants Might Reach One Billion by 2050. Retrieved from http://www.ipsnews.net/2017/08/climate-migrants-might-reach-one-billion-by-2050/.